普通高等教育"十三五"规划教材

# 环境科学与工程专业英语
## ——阅读与写作(第2版)

# English for Environmental Science and Engineering
## ——Reading Materials & Academic Writing (2nd Edition)

胡龙兴 编著



中国石化出版社

## 内 容 提 要

本书分两部分，第一部分为环境学科专业英语阅读材料，第二部分为环境类科技（学术）论文写作。阅读材料选自原版英文资料，涉及环境学科的不同领域。在每篇阅读材料后，均配有词汇表、疑难句子注释等。科技（学术）论文写作部分聚焦环境类英语科技论文摘要和正文的写作，紧扣英语科技（学术）论文文体特征和句子结构特点。同时，又列举了大量实例，进行适当剖析和点评。写作部分具有鲜明的"理论联系实际"和"案例分析"的特点。

本书旨在作为环境学科本科生和研究生的专业英语教材，也可供有关专业科技人员、工程技术人员和管理人员作为环境类英语的学习材料。

**图书在版编目(CIP)数据**

环境科学与工程专业英语:阅读与写作 / 胡龙兴编著 .
—2 版 . —北京:中国石化出版社,2018. 8(2023. 1 重印)
普通高等教育"十三五"规划教材
ISBN 978-7-5114-4981-8

Ⅰ. ①环… Ⅱ. ①胡… Ⅲ. ①环境科学-英语-阅读教学-高等学校-教材 ②环境科学-英语-写作-高等学校-教材 ③环境工程-英语-阅读教学-高等学校-教材 ④环境工程-英语-写作-高等学校-教材 Ⅳ. ①X

中国版本图书馆 CIP 数据核字(2018)第 184749 号



**中国石化出版社出版发行**

地址:北京市东城区安定门外大街 58 号
邮编:100011　电话:(010)57512500
发行部电话:(010)57512575
http://www. sinopec-press. com
E-mail:press@ sinopec. com
北京力信诚印刷有限公司印刷
全国各地新华书店经销

*

787×1092 毫米 16 开本 20. 25 印张 506 千字
2018 年 9 月第 1 版　2023 年 1 月第 2 次印刷
定价:60. 00 元

# 再版前言

本书自出版以来已有七年时间，现因实际需要进行再版。这次再版未对书中内容作大的变动，仅对书中内容进行少量删减和必要的纠错。

在本书再版之际，编者结合教学实践经验，就有关问题提出意见和建议如下：

1. 大学生的专业英语学习仍具必要性和重要性

对于大学本科生，尤其是研究生来说，专业英语学习是大学基础英语学习之后的一种特定用途的英语学习，将为学生所从事的专业工作或进一步的专业深造提供强大工具和技能。事实证明，没有经过较好的专业英语学习和训练，本科毕业生会在科研开发活动中、研究生会在其论文课题研究中遇到很大困难，会导致不能有效阅读大量英文文献、获取有用信息，也不能撰写英文学术论文，让好的学术研究结果成果化。

2. 本书的主要读者

本书编写的主要目的是作为国内环境学科本科生和研究生的专业英语教材，以帮助学生在教师的指导下，进行较系统的专业英语的阅读与写作训练。

3. 本书的使用建议

本书分环境学科专业英语阅读材料和科技论文写作讲解两部分。由于各校设置专业英语课程的课时不尽相同，教材的使用方法也各不相同。建议本科生的教学要求为：通过学习，扩大专业英语词汇量，巩固基础英语语法知识，熟悉科技英语结构及文体，能较顺利阅读专业英语文献资料，并能用英语撰写论文摘要。本书中有大量阅读材料，这部分材料的学习可采用教师课堂讲解及师生互动的教学方式与学生课外自学方式相结合，当然，课外自学有要求、有导向。本科生的英语写作能力培训可聚焦科技论文英文摘要写作，本教材在这方面有充分的安排。建议对研究生的教学要求为：补齐本科生专业英语水平不足的短板，熟悉科技英语结构及文体，能顺利阅读专业英语文献资料，并能用英语正确撰写论文摘要和撰写科技(学术)论文全文。建议研究生阅读教材中的全部阅读材料，并保证很高的阅读理解准确性。研究生应掌握论文写作规范，在教师的帮助下，进行持续、有效的写作练习。

由于环境学科的复杂性和作者水平有限，本教材仍然会有疏漏和不当之处，在此，恳请读者提出批评、意见和建议。

编者

于上海大学

# 前言

环境专业英语是在大学基础英语之后，环境专业本科生和研究生学习的一门特定用途英语。为了了解和掌握国际环境领域的最新发展动态和科技发展现状及趋势，撰写和发表英文科技(学术)论文，以参与国际学术交流和使科研工作成果化，具备较高的专业英语水平是十分必要的。学好专业英语是达到此目标的有效途径。对于本科生和研究生来说，专业英语是基础英语学习之后的另一个英语学习的重要方面，从某种意义上说，专业英语的学习更具实用性和紧迫性。

环境专业英语课程的教学目的主要是使学生在专业内容方面进行英语阅读的系统训练，把学到的基础英语扩展到专业应用，熟悉专业英语的特点，提高阅读英语科技文献的能力，能借助扎实的英语能力获取有用的信息，学习专业英语的基本写作知识，初步掌握专业英语的写作技能。然而，现行的环境学科专业英语教学和教材大多注重于阅读训练，对专业英语的写作关注得不够。培养专业英语的阅读能力的一条有效途径是熟读一定量的原版资料，泛读和精读相结合，更注重精读，强调阅读理解的准确性。培养专业英语的写作能力，首先要培养意识，要懂得写作的基本规范和理论，即要懂“规矩”。然后，要尽可能多学习和模仿地道英美人的科技(专业)英语的表达方式，力戒“中文式英语”的写作。要多阅读，甚至精心研读优秀的专业学术论文，在阅读专业论文时不妨带着两个任务：(1)获取有用信息；(2)评判写作质量。

本书的编著是作者在多年承担本科生和研究生环境专业英语课程教学任务的基础上完成的，在编著过程中参阅了不少有益书籍和资料，吸收了不少同行、专家的智慧和经验。本书由上海大学环境与化学工程学院副院长、教授、博士生导师陈捷博士担任审稿人。在编著过程中得到了上海市环境科学研究院江研因研究员、复旦大学侯惠奇教授、同济大学杨海真教授、上海大学张仲燕教授等的有益指导，在此表示衷心的感谢！本书的编著还得到了上海大学重点教材建设项目的资助。

由于环境学科的复杂性和作者水平有限，本教材在材料选取、理论论述、实例解析等方面难免有疏漏和不当之处，在此，恳请读者不吝指教。

编者

于上海大学

# CONTENTS

## PART ONE Reading Materials

## PART TWO Academic Writing

# PART ONE
# Reading Materials

# Unit 1 General

## 1. Environmental Engineering and Environmental Engineer

### WHAT IS ENVIRONMENTAL ENGINEERING?

Engineering may be defined as the profession in which a knowledge of the mathematical and natural sciences gained by study, experience, and practice is applied with judgment to develop ways to economically utilize the materials and forces of nature for the benefit of mankind.

Environmental engineering in particular has been defined as that branch of engineering which is concerned with (a) the protection of human populations from the effects of adverse environmental factors; (b) the protection of environments, both local and global, from the potentially deleterious effects of human activities; and (c) the improvement of environmental quality for people's health and well being.

### WHAT DO ENVIRONMENTAL ENGINEERS DO?

The common theme of environmental engineering is a basic understanding of environments, how they function, how they can be damaged, what hazards they present to people, and how people and environments can be protected from such effects. Environmental engineers not only design, operate, and manage facilities and systems for environmental protection, but they also measure environmental quality and continually seek ways to improve it at reasonable cost. Most environmental engineers work in one or more specific areas of application of their engineering knowledge and skills. They deal with atmospheric, aquatic, and terrestrial environments, as well as interactions among these environments. A modern practitioner of environmental engineering stays knowledgeable about all these areas, though most specialize in fields such as those described below.

**Water and wastewater engineering** is concerned with the provision of good quality water for cities and industries, the proper disposal of wastewater, and the protection and enhancement of water quality as related to many uses such as sport fishing and recreation(1). This field has been recognized as a major sub-branch of civil engineering for almost 100 years and has been largely responsible for solving the problems of water-borne disease between 1900 and 1940. Prior to 1900, water-borne diseases such as typhoid and cholera were commonplace and thousands of people died as a result of water pollution in the U. S. The advent of modern water and wastewater treatment practices has reduced the threat of water-borne disease, but there are many new and complex water quality problems. Current concerns of this branch of environmental engineering are the development of more efficient treatment processes and the removal of very low levels of toxic materials from drinking waters and wastewaters(2). As population growth places additional demands on water supply in more arid regions, this field will be called upon to help reclaim potable and industrial process water from wastewaters.

**Air pollution control engineering** is concerned with fuels, combustion processes, and cleaning of exhaust gases from combustion and with the transport and fates of these products in the atmosphere(3). Particulate abatement procedures have been generally successful in cleaning up many cities throughout the world. Deaths due to high concentrations of gases such as sulfur oxides in the air are not nearly as common as they used to be. However, photochemical smog from automobile emissions is still a problem in many cities. There are now more automobiles than ever before and more power plants generating more electricity at a time when clean fuels are becoming scarce and expensive. Acid rain from both human and natural sources of atmospheric pollution is a newly recognized problem which poses a threat to lakes. On an even larger scale looms the possibility of climatic alterations from increased levels of carbon dioxide produced in fossil fuel combustion(4). The continuing challenges in this field are substantial.

**Solid waste engineering** is concerned with finding ways for cities and industries to handle and dispose of refuse and other solid wastes(5). In urban areas each person produces between five and eight pounds of solid wastes each day, excluding junked automobiles and appliances or industrial solid wastes. The wastes can sometimes be burned or buried, but burial may result in pollution of ground or surface waters, and burning has the potential of causing air pollution. Some cities, including St. Louis and Chicago, are now burning part of their refuse in power plants to produce electricity. Many cities recover metals, glass, and paper from refuse. However, some municipalities are still dumping most of their solid wastes at sea. A recently recognized problem is the disposal of hazardous liquid and solid wastes from industries, many of which are especially dangerous and require extreme precaution in handling. Much work remains to be done.

**Industrial hygiene engineering** is concerned with the protection of people from physical, chemical, and biological hazards in the work environment(6). Machinery may be dangerous, and chemical vapors may be poisonous in any large industrial plant. The industrial hygiene engineer finds ways to make the work environment safer and to keep working people healthy.

**Radiological health engineering** is concerned with protecting the general public as well as those who work around nuclear installations or with radioactive materials from external and internal radiation dangers(7). The discovery of radiation and of radioactive materials was accompanied by the discovery that too much radiation causes physiological damage. Yet society found need for X-ray machines, radioactive tracers, and nuclear power plants because their benefits to society were, and still are, considered greater than the hazards presented by properly designed and operated systems. The radiological health engineer tries to minimize these hazards.

**Environmental impact assessment** is concerned with evaluating, eliminating, and predicting the effects of human activities upon the environment(8). This is a new field of practice which has developed since 1970. However, the principles of the work go back many years to sanitary engineers who studied the effects of wastes on lakes and streams and air pollution control engineers who worked out the damaging effects of smoke. As a new and developing field it can be expected to be very important in the future.

**Environmental management** is concerned with the development of new and better ways to design and operate facilities and systems which will provide for protection and improvement of environmental quality and the conservation of natural resources(9). This is also a new and developing field, and the emphasis is on conservation and environmental protection. Pollution is not so much a matter of what is done, but how it is done and of where it is done. Engineers working in this field try to help industries and government agencies to find ways of accomplishing what they want to do without causing pollution and without damaging environments in other ways.

## HOW DOES ONE BECOME AN ENVIRONMENTAL ENGINEER?

The first step in becoming an environmental engineer is to obtain a bachelor of science degree in engineering from a school accredited by the Accreditation Board of Engineering and Technology (ABET). There are over 200 such engineering schools in the U. S. In the past most environmental engineers have majored in civil, mechanical, or chemical engineering as undergraduates. Recently, some engineering schools have begun to offer undergraduate programs in environmental engineering. However, most environmental engineers obtain the masters degree at some point in their careers.

The typical undergraduate curriculum is divided into approximately one-third in basic sciences, humanities, and social sciences; one-third in the engineering sciences such as solid mechanics, fluid mechanics, thermodynamics, electrical science, and materials; and one-third in design and other courses related to the area of the student's specialization. Engineering schools which do not have an undergraduate major in environmental engineering will usually offer a "concentration" in environmental engineering within the degree programs of civil engineering or other disciplines.

A number of students who become environmental engineers obtain bachelor's degrees in one of the sciences, and then proceed to a graduate program in environmental engineering. Although this is possible, in many cases such students may be required to take additional mathematics and engineering science courses before commencing their graduate work. In addition, students without an accredited undergraduate engineering degree are prohibited from obtaining professional registration as an engineer in some states.

## WHERE DO ENVIRONMENTAL ENGINEERS WORK?

Environmental engineers work in consulting firms, in industrial corporations, in local, state and federal government, in private research organizations, and in small but increasing numbers with environmental activist and other public-interest groups. In addition, some environmental engineers who obtain doctoral degrees are on university faculties, although many doctoral-level environmental engineers are employed elsewhere.

The environmental engineer working in a consulting engineering firm makes studies and prepares reports, plans and specifications for a client. The clients are usually cities, states, or industries which have a specific environmental problem in need of a solution. Another area of consulting for environmental engineers is environmental sampling and monitoring. A number of industries will hire consulting firms to obtain samples of air, water, vegetation, soil, food, and wastes and analyze them to obtain measurements of radioactivity, pesticides, and other materials present. The industries use the results to determine if they are in compliance with laws and regulations, to design treatment units, and to defend themselves in law suits.

Many larger cities and industries have a director of environmental engineering and a staff which does some of the environmental engineering work in-house as well as contracting part of the work to consultants. The in-house staff takes responsibility for managing and operating pollution control facilities, while the consultants are more involved in design. Both groups engage to some extent in environmental assessments.

Environmental engineering is a small but identifiable branch of engineering. Nearly all past graduates have been able to find employment in their chosen fields after graduation. The opportunities in the field are many and varied. There are always new challenges to be faced, not only in our modern, technological society, but also in less industrially developed societies in other parts of the world. Thus, the long term outlook for employment is very good.

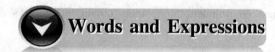

## Words and Expressions

1. branch of engineering 工程分支
2. be concerned with 涉及，与……有关，参与
3. effects of adverse environmental factors 不利的环境因素的影响
4. deleterious effects of human activities 人的活动的不利的影响
5. well being 幸福，福利
6. common theme 共同的主题
7. facility 设施，设备
8. at reasonable cost 以合理的成本[费用]
9. deal with 应付，对付，处理
10. atmospheric environment 大气环境
11. aquatic environment 水环境
12. terrestrial environment 陆地环境
13. be knowledgeable about 知晓，了解
14. a major sub-branch of civil engineering 土木工程的一门主要分支
15. be responsible for 引起，造成，是……的原因，对……负责
16. water-borne disease 水致疾病，以水为媒介传播的疾病
17. prior to 在……以前，早于，居先
18. typhoid 伤寒
19. cholera 霍乱
20. commonplace 平凡的；平凡的事物
21. efficient treatment processes 有效的处理工艺[过程]
22. removal of low levels of toxic materials 低含量有毒物质的去除，去除低含量的有毒物质
23. drinking water 饮用水
24. water supply 供水，水源
25. arid 干旱的，不毛的
26. reclaim 回收，再生，重新使用，收回
27. potable water 饮用水
28. exhaust gas 废气，尾气
29. particulate abatement 颗粒物去除
30. clean up 清除，净化
31. throughout the world 遍及世界各地
32. sulfur oxide 硫氧化物
33. photochemical smog 光化学烟雾
34. automobile emission 汽车排气[废气]

35. power plant 发电厂，发电站
36. clean fuel 清洁燃料
37. scarce 缺乏的，不足的，罕见的
38. acid rain 酸雨
39. pose 引起，造成，提出，使……摆好姿势
40. on a large scale 大规模
41. loom 隐隐出现[呈现]
42. climatic alteration 气候变化
43. carbon dioxide 二氧化碳
44. fossil fuel 矿物燃料，化石燃料
45. refuse 垃圾
46. solid waste 固体废物
47. junked automobile 废汽车
48. appliance 器具，设备，装置
49. burial 埋葬，埋入
50. ground water 地下水
51. surface water 地表水
52. recover metals, glass, and paper from refuse 从垃圾中回收金属，玻璃和纸
53. dump 堆，堆场；倾倒，卸料
54. disposal of hazardous liquid and solid wastes 危险液体和固体废物的处置
55. industrial hygiene engineering 工业卫生工程
56. machinery 机器，机械
57. chemical vapor 化学蒸气
58. poisonous 有毒的
59. radiological health engineering 辐射安全工程；放射性安全工程
60. the general public (the public at large) 公众
61. nuclear installation 核设施，核装置
62. radioactive material 放射性物料
63. radiation 放射，辐射
64. physiological damage 生理伤害[损害]
65. X-ray machine X 射线机
66. radioactive tracer 放射性示踪剂
67. nuclear power plant 核电厂，核电站
68. sanitary engineer 卫生工程师
69. effects of wastes on lakes and streams 废物对湖泊和河流的影响
70. work out 制订出，估计出
71. be expected to be very important 预期是非常重要的
72. conservation of natural resource 自然资源的保护
73. bachelor of science degree in engineering 工学学士
74. accredit 鉴定……为合格，认可，特许，委派，任命
75. the Accreditation Board of Engineering and Technology (ABET) 工程与技术鉴定[认可]委员会
76. be majored in civil, mechanical, or chemical engineering 以土木工程，机械工程或化学工程为专业
77. master degree 硕士学位
78. undergraduate curriculum 本科生课程
79. basic science 基础科学
80. humanities 人文学
81. social science 社会科学
82. solid mechanics 固体力学
83. fluid mechanics 流体力学
84. thermodynamics 热力学
85. electrical science 电气科学
86. discipline 学科
87. commence 开始，开始做
88. professional registration 职业注册
89. consulting firm 咨询公司
90. industrial corporation 工业(股份有限)公司
91. public-interest group 公益团体
92. doctoral degree 博士学位
93. university faculties 大学的系、院
94. specification 详细说明，说明书，明细表
95. client 委托人，买方，顾客
96. environmental sampling and monitoring 环境采样和监测
97. vegetation 植物，植被

98. radioactivity 放射性
99. pesticide 农药，杀虫剂
100. in compliance with 依从，按照，遵照，与……一致
101. law and regulation 法规
102. treatment unit 处理单元[设备]
103. defend themselves in law suits 在法律诉讼中为他们自己辩护
104. contract 合同，契约；签合同
105. consultant 顾问，咨询，请教者，查阅者
106. be involved in 包括在……之内，与……有关，专心地做
107. be engaged in 正从事于，正忙于，正致力于，参加
108. to some extent 在某种程度上
109. environmental assessment 环境评价
110. identifiable branch of engineering 可认同的[可识别的]工程分支
111. challenge to be faced 将面临的困难[挑战]

## Notes

(1) Water and wastewater engineering is concerned with the provision of good quality water for cities and industries, the proper disposal of wastewater, and the protection and enhancement of water quality as related to many uses such as sport fishing and recreation.

水和废水工程涉及为城市和工业界提供高质量的水、适当处理废水、保护和改善与许多用途(如休闲性的垂钓和娱乐)有关的水的质量。

(2) Current concerns of this branch of environmental engineering are the development of more efficient treatment processes and the removal of very low levels of toxic materials from drinking waters and wastewaters.

目前对环境工程这一分支的关注是开发出更有效的处理工艺和从饮用水和废水中去除非常低含量的有毒物质。

(3) Air pollution control engineering is concerned with fuels, combustion processes, and cleaning of exhaust gases from combustion and with the transport and fates of these products in the atmosphere.

空气污染控制工程涉及燃料、燃烧过程及燃烧产生的废气的净化，还涉及这些燃烧产物在大气中的流通和归宿。

(4) On an even larger scale looms the possibility of climatic alterations from increased levels of carbon dioxide produced in fossil fuel combustion

在更大的规模内可以看到在化石燃料燃烧中产生的二氧化碳的量的增加导致气候变化的可能性。本句为倒装句。

(5) Solid waste engineering is concerned with finding ways for cities and industries to handle and dispose of refuse and other solid wastes.

固体废物工程涉及寻找各种方法供城市和工业界来处理和处置垃圾和其他固体废物。

(6) Industrial hygiene engineering is concerned with the protection of people from physical, chemical, and biological hazards in the work environment.

工业卫生工程涉及保护员工在工作环境中不受物理、化学和生物的危害。

(7) Radiological health engineering is concerned with protecting the general public as well as those who work around nuclear installations or with radioactive materials from external and internal radiation dangers.

辐射安全工程涉及保护公众及在核设施周围工作的人员或操作放射性物料人员不受内外部辐射的危害。

(8) Environmental impact assessment is concerned with evaluating, eliminating, and predicting the effects of human activities upon the environment.

环境影响评价涉及评估、消除和预测人的活动对环境的影响。

(9) Environmental management is concerned with the development of new and better ways to design and operate facilities and systems which will provide for protection and improvement of environmental quality and the conservation of natural resources.

环境管理涉及研发新的和更好的设计和操作设施与系统的方法，该设施系统能为保护及改善环境质量和保护自然资源提供保证。

# 2. A Symbiotic Relationship in Environmental Engineering

## Introduction

In response to society's demands, the subject of successfully addressing environmental problems has become extremely complex and specialized. This was not always so(1). Until the midsixties civil engineers predominated in the field of environmental engineering. Since that time we have witnessed a considerable influx of chemical engineers.

A number of questions come to mind "What prompted this change?", "What are the current roles of civil and chemical engineers in environmental engineering?", "How has it impacted on the profession?" and "To what degree do civil and chemical engineers complement each other in addressing and solving the environmental problems of today?". The object of this discussion is to examine this symbiotic relationship; and to understand it requires that we first examine the traditional roles of chemical and civil engineers in relation to environmental matters.

## The Civil Engineer in Environmental Engineering

Civil engineers learn about structures, soils, roads, materials of construction, hydraulics etc. Because of this, it was quite natural to include water supply, distribution and treatment on the one hand and wastewater collection and treatment on the other. This was usually taught in the senior year at the undergraduate level. Until the early sixties the emphasis was on rural and municipal sanitation, vector borne disease control and general public health. Industrial wastes were generally treated jointly with municipal wastes. Seldom was the need to treat industrial waste-waters on a separate basis identified(2).

Our general understanding of, and concern for, the environment was deemed to be adequate to address pollution control requirements with available technologies. This of course pertained to water and wastewater since concerns related to air quality and solids waste management, while in the background, were generally absent.

Thus, civil engineers were generally well prepared to respond to environmental problems. This is in spite of their general aversion to chemistry and things chemical.

## The Chemical Engineer in Environmental Engineering

Chemical engineers are trained to be industry process oriented(3). Their training emphasizes production and the development of processes to produce a product at the lowest cost. As chemical engineers started to respond to environmental problems, subjects such as unit operations, process kinetics, chemical theory, process control, process scale-up etc., were demonstrated to be valuable stepping stones and tools in solving industrial pollution control problems(4). The requirement to respond in this fashion did not occur until the late sixties and early seventies, when society demanded a cleaner environment and were quick to point a finger at industry for not being the good corporate citizen it should be(5).

Chemical engineers are also taught to look at the complete industry with all its facets. The process/systems approach to problem solving is second nature and thus made it much easier for chemical engineers to define opportunities for in-plant process control through the application of mass and materials balances in pin-pointing product leakage(6). This knowledge also went a long way toward initiating better house-keeping procedures within plants, identifying opportunities for raw materials substitution and production process modifications, all resulting in reduced water and waste quantities(7). Because of product process knowledge, chemical engineers also have the training to develop new or change old production processes so that the inevitable waste products are less toxic to the environment.

Society was demanding higher treatment levels of wastes prior to discharge to the environment. Since municipal treatment plants which accepted industrial wastes for joint treatment had difficulty in meeting these demands, more sophisticated and specially designed treatment technologies were required. The chemical engineers' familiarity with technologies such as spray drying, and activated carbon, membranes (reverse osmosis, ultra- and hyper-filtration), electrolysis, electrodialysis, ion exchange, heat recovery and liquid/solid separation all came in handy as the search continued for technologies to process wastes more economically than before(8).

In this search for new processes and their application, chemical engineers in wastewater treatment process research and development, because of their process knowledge, were a jump ahead of civil engineers(9). With few exceptions chemical engineers, because of their process orientation, are generally better equipped than civil engineers to handle problems related to using recycled materials and alternate sources of raw materials as feed stock for industrial processes(10).

These are the problems and opportunities chemical engineers addressed when involved in environmental engineering.

It would appear from the foregoing that chemical engineers have replaced civil engineers in addressing industrial environmental engineering problems. This is not so at all. What has evolved over the past fifteen years is a symbiotic relationship.

## Words and Expressions

1. symbiotic relationship 共生关系
2. in response to society's demands 响应社会的需要
3. subject of successfully addressing environmental problems 成功论述环境问题的学科
4. field of environmental engineering 环境工程领域
5. influx 涌入，流入，涌进，汇集
6. complement each other 互相补充
7. in relation to 关于
8. environmental matter 环境事务
9. materials of construction 建筑材料
10. hydraulics 水力学
11. wastewater collection and treatment 废水收集和处理
12. water distribution 水的分配
13. the senior year at the undergraduate level (大学)本科的四年级
14. rural and municipal sanitation 农村和城市下水道系统，农村和城市卫生设备
15. vector borne disease 媒介传播疾病
16. general public health 公众健康，公共卫生(事业)
17. municipal waste 城市废物
18. deem 认为，相信
19. available technologies 可用技术
20. pertain to 关于，从属于，适合
21. background 背景，本底
22. aversion 厌恶，反感
23. unit operation 单元操作
24. process kinetics 过程动力学
25. chemical theory 化学理论
26. process control 过程控制
27. process scale-up 过程放大
28. facet 方面
29. mass and materials balance 物料平衡[衡算]
30. leakage 漏，泄漏
31. initiate 开始，启动，发动
32. raw materials substitution 原料替代
33. production process modification 生产过程改进，生产工艺改进
34. inevitable 不可避免的，必然发生的
35. spray drying 喷雾干燥
36. activated carbon 活性炭
37. membrane 膜
38. reverse osmosis 逆渗透，反渗透
39. ultrafiltration 超过滤，超滤
40. hyperfiltration 超过滤，超滤
41. electrolysis 电解
42. electrodialysis 电渗析
43. ion exchange 离子交换
44. heat recovery 热回收
45. liquid/solid separation 液固分离
46. orientation 定向，定方位
47. recycled materials 循环物料
48. raw materials 原料
49. feed stock(送入机器或加工厂的)原料
50. stepping stones 踏脚石，达到目的的手段
51. in-plant process control 厂内过程控制
52. house-keeping procedures within plants 厂内的房屋保养措施

## Notes

(1) This was not always so.

情况并不总是这样。

(2) Seldom was the need to treat industrial wastewaters on a separate basis identified.

很少需要分开来单独地处理工业废水。本句为倒装句，以 seldom、little、hardly 等有否定意义的词引导的句子，采用倒装结构。

(3) Chemical engineers are trained to be industry process oriented.

化学工程师的培养是面向工业生产过程的。"名词或名词性词组+oriented"是一种常见的用法，译为"面对一定方向的"或"以……为目标的"。

(4) As chemical engineers started to respond to environmental problems, subjects such as unit operations, process kinetics, chemical theory, process control, process scale-up etc., were demonstrated to be valuable stepping stones and tools in solving industrial pollution control problems.

当化学工程师面对环境问题时，像单元操作、过程动力学、化学理论、过程控制、过程放大等学科知识被证明是解决工业污染控制问题有价值的手段和工具。

(5) The requirement to respond in this fashion did not occur until the late sixties and early seventies, when society demanded a cleaner environment and were quick to point a finger at industry for not being the good corporate citizen it should be.

以这种方式作出响应的要求直到60年代的后期和70年代的早期才出现，那时社会要求有一个更清洁的环境，并很快指责工业界应该是而实际却不是守法的好公民。句中的"point a finger at"意为"对……的指责"，"the good corporate citizen"译为"守法的好公民"。

(6) The process/systems approach to problem solving is second nature and thus made it much casier for chemical engineers to define opportunities for in-plant process control through the application of mass and materials balances in pin-pointing product leakage.

解决问题的过程/系统方法是第二个特性，因此使化学工程师能更容易地通过应用物料衡算极精确地测定产物的泄漏来确定厂内过程控制的机会。

(7) This knowledge also went a long way toward initiating better house-keeping procedures within plants, identifying opportunities for raw materials substitution and production process modifications, all resulting in reduced water and waste quantities.

该知识对引进更好的厂内房屋保养措施和识别原料替代及生产工艺改进的可能性也非常有效，所有这些都导致用水和废物量减少。句中"go a long way toward……"译为"(对……)非常有效"，"(对……)很有帮助"。在 toward 后，有两个并列的动名词短语，分别用 initiating 与 identifying 引出。

(8) The chemical engineers' familiarity with technologies such as spray drying, and activated carbon, membranes (reverse osmosis, ultra- and hyper-filtration), electrolysis, electrodialysis, ion exchange, heat recovery and liquid/solid separation all came in handy as the search continued for technologies to process wastes more economically than before.

随着对能更经济处理废物技术探索的继续，化学工程师对诸如喷雾干燥、活性炭、膜(逆渗透、超滤)、电解、电渗析、离子交换、热回收和液固分离等技术的熟悉迟早会有用处的。

(9) In this search for new processes and their application, chemical engineers in wastewater treatment process research and development, because of their process knowledge, were a jump ahead of civil engineers.

在探索新工艺及其应用的过程中，由于他们的工艺知识，化学工程师在废水处理工艺的研究和开发中，要领先土木工程师一步。

(10) With few exceptions chemical engineers, because of their process orientation, are generally better equipped than civil engineers to handle problems related to using recycled materials and alternate sources of raw materials as feed stock for industrial processes.

几乎没有例外，化学工程师，由于他们的工艺定向，在处理关系到使用循环物料和替换原料作为工业过程的原料问题时，一般比土木工程师更适合。

## 3. The Field of Environmental Science

Environmental science is an interdisciplinary area of study that includes both applied and theoretical aspects of human impact on the world. Since humans are generally organized into groups, environmental science must deal with the areas of politics, social organization, economics, ethics, and philosophy. Thus, environmental science is a mixture of traditional science, societal values, and political awareness.

Environmental science as a field of study is still in the process of evolving, but its beginnings are rooted in the early history of civilization. Many ancient cultures expressed a reverence for the plants, animals, and geographic features that provided them with food, water, and transportation which can still be appreciated by modern people.

The current interest in the state of the environment began with people like Thoreau and other philosophers but received an additional push by the organization of the first Earth Day on 22 April 1970. The second Earth Day on 22 April 1990 reaffirmed this commitment, as have similar Earth Days since then. As a result of this continuing interest in the state of the world and how people affect it, environmental science is now a standard course on many college campuses and is also a part of high school course offerings. Most of the concepts covered by environmental science courses had previously been taught in ecology, conservation, or geography courses. Environmental science incorporates the scientific aspects of these courses with input from the social sciences, such as economics, sociology, and political science, into a new interdisciplinary field(1).

### The Interrelated Nature of Environmental Problems

Environmental science is by nature an interdisciplinary field. The word environment is usually understood to mean the surrounding conditions that affect people and other organisms. In a broader definition, environment is everything that affects an organism during its lifetime. Environmental issues from a human perspective involve concerns about science, nature, health, employment, profits, politics, ethics, economics, and other considerations.

Most social and political decisions are made with respect to political jurisdictions, but environmental problems do not necessarily coincide with these artificial, political boundaries. For example, air pollution problems may involve several local units of government, several states or provinces, and in many situations, different nations. Air pollution problems of Juarez, Mexico, are also air pollution problems of El Paso, Texas. But the issue is more than an air-quality and human-health issue. Lower wage rates and less-strict environmental laws have influenced some industries to move to Mexico for economic reasons. Mexico and many other developing nations are struggling to improve their environmental image and need the money generated by foreign investment to be able to improve the conditions of their people and the environment in which they live.

Similarly, air pollutants produced in the major industrial regions of the United States drift across the border into Canada, where acid rain damages lakes and forests. A long-standing dispute exists between the United States and Canada over this issue. Canada claims that the United States should be doing more to reduce emissions that cause acid rain, and the United States claims it is doing as much as it can. Using water from the Colorado River for irrigation reduces the quality and quantity of water entering Mexico and causes political friction between Mexico and the United States.

Because of all these political, economic, ethical, and scientific linkages, solving environmental problems is a complex task. Such problems seldom have a simple solution. However, successful organizations such as the International Joint Commission have had major bearing on the quality of the environment over broad regions of the world.

The International Joint Commission was established in 1909 when the Boundary Waters Treaty was signed between the United States and Canada. The treaty was established in part to provide that the "boundary waters and waters flowing across the boundary shall not be polluted on either side to the injury of health or property of the other.". The International Joint Commission has been instrumental in identifying areas of concern and encouraging the cleanup of polluted sites that affect the quality of the Great Lakes and other boundary waters. In general, the two governments have listened to the advice of the International Joint Commission and have responded by initiating cleanup activities.

The first worldwide meeting of heads of state directed to concern for the environment took place at the Earth Summit, formally known as the United Nations Conference on Environment and Development (UNCED) in Rio de Janeiro in 1992. Most countries signed agreements on sustainable development and biodiversity. Previous agreements on global warming and depletion of the ozone layer had been signed by many nations. It may be years before we will know if all countries that signed these agreements will meet their commitments to environmental improvement, but they have at least stated their intention to do so.

The United Nations through the United Nations Educational, Scientific, and Cultural Organization (UNESCO) and the United Nations Environment Programme (UNEP), has supported many environmental programs. A recent undertaking is the International Environmental Education Programme(IEEP). This program recognizes the need for environmental education at all levels of society in both the formal education and the informal education that occurs through the media and groups of interested citizens. Conferences on environmental education were first held during the 1970s and continue to the present.

Another international effort is the Trilateral Committee on Environmental Education consisting of representatives from Canada, Mexico, and the United States. The first trilateral project was a conference on education, communication, and the environment held in March 1993. A major project of the Trilateral Committee is collecting and distributing environmental education materials to primary and secondary schools.

## Words and Expressions

1. interdisciplinary 各学科之间的，跨学科的
2. ethics 伦理学
3. reverence 尊重，尊敬
4. geographic features 地势
5. appreciate 意识到，（正确）评价，珍视，感激
6. be consistent with 与……一致
7. current interest in the state of the environment 当前对环境状况的关注
8. reaffirm 重申，再肯定，再证实
9. commitment 所承诺之事，保证，委托，委任，赞成
10. college campus 大学校园
11. ecology 生态学
12. geography 地理学
13. interrelated 互相关联的
14. by nature 本质上，本性上，生性
15. organism 有机体，生物体
16. lifetime 寿命
17. environmental issue 环境问题
18. with respect to 关于，至于
19. political jurisdiction 政治权限，政治管辖权
20. not necessarily 未必
21. be coincide with 与……一致，与……相符
22. drift across the border 漂过边界
23. a long-standing dispute 长期的争论
24. political friction 政治摩擦
25. linkage 联系，连通；联动
26. the International Joint Commission 国际联合委员会
27. have major bearing on 与……有重大关系，对……有重大影响
28. the BoundaryWaters Treaty 边界水协议
29. in part 部分地
30. instrumental 有用的
31. of concern 所关注的
32. the Great Lakes 北美洲五大湖
33. the United Nations Conference on Environment and Development（UNCED）联合国环境与发展大会
34. sustainable development 可持续发展
35. biodiversity 生物多样性
36. global warming 全球变暖
37. depletion of the ozone layer 臭氧层耗竭[变薄]
38. the United Nations Educational, Scientific, and Cultural Organization（UNESCO）联合国教(育)科(学)文(化)组织
39. the United Nations Environment Programme（UNEP）联合国环境规划署
40. undertaking 任务，事业，计划，许诺，承担
41. project 计划，方案，工程，项目

## Notes

Environmental science incorporates the scientific aspects of these courses with input from the social sciences, such as economics, sociology, and political science, into a new interdisciplinary field.

环境科学将这些课程的科学方面与社会科学(如经济学，社会学和政治学)的内容相结合，形成了一个新的跨学科领域。

## 4. Elements of Ecology

Plants and animals in their physical environment make up an ecosystem. The study of such ecosystems is *ecology*. Although we often draw lines around a specific ecosystem in order to be able to study it more fully (e. g., a farm pond) and in so doing assume that the system is totally self-contained, this obviously is not true, and we must remember that one of the tenets of ecology is that "everything is connected with everything else."

Within an ecosystem there exist three broad categories of actors. The *producers* take energy from the sun, nutrients such as nitrogen and phosphorus from the soil, and through the process of photosynthesis produce high-energy chemicals. The energy from the sun is thus stored in the chemical structure of their organic molecules. These organisms are often referred to as being in the first trophic level and are called autotrophs.

A second group of organisms are the *consumers* who use some of this energy by ingesting the high-energy molecules. These organisms are in the second trophic level in that they directly use the energy of the producers. There can be several more trophic levels as the consumers use the level above as a source of energy.

The third group of organisms, the decomposers or decay organisms, use the energy in animal wastes and dead plants and animals, and in so doing convert the organic molecules to stable inorganic compounds. The residual inorganics then become the building blocks for new life, using the sun as the source of energy.

Ecosystems exhibit a flow of both energy and nutrients. Energy flow is in only one direction: from the sun and through each trophic level. Nutrient flow, on the other hand, is cyclic. Nutrients are used by plants to make high-energy molecules, which are eventually decomposed to the original inorganic nutrients, ready to be used again.

The entire food web, or ecosystem, stays in dynamic balance, with adjustments being made as required[(1)]. Such a balance is called homeostasis. For example, a drought one year may produce little grass, thus exposing field mice to predators such as owls. The mice, in turn, spend more time in burrows, thus not eating as much, and allowing the grass to reseed for the following year. External perturbations, however, can upset and even destroy an ecosystem. In the previous example, the use of an herbicide to kill the grass might also destroy the field mouse population, and in turn diminish the number of owls. It must be recognized that although most ecosystems can absorb a certain amount of insult, a sufficiently large perturbation can cause irreparable damage.

The amount of perturbation a system is able to absorb without being destroyed is tied to the concept of the ecological niche. The combination of function and habitat of an organism in an ecological system is its niche. A niche is not a property of a type of organism or species, but is its best accommodation with the environment. In the example above, the grass is a producer which acts as food for the field mouse, which in turn is food for the owl. If the grass were destroyed, both the mice and owls might eventually die out. But suppose there were two types of grass, each equally acceptable as mouse food. Now if one died out, the ecosystem would not be destroyed because the mice would still have food. This simple example demonstrates an important ecological principle: the stability of an ecosystem is proportional to the number of organisms capable of filling various niches[2]. A jungle, for example, is a very stable ecosystem, whereas the tundra in Alaska is extremely fragile. Another fragile system is deep oceans-a fact which should be a consideration in the disposal of hazardous and toxic materials in deep ocean areas. Inland water courses tend to be fairly stable ecosystems, but certainly not totally resistant to destruction by outside forces. Other than the direct effect of toxic materials such as heavy metals and refractory organics, the most serious effect of water pollution for inland waters is the depletion of dissolved (free) oxygen. All higher forms of aquatic life exist only in the presence of oxygen, and most desirable microbiologic life also requires oxygen. Generally, all natural streams and lakes are aerobic (containing dissolved oxygen). If a watercourse becomes anaerobic (absence of oxygen), the entire ecology changes to make the water unpleasant or unsafe.

Problems associated with pollutants which affect the dissolved oxygen levels cannot be appreciated without a fuller understanding of the concept of decomposition or biodegradation, part of the total energy transfer system of life[3].

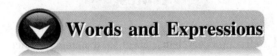

## Words and Expressions

1. elements of ecology 生态学原理[基础]
2. physical environment 自然环境
3. make up 组成，构成
4. ecosystem 生态系统
5. tenet 原理，原则
6. three broad categories of actors 三大类行动者
7. producer 生产者
8. nutrient 养分，营养物
9. photosynthesis 光合作用
10. organic molecule 有机分子
11. be referred to as 被称之为
12. trophic level 营养级
13. autotroph 自养生物
14. consumer 消费者
15. ingest 吞入，咽下，摄取
16. high-energy molecule 高能分子
17. the progressive use of energy 能量的渐进

使用

18. decomposer 分解者

19. stable inorganic compound 稳定的无机化合物

20. residual inorganics 剩余无机物

21. building block 结构单元，构件，标准块，组装的部件

22. cyclic 循环的

23. dynamic balance 动态平衡

24. homeostasis 生物体内平衡

25. drought 干旱

26. expose... to 使……暴露于，接触

27. predator 食肉动物

28. owl 猫头鹰

29. mice 田鼠

30. external perturbation 外界的扰动

31. herbicide 除草剂

32. field mouse population 田鼠种群

33. diminish 减少，减小

34. irreparable 不可弥补的

35. ecological niche 生态小生境

36. habitat 栖息地

37. ecological system 生态系统

38. die out 消亡，消失

39. ecological principle 生态学原理

40. jungle 丛林

41. tundra 冻原

42. fragile 脆弱的，易损的

43. hazardous and toxic materials 危险和有毒物料

44. resistant to destruction by outside forces 抵抗外力的破坏

45. other than 除……之外

46. heavy metal 重金属

47. refractory organics 难降解有机物

48. depletion of dissolved oxygen 溶解氧耗竭

49. aquatic life 水生物

50. in the presence of 存在，有

51. desirable 所希望的，合乎需要[要求]的

52. microbiologic life 微生物

53. aerobic 好氧的，需氧的，有氧的

54. anaerobic 厌氧的，厌气的，无溶解氧的

55. decomposition 分解

56. biodegradation 生物降解(作用)

## Notes

(1) The entire food web, or ecosystem, stays in dynamic balance, with adjustments being made as required.

整个食物网，即生态系统，处于动态平衡，根据要求可作出适当的调整。

(2) This simple example demonstrates an important ecological principle: the stability of an ecosystem is proportional to the number of organisms capable of filling various niches.

这一简单的例子说明了一个重要的生态学原理：生态系统的稳定性与能够给各种小生态环境提供食物的生物体的数目成比例。

(3) Problems associated with pollutants which affect the dissolved oxygen levels cannot be appreciated without a fuller understanding of the concept of decomposition or biodegradation, part of the total energy transfer system of life.

没有较充分理解分解，即生物降解(生命总能量传递系统的一部分)的概念，就不能理解与影响溶解氧含量的污染物有关的问题。

# 5. Biodegradation and Decomposition

## Biodegradation

Plant growth, or photosynthesis, can be represented by the equation

$$CO_2 + H_2O \xrightarrow{\text{sunlight \& nutrients}} HCOH + O_2$$

In this representation formaldehyde (HCOH) and oxygen are produced from carbon dioxide and water, with sunlight the source of energy. If the formaldehyde and oxygen are combined and ignited, an explosion results. The energy which is released during such an explosion is stored in the carbon-hydrogen-oxygen bonds of formaldehyde.

As discussed above, plants (producers) use inorganic chemicals as nutrients and, with sunlight as a source of energy, build high-energy molecules. The animals (consumers) eat these high-energy molecules and during their digestion process some of the energy is released and used by the animals. The release of this energy is quite rapid and the end products of digestion (excrement) consist of partially stable compounds. These compounds become food for other organisms and are thus degraded further but at a slower rate. After several such steps, very low-energy compounds are formed which can no longer be used by microorganisms for food. Plants then use these compounds to build more high-energy molecules and the process starts all over. The process is symbolically shown in Figure 1.

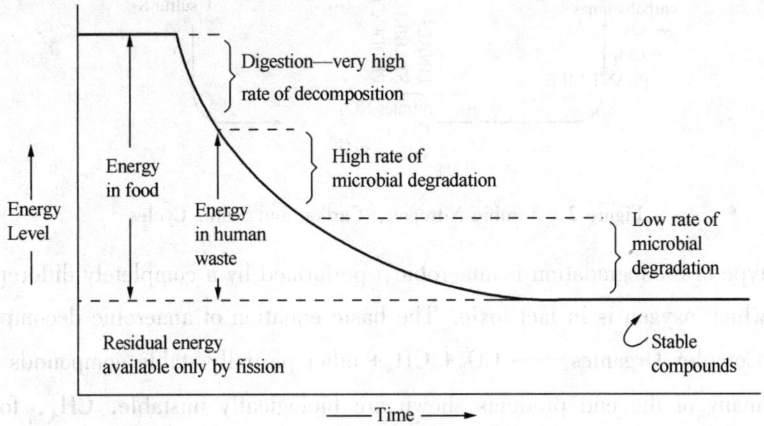


Figure 1 Energy Loss in Biodegradation

It is important to realize that many of the organic materials responsible for water pollution enter watercourses at a high energy level(1). It is the biodegradation, or the gradual use of this energy, by a chain of organisms which causes many of the water pollution problems(2).

## Aerobic and Anaerobic Decomposition

Decomposition, or biodegradation, can take place in one of two distinctly different ways: aerobic (using free oxygen) or anaerobic (in the absence of free oxygen).

The basic equation of aerobic decomposition is

$$\text{Complex Organics} + O_2 \longrightarrow CO_2 + H_2O + \text{Stable Products}$$

Carbon dioxide and water are always two of the end products of aerobic decomposition. Both are stable, low in energy, and are used by plants in the process of photosynthesis. If sulfur compounds are involved in the reaction, the most stable end product is $SO_4^{2-}$, the sulfate ion. Similarly, phosphorus ends up as $PO_4^{3-}$, orthophosphate. Nitrogen goes through a series of increasingly stable compounds, finally ending up as nitrate. The progression is

$$\text{Organic Nitrogen} \longrightarrow NH_3(\text{ammonia}) \longrightarrow NO_2^-(\text{nitrite}) \longrightarrow NO_3^-(\text{nitrate})$$

Because of this distinctive progression, nitrogen has been in the past and to some extent is still used as an indicator of pollution.

A schematic representation of the aerobic cycle for carbon, sulfur and nitrogen compounds is shown as Figure 2. This figure illustrates only the basic facts, and is a gross simplification of the actual steps and mechanisms involved.

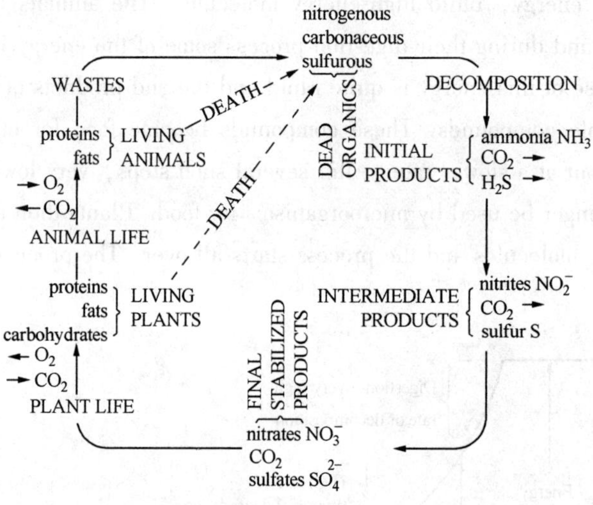


Figure 2 Aerobic Nitrogen, Carbon and Sulfur Cycles

A second type of biodegradation is anaerobic, performed by a completely different set of microorganisms, to which oxygen is in fact toxic. The basic equation of anaerobic decomposition is

$$\text{Complex Organics} \longrightarrow CO_2 + CH_4 + \text{other partially stable compounds}$$

Note that many of the end products shown are biologically unstable. $CH_4$, for example, is methane, a high-energy gas commonly called marsh gas, physically stable but still able to be decomposed biologically. Nitrogen compounds stabilize only to ammonia ($NH_3$), and sulfur ends up as evil-smelling hydrogen sulfide ($H_2S$) gas. Figure 3 is a schematic representation of anaerobic decomposition. Note that the left half of the cycle, the photosynthesis by plants, is identical to the aerobic cycle in Figure 2.

Biologists often speak about various compounds as "hydrogen acceptors". The hydrogen atoms, torn from high-energy organic molecules, must be attached to various compounds. In aerobic decomposition oxygen serves this purpose and is thus known as the hydrogen acceptor. It accepts the hydrogen atoms to form water.

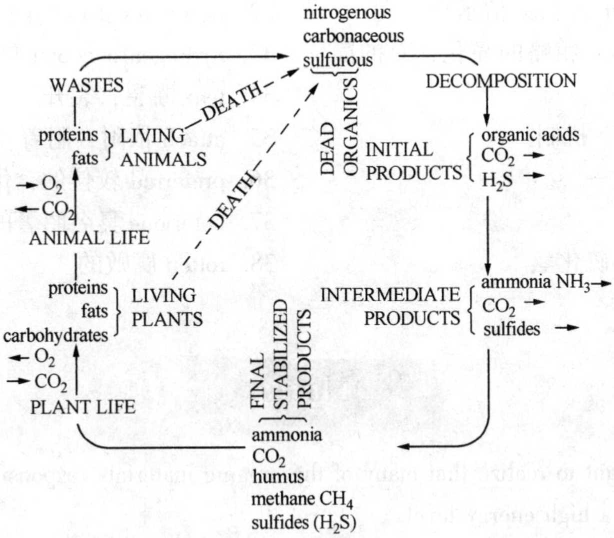


Figure 3  Anaerobic Nitrogen, Carbon and Sulfur Cycles

In anaerobic decomposition free oxygen is not available, and the next preferred hydrogen acceptor is nitrogen, thus forming ammonia, $NH_3$. If free oxygen is not available, ammonia cannot be converted to nitrites or nitrates. If nitrogen is not available, the next preferred hydrogen acceptor is sulfur, thus forming hydrogen sulfide, $H_2S$, the chemical responsible for the notorious rotten egg smell(3).

## Words and Expressions

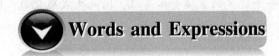

1. photosynthesis 光合作用
2. formaldehyde 甲醛
3. carbon dioxide 二氧化碳
4. ignite 点燃
5. explosion 爆炸
6. carbon-hydrogen-oxygen bond 碳-氢-氧键
7. digestion process 消化过程
8. end product 最终产物
9. excrement 排泄物，粪便
10. consist of 由……组成
11. partially stable compound 部分稳定的化合物
12. a chain of organisms 一系列有机体，一系列生物体
13. two distinctly different ways 两种截然不同的方式
14. free oxygen 游离氧
15. in the absence of 没有，不存在
16. basic equation of aerobic decomposition 好氧分解的基本方程
17. complex organics 复杂的有机物
18. sulfate ion 硫酸根离子
19. orthophosphate 正磷酸盐
20. a series of 一系列
21. organic nitrogen 有机氮
22. ammonia 氨
23. nitrite 亚硝酸盐
24. nitrate 硝酸盐
25. distinctive progression 有区别的渐进(过程)，鉴别性的渐进(过程)

26. indicator of pollution 污染指示物
27. gross simplification 粗略的简化，总的简化
28. mechanism 机理，机制
29. methane 甲烷
30. marsh gas 沼气
31. hydrogen sulfide 硫化氢
32. be identical to 等同于
33. hydrogen acceptor 氢接受体
34. tear 撕裂，裂开
35. attach 依附，附着；接合
36. preferred 较佳的，优先的，优先选用的
37. notorious 臭名昭著的，有名的
38. rotten 腐败的

## Notes

(1) It is important to realize that many of the organic materials responsible for water pollution enter watercourses at a high energy level.

重要的是要意识到许多引起水污染的有机物是以高能级进入水体的。

(2) It is the biodegradation, or the gradual use of this energy, by a chain of organisms which causes many of the water pollution problems.

正是生物降解作用，即由一系列生物体对能量的逐渐使用导致了许多水污染问题。本句为强调句。

(3) If nitrogen is not available, the next preferred hydrogen acceptor is sulfur, thus forming hydrogen sulfide, $H_2S$, the chemical responsible for the notorious rotten egg smell.

如果氮不存在，下一个优先的氢接受体是硫，因此生成硫化氢，$H_2S$，它是能引起腐败蛋臭味的化学物质。

# Unit 2 Water and Wastewater

## 6. Measurement of Water Quality

Quantitative measurements of pollutants are obviously necessary before water pollution can be controlled. Measurement of these pollutants is, however, fraught with difficulties.

The first problem is that the specific materials responsible for the pollution are sometimes not known. The second difficulty is that these pollutants are generally at low concentrations, and very accurate methods of detection are therefore required.

Only a few of the many analytical tests available to measure water pollution are discussed in this chapter. A complete volume of analytical techniques used in water and wastewater engineering is compiled as *Standard Methods*. This volume, now in its 15th edition, is the result of a need for standardizing test techniques. It is considered definitive in its field and has the weight of legal authority.

Many of the pollutants are measured in terms of milligrams of the substance per liter of water (mg/L). This is a weight/volume measurement. In many older publications pollutants are measured as parts per million (ppm), a weight/weight parameter. If the liquid involved is water these two units are identical, since 1 milliliter of water weighs 1 gram. Because of the possibility of some wastes not having the specific gravity of water, the ppm measure has been scrapped in favor of mg/L(1).

A third commonly used parameter is percent, a weight/weight relationship. Obviously 10,000 ppm = 1 percent and this is equal to 10,000 mg/L only if 1 mL = 1g.

### Sampling

Some tests require the measurement to be conducted in the stream since the process of obtaining a sample may change the measurement. For example, if it is necessary to measure the dissolved oxygen in a stream, the measurement should be conducted right in the stream, or the sample must be extracted with great care to assure that no transfer of oxygen between the air and water (in or out) has occurred(2).

Most tests can be performed on a water sample taken from the stream. The process by which that sample is obtained, however, can greatly influence the result.

There are basically three types of samples:

1. grab.
2. composite.
3. flow weighed composite.

The grab, as the name implies, simply measures a point. Its value is that it represents accurately the water quality at the moment of sampling, but obviously says nothing about the quality before or after the sampling.

The composite sample is obtained by taking a series of grab samples and mixing them together. The flow weighed composite is obtained by taking each sample so that the volume of the sample is proportional to the flow at that time. The last method is especially useful when daily loadings to wastewater treatment plants are calculated.

Whatever the technique or method, however, it is necessary to recognize that the analysis can only be as accurate as the sample, and often the sampling methodology is far more sloppy than the analytical determination.

## Dissolved Oxygen

Probably the most important measure of water quality is the dissolved oxygen. Oxygen, although poorly soluble in water, is fundamental to aquatic life. Without free dissolved oxygen, streams and lakes become uninhabitable to most desirable aquatic life. Yet, the maximum oxygen that can possibly be dissolved in water at normal temperatures is about 9 mg/L, and this saturation value decreases rapidly with increasing water temperature, as shown in Table 1. The balance between saturation and depletion is therefore tenuous.

**Table 1 Solubility of Oxygen**

| Temperature of Water/℃ | Saturation Concentration of Oxygen in Water /(mg/L) |
|---|---|
| 0 | 14.6 |
| 2 | 13.8 |
| 4 | 13.1 |
| 6 | 12.3 |
| 8 | 11.9 |
| 10 | 11.3 |
| 12 | 10.8 |
| 14 | 10.4 |
| 16 | 10.0 |
| 18 | 9.5 |
| 20 | 9.2 |
| 22 | 8.8 |
| 24 | 8.5 |
| 26 | 8.2 |
| 28 | 8.0 |
| 30 | 7.6 |

The amount of oxygen dissolved in water is usually measured either by an oxygen probe or the old standard wet technique, the Winkler Dissolved Oxygen Test. The Winkler test for dissolved oxygen, developed more than 80 years ago, is the standard to which all other methods are compared.

Chemically simplified reactions in the Winkler test are as follows:

1. Manganese ions added to the samples combine with the available oxygen forming a precipitate

$$Mn^{2+} + O_2 \longrightarrow MnO_2 \downarrow$$

2. Iodide ions are added, and the manganous oxide reacts with the iodide ions to form iodine

$$MnO_2 + 2I^- + 4H^+ \longrightarrow Mn^{2+} + I_2 + 2H_2O$$

3. The quantity of iodine is measured by titrating with sodium thiosulfate, the reaction being

$$I_2 + 2S_2O_3^{2-} \longrightarrow S_4O_6 + 2I^-$$

Note that all of the dissolved oxygen combines with $Mn^{2+}$, so that the quantity of $MnO_2$ is directly proportional to the oxygen in solution. Similarly, the amount of iodine is directly proportional to the manganous oxide available to oxidize the iodide. Although the titration measures iodine, the quantity of iodine is thereby directly related to the original concentration of oxygen.

The Winkler test has obvious disadvantages, such as chemical interference and the necessity to either carry a wet laboratory to the field or bring the samples to the laboratory and risk the loss (or gain) of oxygen during transport. All of these disadvantages are overcome by using a dissolved oxygen electrode, often called a probe.

The simplest (and historically the first) probe is shown in Figure 1. The principle of operation is that of a galvanic cell. If lead and silver electrodes are put in an electrolyte solution with a microammeter between, the reaction at the lead electrode would be

$$Pb + 2OH^- \longrightarrow PbO + H_2O + 2e^-$$

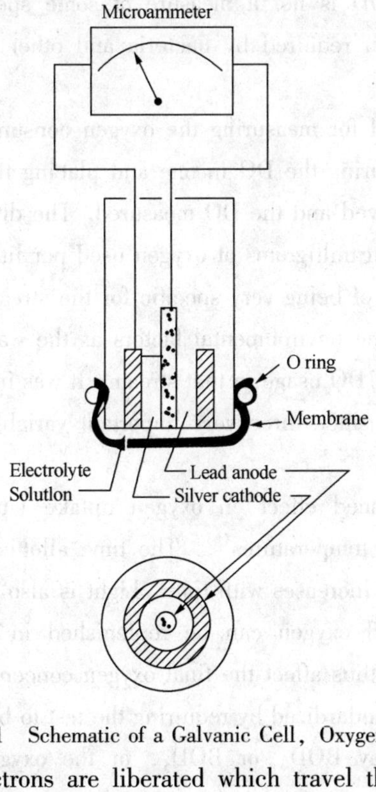


Figure 1 Schematic of a Galvanic Cell, Oxygen Probe

At the lead electrode, electrons are liberated which travel through the microammeter to the silver electrode where the following reaction takes place:

$$2e^- + \frac{1}{2} O_2 + H_2O \longrightarrow 2OH^-$$

The reaction would not go unless free dissolved oxygen is available, and the microammeter would not register any current. The trick is to construct and calibrate a meter in such a manner that the electricity recorded is proportional to the concentration of oxygen in the electrolyte solution(3).

In the commercial models the electrodes are insulated from each other with nonconducting plastic and are covered with a permeable membrane with a few drops of an electrolyte between the membrane and electrodes(4). The amount of oxygen that travels through the membrane is proportional to the DO concentration. A high DO in the water creates a strong push to get through the membrane, while a low DO would force only limited $O_2$ through to participate in the reaction and thereby create electrical current. Thus the current registered is proportional to the oxygen level in solution.

## Biochemical Oxygen Demand

Perhaps even more important than the determination of dissolved oxygen is the measurement of the rate at which this oxygen is used(5). A very low rate of use would indicate either clean water or that the available microorganisms are uninterested in consuming the available organics. A third possibility is that the microorganisms are dead or dying. (Nothing decreases oxygen consumption by aquatic microorganisms quite so well as a healthy slug of arsenic.)

The rate of oxygen use is commonly referred to as *biochemical oxygen demand* (BOD). It is important to understand that BOD is not a measure of some specific pollutant. Rather, it is a measure of the amount of oxygen required by bacteria and other microorganisms while stabilizing decomposable organic matter.

The BOD test was first used for measuring the oxygen consumption in a stream by filling two bottles with stream water, measuring the DO in one and placing the other in the stream. In a few days the second bottle was retrieved and the DO measured. The difference in the oxygen levels was the BOD, or oxygen demand, in milligrams of oxygen used per liter of sample.

This test had the advantage of being very specific for the stream in question since the water in the bottle is subjected to the same environmental factors as the water in the stream, and thus the result was an accurate measure of DO usage in that stream. It was impossible, however, to compare the results in different streams, since three very important variables were not constant: temperature, time and light.

Temperature has a pronounced effect on oxygen uptake (usage), with metabolic activity increasing significantly at higher temperatures(6). The time allotted for the test is also important, since the amount of oxygen used increases with time. Light is also an important variable since most natural waters contain algae and oxygen can be replenished in the bottle if light is available. Different amounts of light would thus affect the final oxygen concentration.

The BOD test was finally standardized by requiring the test to be run in the dark at 20℃ for five days. This is defined as five-day BOD, or $BOD_5$, or the oxygen used in the first five days. Although there appear to be some substantial scientific reasons why five days was chosen(7), it has been suggested that the possibility of preparing the samples on a Monday and taking them out on

Friday, thus leaving the weekend free, was not the least important of these reasons.

The BOD test is almost universally run using a standard BOD bottle (about 300 ml volume) as shown in Figure 2. It is of course also possible to have a 2-day, 10-day, or any other day BOD. One measure used in some cases is ultimate BOD or the $O_2$ demand after a very long time.

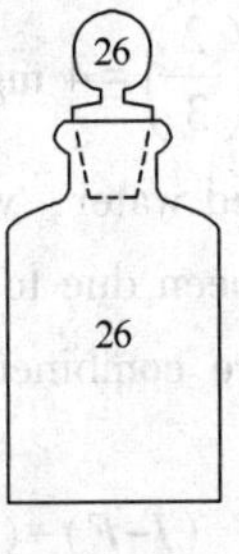


Figure 2 A BOD Bottle

If we measure the oxygen in several samples every day for five days, we may obtain curves such as Figure 3. Referring to this figure, sample A had an initial DO of 8 mg/L, and in five days this has dropped to 2 mg/L. The BOD therefore is 8-2 =6 mg/L.

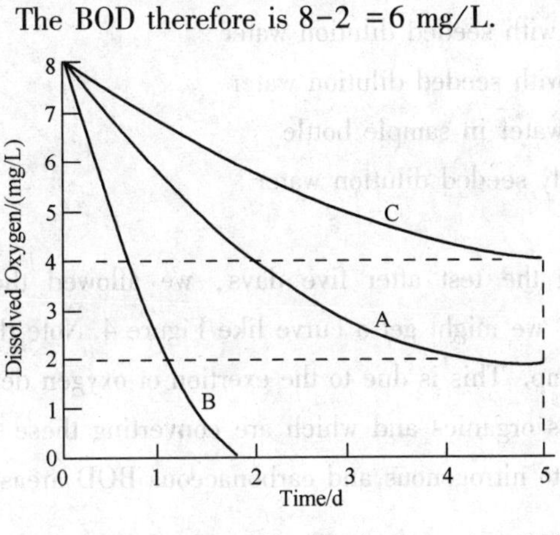


Figure 3 Typical Oxygen Uptake (Use) Curves in a BOD Test.

Sample B also had an initial DO of 8 mg/L, but the oxygen was used up so fast that it dropped to zero. If after five days we measure zero DO, we know that the BOD of sample B was more than 8 -0=8 mg/L, but we don't know how much more since the organisms might have used more DO if it were available. For samples containing more than about 8 mg/L, dilution of the sample is therefore necessary.

Suppose sample C shown on the graph is really sample B diluted by 1 :10. The BOD of sample B is therefore

$$\frac{8-4}{0.1}=40 \text{ mg/L}$$

It is possible to measure the BOD of any organic material (e. g., sugar) and thus estimate its influence on a stream, even though the material in its original state might not contain the necessary organisms. Seeding is a process in which the microorganisms which create the oxygen intake are added to the BOD bottle.

Suppose we used the water previously described by the A curve as seed water since it obviously contains microorganisms (it has a 5-day BOD of 6 mg/L) We now put 100 mL of an unknown solution into a bottle and add 200 mL of seed water, thus filling the 300-mL bottle. Assuming that the initial DO of this mixture was 8 mg/L and the final DO was 1 mg/L, the total oxygen used was 7 mg/L. But some of this was due to the seed water, since it also has a BOD, and only a portion was due to the decomposition of the unknown material. The DO uptake due to the seed water was

$$6\times\left(\frac{2}{3}\right)=4\ \mathrm{mg/L}$$

since only 2/3 of the bottle was the seed water, which has a BOD of 6 mg/L. The remaining oxygen uptake (7−4=3 mg/L) must have been due to the unknown material.

If the seeding and dilution methods are combined, the following general formula is used to calculate the BOD:

$$\mathrm{BOD\ (mg/L)}=\frac{(I-F)-(I'-F')(X/Y)}{D}$$

where $I$ = initial DO of bottle with sample and seeded dilution water

$F$ = final DO of bottle with sample and seeded dilution water

$I'$ = initial DO of bottle with seeded dilution water

$F'$ = final DO of bottle with seeded dilution water

$X$ = ml seeded dilution water in sample bottle

$Y$ = ml in bottle with only seeded dilution water

$D$ = dilution of sample.

If, instead of stopping the test after five days, we allowed the reactions to proceed and measured the DO each day, we might get a curve like Figure 4. Note that some time after five days the curve takes a sudden jump. This is due to the exertion of oxygen demand by the microorganisms which decompose nitrogenous organics and which are converting these to the stable nitrate, $NO_3^-$. The curve is thus divided into nitrogenous and carbonaceous BOD areas. Note also the definition of the ultimate BOD.

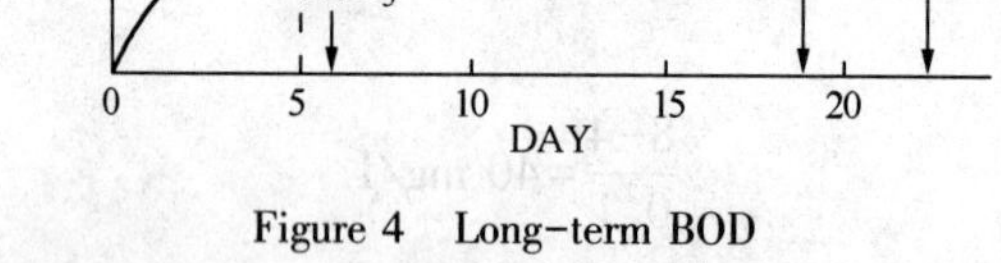

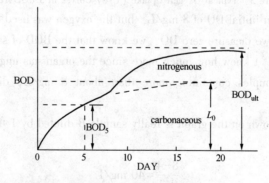


Figure 4 Long-term BOD

Note that BOD$_{ult}$ includes nitrogenous as well as the ultimate carbonaceous BOD ($L_0$)

It is important to remember that BOD is a measure of oxygen use, or potential use. An effluent with a high BOD can be harmful to a stream if the oxygen consumption is great enough to cause anaerobic conditions. Obviously, a small trickle going into a great river will have negligible effect, regardless of the mg/L of BOD involved. Similarly, a large flow into a small stream can seriously affect the stream even though the BOD might be low. Accordingly, engineers often talk of "pounds of

BOD" a value calculated by multiplying the concentration by the flow rate, with a conversion factor, so that

$$\text{lb BOD/day} = [\text{mg/L BOD}] \times [\text{flow in million gallons per day}] \times 8.34$$

The BOD of most domestic sewage is about 250 mg/L. Many industrial wastes run as high as 30,000 mg/L. The potential detrimental effect of an untreated dairy waste which might have a BOD of 20,000 mg/L is quite obvious.

As discussed in Chapter 2, the carbonaceous part of the BOD curve can be modeled using the equation

$$y = L_0(1 - e^{-k'_1 t})$$

where $y = \text{BOD}_t$, or the amount of oxygen demanded by the microorganisms at some time t, mg/L

$L_0$ = the ultimate demand for oxygen, mg/L

$k'_1$ = rate constant, previously termed the deoxygenation constant, $\text{days}^{-1}$

$t$ = time, days.

It should be again emphasized that this equation is applicable only to the carbonaceous BOD curve, and $L_0$ is defined as the ultimate carbonaceous BOD.

It is often necessary, such as when modeling the DO profile in a stream, to know both $k'_1$ and $L_0$. This must be done using laboratory BOD tests.

There are a number of techniques for calculating $k'_1$ and $L_0$. One of the simplest is a method devised by Thomas. Starting with the equation

$$y = L_0(1 - 10^{-k'_1 t})$$

we can rearrange it to read

$$\left(\frac{t}{y}\right)^{1/3} = (2.3k_1 L_0)^{-1/3} + \left(\frac{k_1^{2/3}}{3.43 L_0^{1/3}}\right) t$$

This equation is in the form of a straight line

$$x = a + bt$$

where $x = (t/y)^{1/3}$

$a = (2.3kL_0)^{-1/3}$

$b = k_1^{2/3} / (3.43 L_0^{1/3})$

Thus plotting $x$ vs $t$, the slope ($b$) and intercept ($a$) can be obtained, and

$$k_1 = 2.61(b/a)$$

$$L_0 = 1/(2.3 k_1 a^3)$$

For streams and rivers with travel times greater than about five days, the ultimate demand for oxygen must include the nitrogenous demand. Although the use of $\text{BOD}_{\text{ult}}$ (carbonaceous plus nitrogenous) in dissolved oxygen sag calculations is not strictly accurate, it is often assumed that the ultimate BOD can be calculated as

$$\text{BOD}_{\text{ult}} = a(\text{BOD}_5) + b(\text{KN})$$

where KN = Kjeldahl nitrogen (organic plus ammonia mg/L)

$a$ and $b$ = constants

The state of North Carolina, for example, uses $a = 1.2$ and $b = 4.0$ for calculating the ultimate

BOD, which is then substituted for $L_0$ in the dissolved oxygen sag equation (Figure 5).

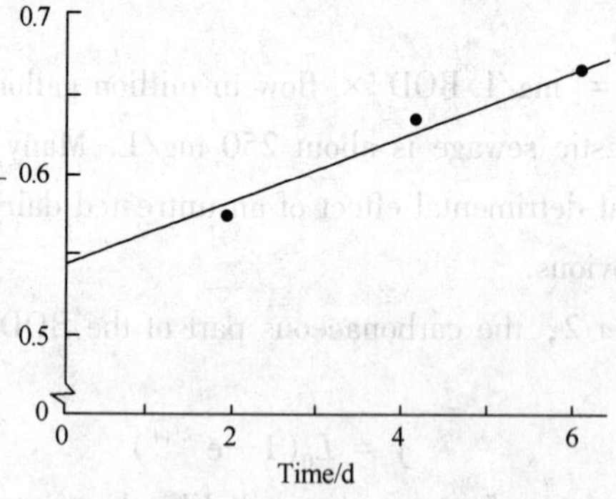


Figure 5 Calculation of $k_1$ and $L_0$ for Example

## Chemical Oxygen Demand

Among the many drawbacks of the BOD test the most important is that it takes five days to run[8]. If the organics were oxidized chemically instead of biologically, the test could be shortened considerably. Such oxidation is accomplished with the chemical oxygen demand (COD) test. Because nearly all organics are oxidized in the COD test and only some are decomposed during the BOD test, COD values are always higher than BOD values. One example of this is wood pulping wastes where compounds such as cellulose are easily oxidized chemically (high COD) but are very slow to decompose biologically (low BOD).

Potassium dichromate is generally used as an oxidizing agent. It is an inexpensive compound which is available in very pure form. A known amount of this compound is added to a measured amount of sample and the mixture boiled. The reaction, in unbalanced form is

$$\underset{\text{(organic)}}{C_xH_yO_z} + \underset{\text{(dichromate)}}{Cr_2O_7^{2-}} + H^+ \xrightarrow{\triangle} CO_2 + H_2O + Cr^{3+}$$

After boiling with an acid, the excess dichromate (not used for oxidizing) is measured by adding a reducing agent, usually ferrous ammonium sulfate. The difference between the chromate originally added and the chromate remaining is the chromate used for oxidizing the organics. The more chromate used, the more organics were in the sample, and hence the higher the COD[9].

## Phosphates

The importance of phosphorus compounds in the aquatic environment is discussed in Chapter 2. Phosphorus in wastewater can be either inorganic or organic. Although the greatest single source of inorganic phosphorus is synthetic detergents, organic phosphorus is found in food and human waste as well. All phosphates in nature will, by biological action, eventually revert to inorganic forms to be again used by the plants in making high-energy material[10].

Ever since phosphates were indicted as one culprit in lake eutrophication, the measurement of total phosphate has assumed considerable importance[11]. Total phosphates can be measured by first boiling the sample in acid solution, which converts all the phosphates to the inorganic forms. From that point the test is colorimetric, using a chemical which when combined with phosphates produces a color directly proportional to the phosphate concentration.

## Bacteriological Measurements

From the public health standpoint the bacteriological quality of water is as important as the chemical quality. A number of diseases can be transmitted by water, among them typhoid and cholera. However, it is one thing to declare that water must not be contaminated by pathogens (disease-causing organisms) and another to determine the existence of these organisms[12]. First, there are many pathogens. Each has a specific detection procedure and must be screened individually. Second, the concentration of these organisms can be so small as to make their detection impossible. It is a perfect example of the proverbial needle in a haystack[13]. And yet only one or two organisms in the water might be sufficient to cause an infection.

How then can we measure for bacteriological quality? The answer lies in the concept of indicator organisms. The indicator most often used is a group of microbes called coliforms which are organisms normal to the digestive tracts of warm-blooded animals. In addition to that attribute, coliforms are:

- plentiful, hence not difficult to find
- easily detected with a simple test
- generally harmless except in unusual circumstances
- hardy, surviving longer than most known pathogens.

Coliforms have thus become universal indicator organisms. But the presence of coliforms does not prove the presence of pathogens. If a large number of coliforms are present, there is a good chance of recent pollution by wastes from warm-blooded animals, and therefore the water may contain pathogenic organisms.

This last point should be emphasized. The presence of coliforms does not mean that there are pathogens in the water. It simply means that there might be. A high coliform count is thus suspicious and the water should not be consumed (although it may be perfectly safe).

There are three ways of measuring for coliforms. The simplest is to filter a sample through a sterile filter, thus capturing any coliforms. The filter is then placed in a petri dish containing a sterile agar which soaks into the filter and promotes the growth of coliforms while inhibiting other organisms. After 24 or 48h of incubation the number of shiny black dots, indicating coliform colonies, is counted. If we know how many milliliters were poured through the filter, the concentration of coliforms can be expressed as coliforms/mL.

The second method of measuring for coliforms is called the most probable number (MPN), a test based on the fact that in a lactose broth coliforms will produce gas and make the broth cloudy. The production of gas is detected by placing a small tube upside down inside a larger tube (Figure 6) so as not to have air bubbles in the smaller tube. After incubation, if gas is produced, some of it will become trapped in the smaller tube and this, along with a cloudy broth, will indicate that the tube had been inoculated with at least one coliform.

And here is the trouble. One coliform can cause a positive tube just as easily as a million coliforms can. Hence it is not possible to ascertain the concentration from just one tube. We get around[13] this problem by inoculating a series of tubes with various quantities of sample, the reasoning being that a 10-m/L sample would be more likely to contain a coliform than a 1-mL sample.

For example, if we take three different inoculation amounts, 10, 1 and 0.1 mL of sample, and inoculate three tubes with each amount, after incubation we might have an array such as follows. The plus signs indicate a positive test (cloudy broth with gas formation) and the minus sign represent tubes where no coliforms were found.

| Amount of Sample, | Tube Number | | |
|---|---|---|---|
| mL put in test tube | 1 | 2 | 3 |
| 10 | + | + | + |
| 1 | − | + | − |
| 0.1 | − | − | + |

Based on these data we would suspect that there is at least 1 coliform per 10 mL, but we still have no firm number.

The solution to this dilemma lies in statistics. It can be proven statistically that such an array of positive and negative results will occur most probably if the coliform concentration was 75 coli/100mL. A higher concentration would most probably have resulted in more positive tubes while a lower concentration would most probably have resulted in more negative tubes. This is how MPN is established.

A third way of measuring coliforms is by a proprietary device called a "Coli-Count". A sterile pad with all the necessary nutrients is dipped into the water sample, incubated, and the colonies counted. The pad is designed to adsorb exactly 1mL of sample water so that the colonies counted give a coliform concentration per mL.

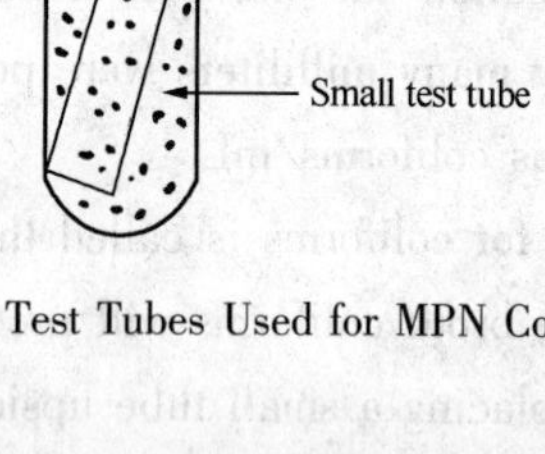

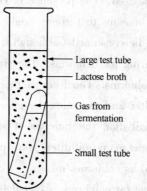


Figure 6 Test Tubes Used for MPN Coliform Test

## Words and Expressions

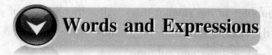

1. water quality 水质
2. quantitative measurement 定量测定
3. be fraught with 充满
4. weight of legal authority 合法权威的分量

5. in terms of milligrams of the substance per liter of water 以每升水中物质的毫克数计
6. parts per million (ppm) 百万分之几
7. milliliter 毫升
8. specific gravity 相对密度，比重
9. scrap 废弃；废料
10. in favor of 采用，废弃(……)而采用
11. grab sample 临时样品，单一样品
12. composite sample 混合样品
13. flow weighed composite sample 考虑流量的混合样品，流量加权的混合样品
14. be proportional to 与……成比例
15. be directly proportional to 与……成正比
16. daily loadings to wastewater treatment plant 废水处理厂的日负荷
17. sloppy 粗糙的，马虎的
18. be fundamental to 对……基本的
19. solubility 溶解度，溶解性
20. saturation concentration 饱和浓度
21. at normal temperature 常温下
22. saturation value 饱和值
23. decrease rapidly with increasing water temperature 随水温的升高而迅速下降
24. balance between saturation and depletion 饱和值和亏空值之间的差额
25. tenuous 细微的，薄弱的
26. oxygen probe 氧探头，氧试探电极
27. standard wet technique 标准湿法技术
28. Winkler Dissolved Oxygen Test Winkler 溶解氧测试法
29. manganese ion 锰离子
30. combine with the available oxygen forming a precipitate 与游离氧化合生成沉淀物
31. iodide ion 碘离子
32. manganous oxide 一氧化锰
33. iodine 碘
34. titrate 滴定
35. titration 滴定
36. sodium thiosulfate 硫代硫酸钠
37. original concentration of oxygen 氧的初始浓度
38. chemical interference 化学干扰
39. dissolved oxygen electrode 溶解氧电极
40. principle of operation 操作原理，工作原理
41. galvanic cell 原电池
42. electrolyte solution 电解质溶液
43. microammeter 微安计
44. between 在其间，在其中
45. register 记录，指示
46. calibrate 标刻度，校准，标定
47. commercial model 商品型号
48. permeable membrane 可透膜
49. participate in 参与
50. electrical current 电流
51. biochemical Oxygen Demand (BOD) 生化需氧量
52. decomposable organic matter 可分解有机物
53. retrieve 收回，取回
54. in milligrams of oxygen used per liter of sample 以每升样品中耗用的氧的毫克数计
55. in question 待测的，讨论中的
56. be subjected to 遭受到，经受，承受
57. same environmental factors as the water in the stream 与河水同样的环境因素
58. variable 可变的；变量
59. pronounced effect on oxygen uptake 对氧吸收的显著影响
60. metabolic activity 新陈代谢活性
61. time allotted for the test 分配给测试的时间，测试所用的时间
62. algae 藻类
63. replenish 补充
64. ultimate BOD 最终 BOD
65. an initial DO of 8 mg/L 初始 8 mg/L 的 DO
66. dilution 稀释
67. seeding 播种，接种
68. oxygen intake 氧的吸收

69. general formula 一般式
70. nitrogenous organics 含氮有机物
71. nitrogenous BOD 含氮 BOD
72. carbonaceous BOD 含碳 BOD
73. effluent 出水，出流，流出物
74. anaerobic condition 厌氧状况，厌氧条件
75. trickle 滴流
76. negligible effect 可忽略不计的影响
77. regardless of 不管，不顾
78. multiply the concentration by the flow rate, with a conversion factor 将浓度与流量及转换因子相乘
79. domestic sewage 生活污水，家庭污水
80. dairy waste 牛奶场污水[废物]
81. model 模式，模型
82. ultimate demand for oxygen 最终需氧量
83. rate constant 速率常数
84. deoxygenation constant 消氧常数，脱氧常数
85. model the DO profile in a stream 模拟河流中的 DO 分布
86. start with the equation 从方程着手
87. in the form of 以……的形式
88. straight line 直线
89. plot Y vs X 将 Y 对 X 作(曲线)图
90. slope 斜率
91. intercept 截距
92. travel time 流行时间
93. dissolved oxygen sag curve 溶解氧下垂曲线，氧垂曲线
94. dissolved oxygen sag equation 溶解氧下垂方程，氧垂方程
95. Kjeldahl nitrogen 凯氏氮
96. substitute A for B 用 A 代替 B
97. Chemical Oxygen Demand (COD) 化学需氧量
98. drawback 缺陷，缺点
99. wood pulping waste 木浆废水
100. cellulose 纤维素
101. potassium dichromate 重铬酸钾
102. oxidizing agent 氧化剂
103. reducing agent 还原剂
104. ferrous ammonium sulfate 硫酸亚铁胺
105. chromate 铬酸盐
106. phosphate 磷酸盐
107. phosphorus compound 含磷化合物
108. synthetic detergent 合成洗涤剂
109. revert to 回复到
110. culprit 肇事者，肇事原因
111. eutrophication 富营养化
112. colorimetric 比色的
113. bacteriological measurement 细菌学测定
114. public health standpoint 公共卫生观点
115. bacteriological quality 细菌学质量
116. transmit 传播
117. pathogen 致病体，病原体
118. screen 筛分
119. proverbial needle in a haystack 大海捞针(谚语)
120. infection 传染
121. indicator organism 指示生物体
122. microbe 微生物
123. coliform 大肠杆菌
124. digestive tract 消化道，消化系统
125. warm-blooded animal 温血动物
126. attribute 特性，性质；把……归因于
127. pathogenic organism 致病有机体
128. coliform count 大肠菌计数
129. suspicious 可疑的，(引起)怀疑的
130. filter a sample through a sterile filter 用无菌过滤器过滤样品
131. capture 截获，捕集
132. petri dish 培养皿
133. sterile agar 无菌的琼脂
134. soak into the filter 渗入到滤层中
135. incubation 培养，孵化，保温
136. shiny black dot 发光的黑点
137. coliform colony 大肠杆菌菌落
138. Most Probable Number (MPN) 最概然数，最可能的数(目)

139. lactose broth 乳糖肉汤
140. make the broth cloudy 使肉汤变混浊
141. trap 截获，捕集
142. positive tube 阳性试管
143. get around 克服(困难)
144. ascertain 确定，弄清，查明
145. inoculate 给……接种，移植，嫁接
146. positive test 阳性试验
147. suspect 怀疑
148. dilemma 困境，困难
149. statistics 统计学
150. proprietary device 专利装置
151. sterile pad 无菌垫片[薄片]

## Notes

(1) Because of the possibility of some wastes not having the specific gravity of water, the ppm measure has been scrapped in favor of mg/L.

由于某些废水可能与水的相对密度不同，因此，倾向于用 mg/L 而不用 ppm 的测定。

(2) For example, if it is necessary to measure the dissolved oxygen in a stream, the measurement should be conducted right in the stream, or the sample must be extracted with great care to assure that no transfer of oxygen between the air and water (in or out) has occurred.

例如，如果必须测定河流中的溶解氧，测定应当就在河流中进行，或者必须非常谨慎地采样以保证空气和水之间没有氧的转移(溶入或逸出)。

(3) The trick is to construct and calibrate a meter in such a manner that the electricity recorded is proportional to the concentration of oxygen in the electrolyte solution.

以这样的方式构造和校准仪表，使其记录的电流与电解质溶液中的氧浓度成正比。

(4) In the commercial models the electrodes are insulated from each other with nonconducting plastic and are covered with a permeable membrane with a few drops of an electrolyte between the membrane and electrodes.

在商品仪表中，电极用不导电的塑料彼此绝缘，并用可透膜覆盖，在膜和电极间加入几滴电解质溶液。

(5) Perhaps even more important than the determination of dissolved oxygen is the measurement of the rate at which this oxygen is used.

或许比测定溶解氧更重要的是测定溶解氧耗用的速率。本句为倒装句。

(6) Temperature has a pronounced effect on oxygen uptake (usage), with metabolic activity increasing significantly at higher temperatures.

温度对氧的吸收(耗用)有显著的影响，在较高温度下新陈代谢活性显著增加。

(7) Although there appear to be some substantial scientific reasons why five days was chosen, it has been suggested that the possibility of preparing the samples on a Monday and taking them out on Friday, thus leaving the weekend free, was not the least important of these reasons.

尽管选择五日似乎有一些重要的科学理由，但在星期一准备样品，在星期五取出样品，从而留出空闲的周末可能是重要的原因。

(8) Among the many drawbacks of the BOD test the most important is that it takes five days to run.

在 BOD 测试法的许多缺陷中，最大的缺陷是它需要五天时间完成。

(9) The more chromate used, the more organics were in the sample, and hence the higher the COD.

铬酸盐用得越多，样品中的有机物就越多，因此，COD 就会越高。

(10) All phosphates in nature will, by biological action, eventually revert to inorganic forms to be again used by the plants in making high-energy material.

实质上所有的磷酸盐通过生物作用最终都会回复到无机磷形式，在构造高能物质中为植物重新使用。

(11) Ever since phosphates were indicted as one culprit in lake eutrophication, the measurement of total phosphate has assumed considerable importance.

自从磷酸盐被认为是湖泊富营养化的一个肇事原因，总磷酸盐的测定就显得相当重要。

(12) However, it is one thing to declare that water must not be contaminated by pathogens (disease-causing organisms) and another to determine the existence of these organisms.

然而，声称水不得被病原体(引起疾病的生物体)污染是一回事，确定这些生物体的存在是另一回事。

(13) It is a perfect example of the proverbial needle in a haystack.

这是谚语“大海捞针”的极好例子。

## 7. Water Treatment

Many aquifers and isolated surface waters are high in water quality and may be pumped from the supply and transmission network directly to any number of end uses, including human consumption, irrigation, industrial processes or fire control. However, such clean water sources are the exception to the rule in many regions of the nation, particularly regions with dense populations or regions that are heavily agricultural. Here, the water supply must receive varying degrees of treatment prior to distribution.

Impurities enter the water as it moves through the atmosphere, across the earth's surface, and between soil particles in the ground(1). These background levels of impurities are often supplemented by man's activities. Chemicals from industrial discharges and pathogenic organisms of human origin, if allowed to enter the water distribution system, can cause health problems. Excessive silt and other solids can make the water both unsightly and aesthetically unpleasing. Water can be contaminated by many routes. For example, heavy metal pollution, including lead, zinc and copper, can be caused by corrosion of the very pipes which carry the water from its source to the consumer(2).

The method and degree of water treatment are important considerations for environmental engineers. Generally speaking, the characteristics of raw water determine the method of treatment. Because most public supply systems are relied on for drinking water, as well as industrial and fire consumption, the highest level of use, human consumption, defines the degree of treatment(3). Thus, we focus only on treatment technologies which produce potable water.

A typical water treatment plant is diagrammed in Figure 1. These plants are designed to remove odors, color and turbidity as well as bacteria and other contaminants from surface water. Raw surface water entering a water treatment plant usually has significant turbidity caused by tiny colloidal clay and silt particles. These particles have a natural electrostatic charge which keeps them continually in motion and prevents them from colliding and sticking together. Chemicals such as alum (aluminum sulfate) are added to the water, first to neutralize the charge on the particles and then to aid in making the tiny particles "sticky" so they can coalesce and form large particles called flocs. This process is called *coagulation and flocculation*.

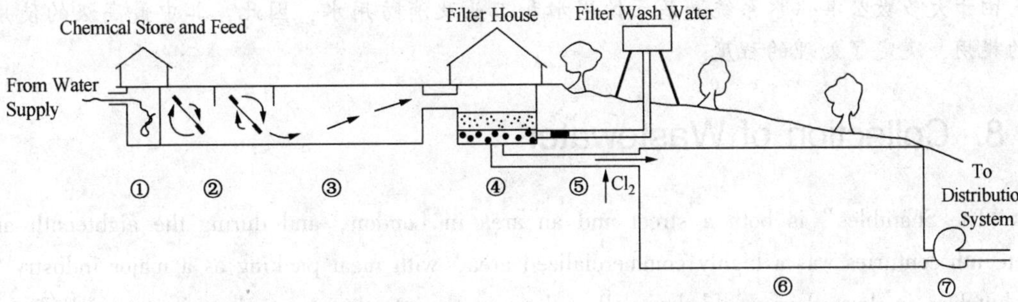


Figure 1　Movement of Water through a Typical Water Treatment Plant

1—Chemical mixing basin; 2—Flocculation basin; 3—Settling tank; 4—Rapid sand filter

5—Disinfection with chlorine; 6—Clean water storage basin (Clear well); 7—Pump

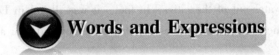

## Words and Expressions

1. aquifer 含水层，蓄水层
2. any number 许多
3. end use 终端用户，终端耗用
4. irrigation 灌溉
5. fire control 消防，防火，火灾控制
6. impurity 杂质
7. background level 本底含量
8. industrial discharge 工业排放物
9. odor 味，气味
10. color 色，色度
11. turbidity 混浊度
12. colloidal clay 胶态黏土
13. silt particle 泥滓颗粒，泥沙颗粒
14. electrostatic charge 静电荷
15. coalesce 黏结，黏合
16. floc 絮体，矾花
17. coagulation 混凝，凝聚
18. flocculation 絮凝
19. chemical mixing basin 化学混合池
20. flocculation basin 絮凝池
21. settling tank 沉淀池，沉降池
22. rapid sand filter 快速砂滤池
23. disinfection with chlorine 用氯消毒
24. clean water storage basin 清水储槽

## Notes

(1) Impurities enter the water as it moves through the atmosphere, across the earth's surface, and between soil particles in the ground.

当水穿过大气、流经地表和穿梭在地下土壤颗粒之间时，杂质会进入水体。

(2) For example, heavy metal pollution, including lead, zinc and copper, can be caused by corrosion of the very pipes which carry the water from its source to the consumer.

例如，重金属污染，包括铅、锌和铜，可能正是由将水从源头输送到用户的管道的腐蚀引起的。

(3) Because most public supply systems are relied on for drinking water, as well as industrial and fire consumption, the highest level of use, human consumption, defines the degree of treatment.

由于大多数公共供水系统是基于饮用水和工业及消防用水，因此，其中最高级的使用，人的耗用，决定了处理的程度。

## 8. Collection of Wastewater

"The Shambles" is both a street and an area in London, and during the eighteenth and nineteenth centuries was a highly commercialized area, with meat packing as a major industry(1). The butchers in those days would throw all of their wastes into the street where it was washed away by rainwater into drainage ditches. The condition of the area was so bad that it contributed its name to the English language.

In old cities, drainage ditches like the ones at the Shambles were constructed for the sole purpose of moving stormwater out of the cities. In fact, it was illegal in London to discard human excrement into these ditches. Eventually, these ditches were covered over and became what we now know as *storm sewers*.

As water supplies developed and the use of the indoor water closet increased, the need for transporting domestic wastewaters, called sanitary wastes, became obvious. This was accomplished in one of two ways: (1) discharge of the sanitary wastes into the storm sewers, which then carried both sanitary wastes and stormwater, and were known as *combined sewers*, and (2) construction of a new system of underground pipes for removing the wastewater, which became known as *sanitary sewers*.

Newer cities, and more recently built (post-1900) parts of older cities almost all have separate sewers for sanitary wastes and stormwater. In this chapter, storm sewer design is not covered in detail, since this is discussed in Chapter 9. Emphasis here is on estimating the quantities of domestic and industrial wastewaters, and in the design of the sewerage systems to handle these flows.

### Estimating Wastewater Quantities

The term sewage is used here to mean only domestic wastewater. In addition to sewage, however, sewers also must carry

- industrial wastes
- infiltration
- inflow

The quantity of industrial wastes can usually be established by water use records. Alternatively, the flows can be measured in manholes which serve only a specific industry, using a small flow meter in a manhole. Typically, a parshall flume is used, and the flow rate is calculated as a direct proportion of the flow depth[2]. Industrial flows often vary considerably throughout the day and continuous recording is mandatory.

Infiltration is the flow of groundwater into sanitary sewers. Sewers are often placed under the groundwater table and any cracks in the pipes will allow water to seep in. Infiltration is the least for new, well-constructed sewers, and can go as high as 500 $m^3$/km-day (200, 000 gal/mi-day) Commonly, for older systems, 700 $m^3$/km-day (300, 000 gal/mi-day) is used in estimating infiltration. This flow is of course detrimental since the extra volume of water must go through the sewers and the wastewater treatment plant. It thus makes sense to reduce this as much as possible by maintaining and repairing sewers, and keeping sewerage easements clear of large trees which could send roots into the sewers and cause severe damage.

The third source of flow in sanitary sewers is called *inflow*, and represents stormwater which is collected unintentionally by the sanitary sewers. A common source of inflow is a perforated manhole cover placed in a depression, so that stormwater flows into the manhole. Sewers laid next to creeks and drainageways which rise up higher than the manhole elevation, or where the manhole is broken, are also a major source. Lastly, illegal connections to sanitary sewers, such as roofdrains, can substantially increase the wet weather flow over the dry weather flow. Commonly, the ratio of dry weather to wet weather flow is between 1 :1. 2 and 1 :4.

Domestic wastewater flows vary with season, day of the week and the hour of the day. Figure 1 shows a typical daily flow for a residential area.

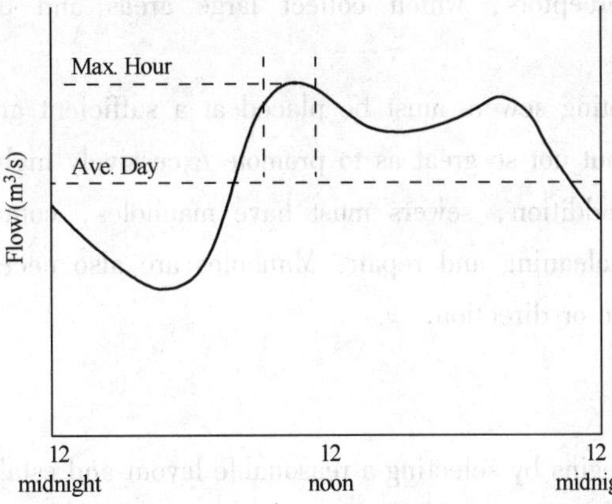


Figure 1 Typical Dry-weather Wastewater Flow for a Residential Area

The three flows of concern when designing sewers are the average flow, the peak or maximum flow, and the extreme minimum[3]. The ratios of average to both the maximum and minimum flows is a function of the total flow, since a higher average daily discharge implies a larger community in which the extremes are evened out. Figure 2 is a plot showing commonly experienced ratios of average to the extremes as a function of the average daily discharge.

## System Layout

Sewers that collect wastewater from residences and industrial establishments almost always operate as open channels, or gravity flow conduit. Pressure sewers are used in a few places, but

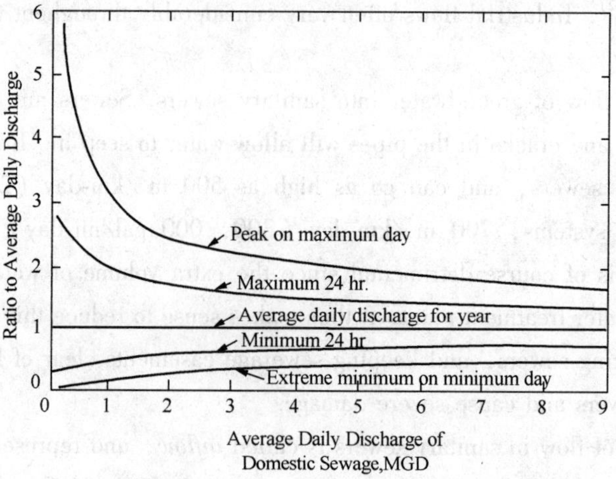


Figure 2 Relation of the Average Daily Flow of Domestic Wastewater to the Extremes of Flow

these are expensive to maintain and are useful only when there either are severe restrictions on water use, or the terrain is such that gravity flow conduits can not be efficiently constructed.

Building connections are usually made with clay or plastic pipe, 6 in. in diameter, to the collecting sewers which commonly run under the street(4). Collecting sewers are sized to carry the maximum anticipated peak flows without surcharging (filling up) and are commonly made of clay, asbestos, cement, concrete or cast iron pipe. They discharge in turn into intercepting sewers, known colloquially as interceptors, which collect large areas and discharge finally into the wastewater treatment plant.

Collecting and intercepting sewers must be placed at a sufficient grade to allow for adequate velocity during low flows, but not so great as to promote excessively high velocities when the flows are at their maximum. In addition, sewers must have manholes, commonly every 120–180 m (400–600 ft) to facilitate cleaning and repair. Manholes are also necessary whenever the sewer changes grade (slope), size or direction.

## Sewer Hydraulics

The design of sewers begins by selecting a reasonable layout and establishing the expected flows within each pipe linking the manholes. For large systems, this often involves the use of economic analysis to determine exactly what routing provides the optimal system. For most smaller systems, this is an unnecessary refinement, and a reasonable system can be drawn by eye.

The average discharge is estimated on the basis of the population served in the drainage area, and the maximum and minimum flows are calculated using Figure 2. Once this is done, the design is a search for the right pipe diameter and grade (slope) which will allow the minimum flow to exceed a velocity necessary for the conveyance of solids, while keeping the velocity at maximum flow less than a limit at which undue erosion and structural damage can occur to the pipes(5). Commonly, the velocity should be held between these limits:

- minimum–0.6 m/sec (~2 ft/sec)
- maximum– 3.0 m/sec ( ~10 ft/sec)

The velocity in sewers is usually calculated using the Manning equation, derived originally from the Chezy open channel flow equation.

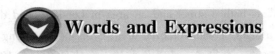

## Words and Expressions

1. butcher 屠夫
2. in those days 在那时
3. drainage ditch 排水沟，排水渠
4. move stormwater out of the city 将雨水从城中移出
5. discard 放弃，丢弃，抛弃
6. human excrement 人的排泄物
7. storm sewer 雨水管
8. domestic wastewater 生活污水，家庭废水
9. sanitary waste 生活污水[废物]
10. combined sewer 合并污水管道，合流污水管道
11. sanitary sewer 生活污水管道
12. infiltration 渗入流
13. inflow 入流
14. manhole 检查井，窨井，人孔
15. flow meter 流量计
16. parshall flume 巴氏水槽，细腰槽
17. flow rate 流量
18. throughout the day 整天，整日
19. mandatory 强制性的，必须遵循的，命令的，指示的
20. groundwater 地下水
21. groundwater table 地下水面
22. crack 缝隙，裂缝
23. seep 渗透
24. make sense 合理，有意义
25. maintain and repair sewer 维修污水管
26. sewerage easement 污水管附属设施
27. perforated manhole cover 开孔的人孔盖
28. creek 小溪流，小河
29. drainageway 排水道
30. wet weather flow 雨天水流[流量]
31. dry weather flow 晴天水流[流量]
32. system layout 系统布局
33. residence 住宅区
34. industrial establishment 工业企业
35. open channel 明渠
36. gravity flow conduit 自流管道，重力流管道
37. pressure sewer 压力管道
38. severe restriction 严格的限制
39. terrain 地域，地形
40. asbestos 石棉
41. cement 水泥
42. concrete 混凝土
43. cast iron 铸铁
44. intercepting sewer 截流污水管
45. at a sufficient grade 以足够的坡度
46. allow for adequate velocity 考虑足够的流速
47. facilitate 使容易，使便利；促进
48. typical packaged pumping station 典型的一体化泵站
49. sewer hydraulics 污水管水力学
50. economic analysis 经济分析
51. refinement 精加工，精处理
52. on the basis of 根据
53. population served in the drainage area 排水区服务人口
54. derive... from 从……导出，从……衍生出

## Notes

(1)"The Shambles" is both a street and an area in London, and during the eighteenth and nineteenth centuries was a highly commercialized area, with meat packing as a major industry.

"Shambles"既是伦敦的一个街名，又是伦敦的一个地区名，在18世纪和19世纪是一个高度商业化地区，以肉类包装加工为主要产业。

(2) Typically, a parshall flume is used, and the flow rate is calculated as a direct proportion of the flow depth.

一般采用一个细腰槽(巴氏槽)，流量通过与水流深度成正比来计算。

(3) The three flows of concern when designing sewers are the average flow, the peak or maximum flow, and the extreme minimum.

设计污水管时所关注的三个流量是平均流量，最大(或峰值)流量和最低流量。

(4) Building connections are usually made with clay or plastic pipe, 6 in. in diameter, to the collecting sewers which commonly run under the street.

建筑物(大楼)与收集污水管的连接管通常是直径为6英寸的陶土管或塑料管，收集污水管通常铺设在街道下。

(5) Once this is done, the design is a search for the right pipe diameter and grade (slope) which will allow the minimum flow to exceed a velocity necessary for the conveyance of solids, while keeping the velocity at maximum flow less than a limit at which undue erosion and structural damage can occur to the pipes.

一旦这一工作完成，设计就是寻求适当的管径和坡度，以在最低流量时有传输固体必须的最低流速，在最大流量时，流速小于会产生管道过分腐蚀和结构损坏的极限流速。

# 9. Wastewater Treatment Options

### Onsite Wastewater Disposal

Environmental engineers have been severely (and sometimes rightly) criticized for having a "sewer syndrome"-they want to collect all wastewater and provide the treatment at a central location. Often this approach does not make much sense.

Consider a situation depicted in Figure 1, where two wastewater treatment options are shown—a centralized treatment plant, and several smaller plants—all discharging their effluents into the same river. The single large plant obviously must provide extremely good treatment in order to attain acceptable dissolved oxygen levels downstream. On the other hand, the smaller plants could take advantage of the assimilative capacity of the river, and would not necessarily have to provide the same high degree of treatment. The logical extension of this idea is to not have any treatment plants at all, but dispose of the wastewater onsite, with each house or building having its own treatment system(1).

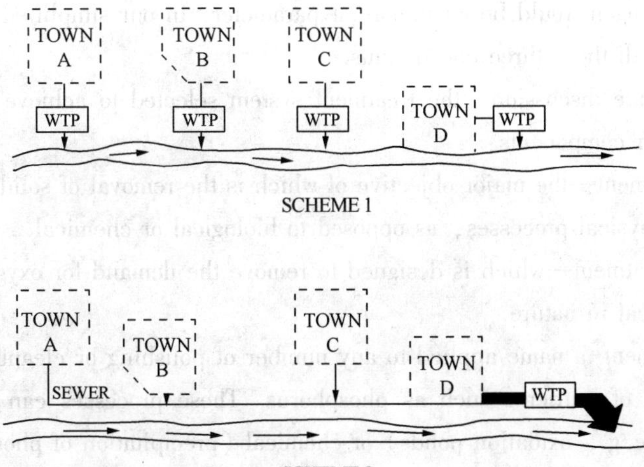


Figure 1 Two Wastewater Treatment Options for Several Small Communities

## Central Wastewater Treatment

The objective of wastewater treatment is to reduce the concentrations of specific pollutants to the level where the discharge of the effluent will not adversely affect the environment. Note that there are two important aspects of this objective. First, wastewater is treated only to reduce the concentrations of selected constituents which would cause harm to the environment or pose a health hazard. Not everything in wastewater is troublesome, and thus is not removed. Secondly, the reduction of these constituents is only to some required level. It is obviously technically possible to produce distilled and deionized $H_2O$ from wastewater, but this is not necessary, and can in fact be detrimental to the watercourse. Fish and other aquatic organisms cannot survive in distilled water.

For any given wastewater in a specific location, the *degree* and *type* of treatment therefore are variables which require engineering decisions.

Often, the degree of treatment is dictated by the assimilative capacity of the recipient(2). The procedure by which dissolved oxygen sag curves are drawn is reviewed in Chapter 2. As noted there, the amount of oxygen-demanding materials (BOD) discharged determines how far the dissolved oxygen level will be depressed. If this depression (deficit) is too large, some BOD must be removed in the treatment plant. Thus a certain plant on a given watercourse is required to produce a given quality of effluent. Such an *effluent standard* (discussed more fully in Chapter 10) dictates in large part the type of treatment required.

In order to facilitate the discussion of wastewater treatment, a "typical wastewater" will be assumed, and it will be further assumed that the effluent from this wastewater treatment must meet the following effluent standards:

BOD ≤ 15mg/L

SS ≤ 15mg/L

P ≤ 1mg/L

Obviously, other criteria might, in given situations, be important. For example, nitrogen is thought to be the limiting nutrient in estuarine waters, and if the discharge was to be a brackish estuary, the total nitrogen would be an important parameter. In our simplified case, however, we are concerned only with these three constituents.

To further facilitate discussion, the treatment system selected to achieve these effluent levels consists of three major components:

- primary treatment—the major objective of which is the removal of solids. Primary treatment systems are always physical processes, as opposed to biological or chemical.
- secondary treatment—which is designed to remove the demand for oxygen. These processes are commonly biological in nature.
- tertiary treatment-a name applied to any number of polishing or cleanup processes, one of which is the removal of nutrients such as phosphorus. These processes can be physical (e. g., filters), biological (e. g., oxidation ponds) or chemical (precipitation of phosphorus).

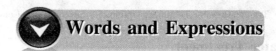

## Words and Expressions

1. onsite wastewater disposal 现场废水处置，就地废水处置
2. syndrome 综合症
3. centralized treatment plant 集中处理厂，中央处理站
4. discharge the effluents into the river 将出水排入河流
5. attain acceptable dissolved oxygen levels downstream 在下游达到可接受的溶解氧含量
6. take advantage of 利用
7. assimilative capacity 同化能力
8. objective of wastewater treatment 废水处理的目的
9. reduce the concentrations of specific pollutant 降低特定污染物的浓度
10. pose a health hazard 引起对健康的危害
11. troublesome 令人讨厌的，不好的
12. distilled water 蒸馏水
13. deionized water 去离子水
14. be detrimental to 对……有害的
15. be dictated by 受……支配，受……控制
16. recipient 接受体，受纳水体
17. depress 减少，减小，使……亏空
18. depression 减少，减小
19. deficit 亏空值，亏空
20. meet the effluent standards 符合出水标准，满足出水标准
21. criteria 标准，准则
22. estuarine water 河口水体
23. brackish estuary 含盐河口水体
24. primary treatment 一级处理
25. secondary treatment 二级处理
26. tertiary treatment 三级处理
27. polishing 精处理，深度处理
28. cleanup process 净化过程
29. oxidation pond 氧化塘

## Notes

(1) The logical extension of this idea is to not have any treatment plants at all, but dispose of

the wastewater onsite, with each house or building having its own treatment system.

这个思想的合理延伸就是不需要任何的处理厂，而是就地处置废水，每个住宅或建筑物都有其自己的处理系统。

(2) Often, the degree of treatment is dictated by the assimilative capacity of the recipient.

处理的程度常常受到接受水体的同化能力控制。

## 10. Primary Treatment

The most objectionable aspect of discharging raw sewage into watercourses is the floating material. It is only logical, therefore, that screens were the first form of wastewater treatment used by communities, and even today, screens are used as the first step in treatment plants. The purpose of a screen in modern treatment plants is the removal of materials which might damage equipment or hinder further treatment. In some older treatment plants screens are cleaned by hand, but mechanical cleaning equipment is used in almost all new plants. The cleaning rakes are automatically activated when the screens get sufficiently clogged to raise the water level in front of the bars(1).

In many plants, the next treatment step is *a comminutor*, a circular grinder designed to grind the solids coming through the screen into pieces about 0.3 cm (1/8 in.) or smaller. Many designs are in use; one common design is shown in Figure 1.

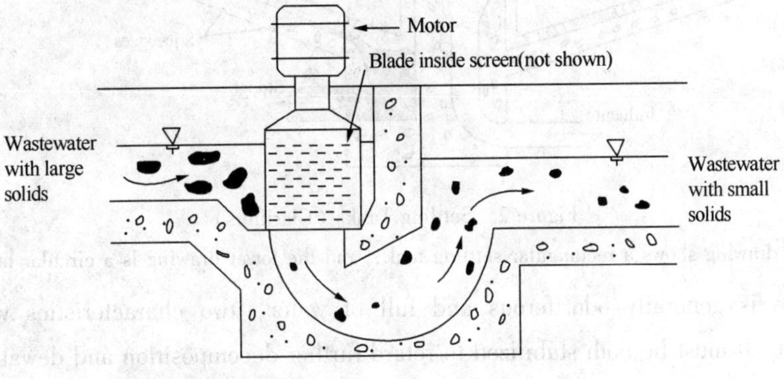


Figure 1 A Comminutor Used to Grind up Large Solids

The third treatment step involves the removal of grit or sand. This is necessary because grit can wear out and damage such equipment as pumps and flow meters. The most common *grit chamber* is simply a wide place in the channel where the flow is slowed down sufficiently to allow the heavy grit to settle out. Sand is about 2.5 times as heavy as most organic solids and thus settles much faster than the light solids. The objective of a grit chamber is to remove sand and grit without removing the organic material. The latter must be further treated in the plant, but the sand can be dumped as fill without undue odor or other problems.

Following the grit chamber most wastewater treatment plants have *a settling tank* (Figure 2) to settle out as much of the solid matter as possible. Accordingly, the retention time is kept long and turbulence is kept to a minimum. The solids settle to the bottom and are removed through a pipe while the clarified liquid escapes over a V-notch weir, a notched steel plate over which the water flows, promoting equal distribution of liquid discharge all the way around a tank. Settling tanks are also known as *sedimentation tanks* and often as *clarifiers*. The settling tank which follows preliminary treatment such as screening and grit removal is known as a *primary clarifier*. The solids which drop to the bottom of a primary clarifier are removed as *raw sludge*, a name which doesn't do justice to the undesirable nature of this stuff (2).

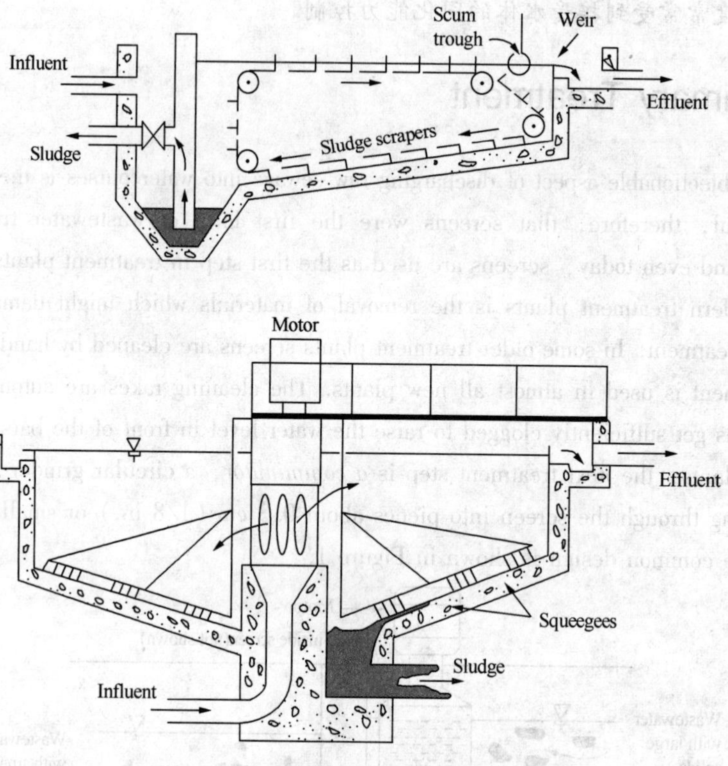


Figure 2 Settling Tanks (Clarifiers)

Top drawing shows a rectangular settling tank, and the lower drawing is a circular tank.

Raw sludge is generally odoriferous and full of water, two characteristics which make its disposal difficult. It must be both stabilized to retard further decomposition and dewatered for ease of disposal. In addition to the solids from the primary clarifier, solids from other processes must similarly be treated and disposed.

Since settling tanks are both ubiquitous and effective, their operation and design is reviewed in greater detail below.

## Settling Tank Design and Operation

As noted above, one of the most efficient means of separating solids from the surrounding liquid is to allow the solids to settle out under the force of gravity. As long as the density of the solids exceeds that of the liquid, it should be possible to achieve solid/liquid separation. Unfortunately, this is not always true, since other forces come into play, as we shall see later.

Particles settle in one of three general ways:

- class I: as discrete particles, unhindered by the container walls or neighboring particles. This is called *discrete particle settling*.

• class II: as discrete particles but hindered by their neighbors and changing in size due to particle contact due to close proximity. This is called *flocculent settling*.

• class III: as a mass of particles, with no interparticle movement so that all particles have the same settling velocity, called *thickening*.

Primary treatment then is mainly a removal of solids, although some BOD is removed as a consequence of the removal of decomposable solids. Typically, the wastewater which was described earlier might now have these characteristics:

| | *Raw Wastewater* | *Following Primary Treatment* |
|---|---|---|
| BOD/(mg/L) | 250 | 175 |
| SS/(mg/L) | 220 | 60 |
| P/(mg/L) | 8 | 7 |

A substantial fraction of the solids has been removed, as well as some BOD and a little P (as a consequence of the removal of raw sludge).

In a typical wastewater treatment plant, this would now move on to secondary treatment.

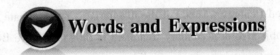

## Words and Expressions

1. objectionable 讨厌的，不适合的，不好的，有害的，不能采用的
2. discharge raw sewage into watercourse 将原污水排入水道
3. floating material 漂浮物
4. screen 粗筛
5. steel bar 钢棒
6. cleaning rake 清洁耙
7. get clogged 堵塞
8. raise the water level in front of the bars 抬高钢棒前的水位
9. comminutor 粉碎机，磨碎机
10. circular grinder 圆型磨碎机
11. wear out 磨损
12. grit chamber 沉砂池
13. fill 填充料
14. settling tank 沉降池，沉淀池
15. retention time 停留时间
16. turbulence 湍流，紊流
17. weir 堰
18. sedimentation tank 沉淀池
19. primary clarifier 一次[初级]澄清池
20. raw sludge 原污泥
21. do justice to 正确对待，正确反映
22. undesirable nature of this stuff 该物质不良的性质
23. odoriferous 散发气味的
24. dewater 脱水
25. ubiquitous 普遍存在的，普通的
26. in detail 详细地
27. discrete particle 离散颗粒
28. flocculent settling 絮凝沉降
29. a mass of particles 一批颗粒，许多颗粒
30. thickening 浓缩，稠化
31. as a consequence of 由于
32. a substantial fraction of 大部分

## Notes

(1) The cleaning rakes are automatically activated when the screens get sufficiently clogged to raise the water level in front of the bars.

当粗筛严重堵塞以至于抬高了筛棒前的水位时，清洁耙就会自动开启。

(2) The solids which drop to the bottom of a primary clarifier are removed as raw sludge, a name which doesn't do justice to the undesirable nature of this stuff.

沉降到初级澄清池底部的固体被作为原污泥去除，但原污泥这一名称没有恰如其分地反映该物质不良的特性。

# 11. Secondary Treatment

The water leaving the primary clarifier has lost much of the solid organic matter but still contains a high demand for oxygen; i. e., it is composed of high energy molecules which will decompose by microbial action, thus creating a biochemical oxygen demand (BOD). This demand for oxygen must be reduced (energy wasted) if the discharge is not to create unacceptable conditions in the watercourse. The objective of secondary treatment is thus to remove BOD while, by contrast, the objective of primary treatment is to remove solids.

Almost all secondary methods use microbial action to reduce the energy level (BOD) of the waste. Although there are many ways the microorganisms can be put to work, the first really successful modern method of secondary treatment was the *trickling filter*.

The trickling filter, shown in Figure 1, consists of a filter bed of fist-sized rocks over which the waste is trickled. A very active biological growth forms on the rocks, and the organisms obtain their food from the waste stream dripping through the bed of rocks. Air is either forced through the rocks or, more commonly, air circulation is obtained automatically by a temperature difference between the air in the bed and ambient temperature. In the older filters the waste is sprayed onto the rocks from fixed nozzles. The newer designs utilize a rotating arm which moves under its own power, like a lawn sprinkler, distributing the waste evenly over the entire bed. Often the flow is recirculated, thus obtaining a higher degree of treatment. The name trickling filter is obviously a misnomer since no filtration takes place.

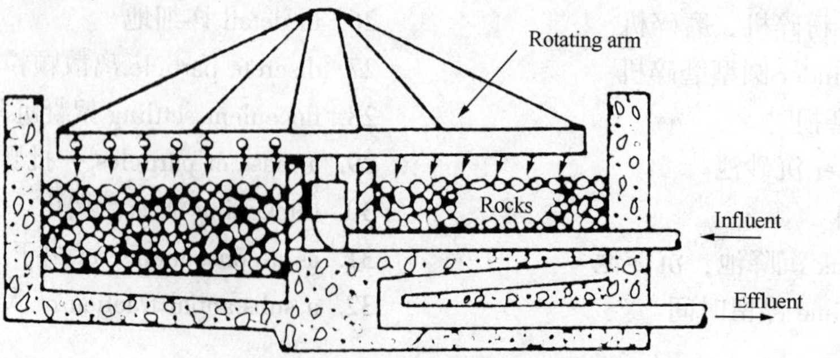


Figure 1 Schematic of a Trickling Filter

Around the turn of the century when trickling filtration was already firmly established, some researchers began musing about the wasted space in a filter taken up by the rocks. Could the microorganisms not be allowed to float free and could they not be fed oxygen by bubbling in air? Although this concept was quite attractive, it was not until 1914 that the first workable pilot plant was constructed. It took some time before this process became established as what we now call the *activated sludge system*.

The key to the activated sludge system is the reuse of microorganisms. The system, shown as a block diagram in Figure 2, consists of a tank full of waste liquid (from the primary clarifier) and a mass of microorganisms. Air is bubbled into this tank (called the *aeration tank*) to provide the necessary oxygen for the survival of the aerobic organisms. The microorganisms come in contact with the dissolved organics and rapidly adsorb these organics on their surface. In time, the microorganisms decompose this material to $CO_2$, $H_2O$, some stable compounds and more microorganisms. The production of new organisms is relatively slow, and most of the aeration tank volume is in fact used for this purpose.

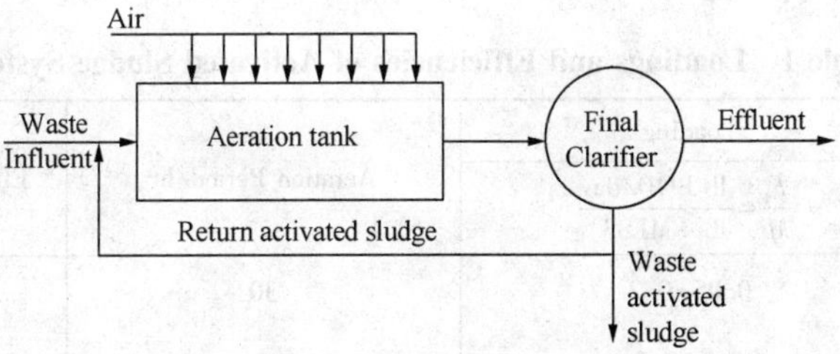


Figure 2 Block Diagram of the Activated Sludge System

Once most of the food has been utilized, the microorganisms are separated from the liquid in a settling tank, sometimes called a *secondary or final clarifier*. The liquid escapes over a weir and can be discharged into the recipient. The separation of microorganisms is an important part of the system. In the settling tanks, the microorganisms exist without additional food and become hungry. They are thus activated; hence the term *activated sludge*.

The settled microorganisms, now known as *return activated sludge*, are pumped to the head of the aeration tank where they find more food (organic in the effluent from the primary clarifier) and the process starts all over again. The activated sludge process is a continuous operation, with continuous sludge pumping the clean water discharge(1).

As mentioned earlier, one of the end products of this process is more microorganisms. If none of the microorganisms are removed, their concentration will soon increase to the point where the system is clogged with solids. It is therefore necessary to waste some of the microorganisms, and this *waste activated sludge* must be processed and disposed of. Its disposal is one of the most difficult aspects of waste treatment.

Activated sludge systems are designed on the basis of loading, or the amount of organic matter (food) added relative to the microorganisms available. This ratio is known as the food-to-microorganisms ratio ($F/M$) and is a major design parameter. Unfortunately it is difficult to measure either $F$ or $M$ accurately, and engineers have approximated these by BOD and the suspended solids in the aeration tank respectively. The combination of the liquid and microorganisms undergoing aeration is known (for some unknown reason) as *mixed liquor*, and thus the suspended solids are called *mixed liquor suspended solids* (MLSS). The ratio of incoming BOD to MLSS, the *F/M* ratio, is also known as the *loading* on the system, calculated as pounds of BOD/day per pound of MLSS in the aeration tank.

If this ratio is low (little food for lots of microorganisms) and the aeration period (retention time in the aeration tank) is long, the microorganisms make maximum use of available food, resulting in a high degree of treatment. Such systems are known as *extended aeration*, and are widely used for isolated sources (e. g., motels, small developments). An added advantage of extended aeration is that the ecology within the aeration tank is quite diverse and little excess biomass is created, resulting in little or no waste activated sludge to be disposed of–a significant saving in operating costs and headaches.

At the other extreme is the "high-rate" system where the aeration periods are very short (thus saving money by building smaller tanks) and the treatment efficiency is lower. The efficiencies and *F/M* ratios for the three types of activated sludge systems are shown in Table 1.

**Table 1 Loadings and Efficiencies of Activated Sludge Systems**

| Process | Loading $\frac{F}{M}=\frac{\text{lb BOD/day}}{\text{lb MLSS}}$ | Aeration Period/hr | Efficiency of BOD Removal/% |
|---|---|---|---|
| Extended Aeration | 0. 05–0. 2 | 30 | 95 |
| Conventional | 0. 2–0. 5 | 6 | 90 |
| High Rate | 1–2 | 4 | 85 |

When the microorganisms first come in contact with the food, the process requires a great deal of oxygen. Accordingly, the dissolved oxygen level in the aeration tank drops immediately after the point at which the waste is introduced. If DO levels are measured over the length of a tank, extremely low concentrations are often found at the influent end of the aeration tank. These low levels of DO can be detrimental to the microbial population. Accordingly, two variations of the activated sludge treatment have found some use: *tapered aeration* and *step aeration* (Figure 3)[2]. The former method consists of blasting additional air where needed, while step aeration involves the introduction of the waste at several locations, thus evening out the initial oxygen demand.

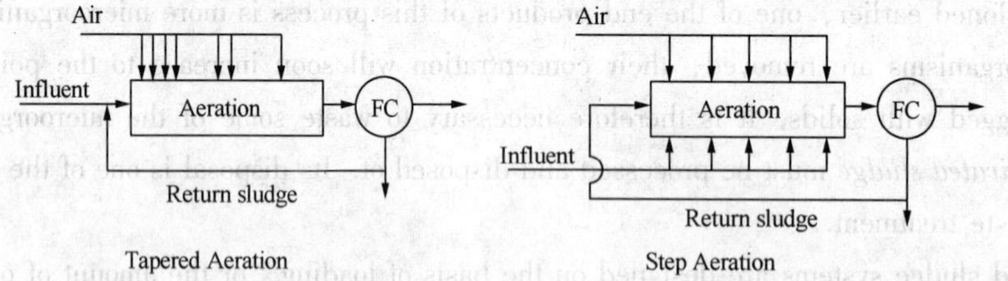


Figure 3 Tapered and Step Aeration Schematics

The third modification is *contact stabilization*, or *biosorption*, a process in which the sorption and bacterial growth phases are separated by a settling tank. The advantage is that the growth can be achieved at high solids concentrations, thus saving tank space. Many existing activated sludge plants can be converted to biosorption plants when tank volume limits treatment efficiency. Figure 4 is a diagram of the biosorption process.

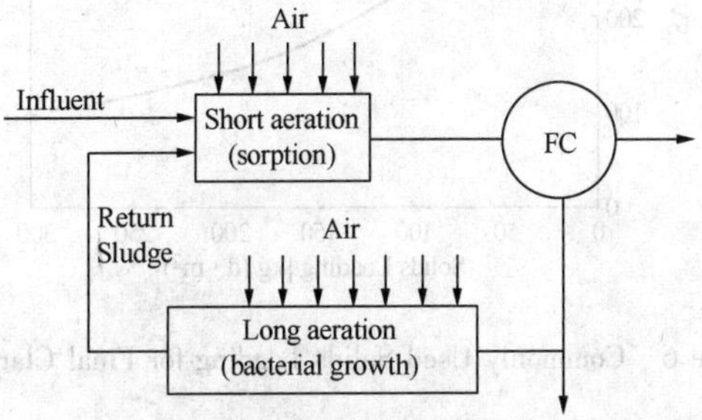


Figure 4 The Biosorption Modification of the Activated Sludge Process

The two principal means of introducing sufficient oxygen into the aeration tank are by bubbling compressed air through porous diffusers or beating air in mechanically[3]. Both diffused air and mechanical aeration are shown in Figure 5.

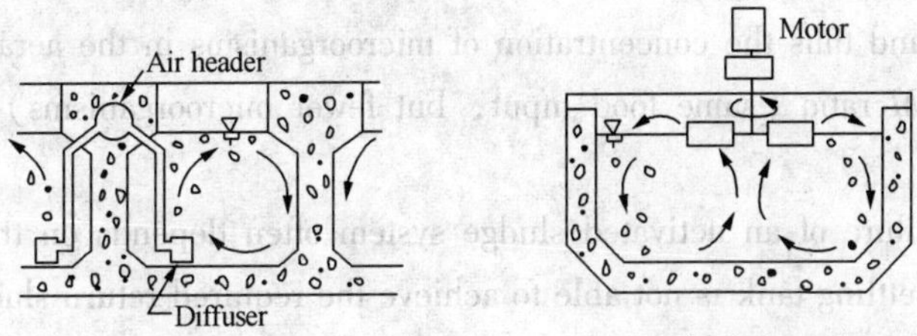


Figure 5 Activated System with Diffused Aeration and Mechanical (surface) Aeration

The success of the activated sludge system depends on many factors. Of critical importance is the separation of the microorganisms in the final clarifier[4]. The microorganisms in the system are sometimes very difficult to settle out, and the sludge is said to be a *bulking sludge*. Often this condition is characterized by a biomass comprised almost totally of filamentous organisms which form a kind of lattice structure with the filaments and refuse to settle[5].

Treatment plant operators should keep a close watch on settling characteristics because a trend toward poor settling can be the forerunner of a badly upset (and hence ineffective) plant. The settleability of activated sludge is most often described by the sludge volume index (SVI), which is determined by measuring the ml of volume occupied by a sludge after settling for 30 minutes in a 1-liter cylinder, and calculated as

$$\mathrm{SVI} = \frac{(\text{volume of sludge after 30 min, in mL}) \times 100}{\text{mg/L of suspended solids}}$$

SVI values below 100 are usually considered acceptable, with SVI greater than 200 defined as badly bulking sludges. Some common loadings, as a function of the SVI, for final clarifiers are shown in Figure 6.

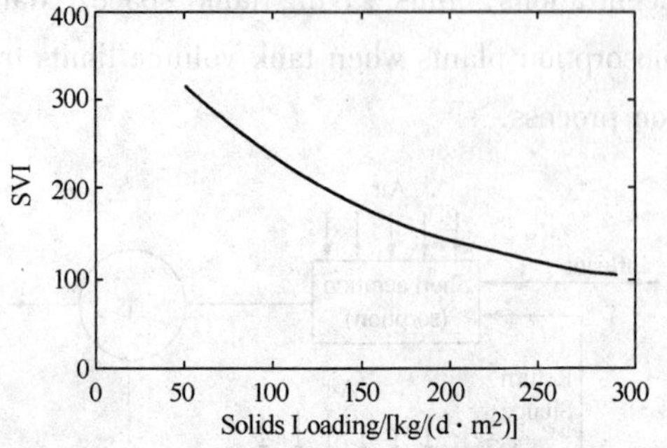


Figure 6 Commonly Used Solids Loading for Final Clarifiers

The causes for poor settling (high SVI) are not always known, and hence the solutions are elusive. Wrong or variable *F/M* ratios, fluctuations in temperature, high concentrations of heavy metals, and deficiencies in nutrients in the incoming wastewater have all been blamed for bulking. Cures include chlorination, changes in air supply, and dosing with hydrogen peroxide ($H_2O_2$) to kill the filamentous microorganisms.

When the sludge does not settle, the return activated sludge become thin (low suspended solids concentration) and thus the concentration of microorganisms in the aeration tank drops. This results in a higher *F/M* ratio (same food input, but fewer microorganisms) and a reduced BOD removal efficiency.

The success or failure of an activated sludge system often depends on the performance of the final clarifier. If this settling tank is not able to achieve the required return sludge solids, the MLSS will drop, and of course the treatment efficiency will be reduced.

Final clarifiers act as both settling tanks (flocculent settling, Class II) and thickeners (Class III). The design of these tanks therefore requires that both the overflow rate and solids loading be considered. The latter, more fully explained in the next chapter, is expressed in terms of kg solids/day applied to a surface area of $m^2$. Figure 6 shows some commonly used solids loadings for final clarifiers as a function of the sludge volume index.

Secondary treatment of wastewater then usually consists of a biological step such as activated sludge, which removes a substantial part of the BOD and the remaining solids. Looking once again at the typical wastewater, we now have the following approximate water quality:

| | *Raw Wastewater* | *Following Primary Treatment* | *Following Secondary Treatment* |
|---|---|---|---|
| BOD/(mg/L) | 250 | 175 | 15 |
| SS/(mg/L) | 220 | 60 | 15 |
| P/(mg/L) | 8 | 7 | 6 |

The effluent, in fact, meets our previously established effluent standards for BOD and SS. Only the phosphorus remains high. The removal of inorganic chemicals like phosphorus is accomplished in tertiary (or advanced) wastewater treatment.

## Words and Expressions

1. be composed of 由……构成，由……组成
2. microbial action 微生物作用
3. trickling filter 滴滤池，生物滤池
4. active biological growth 活性生物生长(物)
5. air circulation 空气流通[环流]
6. filter bed 滤床
7. fist-sized rock 拳头大小的石块
8. ambient temperature 环境温度
9. spray 喷洒
10. fixed nozzle 固定的喷嘴
11. rotating arm 旋转臂
12. under its own power 靠自力
13. lawn sprinkler 草坪洒水机
14. misnomer 误称，名称使用不当
15. around the turn of the century 上世纪末、本世纪初
16. muse about 考虑，思索
17. take up 占据
18. workable pilot plant 可行的试验设备
19. activated sludge system 活性污泥系统
20. block diagram 块状图
21. aeration tank 曝气池
22. secondary clarifier 二次澄清池
23. final clarifier 终沉池
24. dispose of 处置
25. come in contact with 与……接触
26. activated sludge process 活性污泥法
27. approximate A by B 用 B 近似表示 A
28. mixed liquor 混合液
29. mixed liquor suspended solids (MLSS) 混合液悬浮固体
30. loading on the system 系统的负荷
31. make maximum use of 最大程度利用
32. extended aeration 延时曝气
33. tapered aeration 逐减曝气
34. step aeration 逐步曝气；分段曝气
35. contact stabilization 接触稳定化
36. biosorption 生物吸附，生物吸收
37. filamentous organism 丝状生物体
38. keep a close watch on 密切关注
39. forerunner 先兆，预兆
40. badly upset plant 严重失常的设备
41. settleability 可沉降性
42. sludge volume index (SVI) 污泥体积指数
43. ml of volume occupied by a sludge after settling for 30 minutes in a 1-liter cylinder 污泥在 1 升量筒中沉降 30 分钟后所占据体积的毫升数
44. causes for poor settling 不良沉降的原因
45. elusive 难以捉摸的
46. fluctuations in temperature 温度的波动
47. high concentrations of heavy metals 高浓度重金属
48. deficiencies in nutrients 营养物的缺乏
49. be blamed for bulking 因膨胀而受指责
50. cure 对策，治疗的方法
51. chlorination 氯化
52. dose 投配，配加，配料
53. hydrogen peroxide 过氧化氢
54. filamentous microorganisms 丝状微生物
55. removal efficiency 去除效率
56. performance of the final clarifier 终沉池性能
57. overflow rate 溢流速率
58. solids loading 固体负荷
59. advanced wastewater treatment 废水的高级[三级]处理

## Notes

(1) The activated sludge process is a continuous operation, with continuous sludge pumping the clean water discharge.

活性污泥法是一个连续的操作过程，连续不断地产出污泥和清澈的出水。

(2) Accordingly, two variations of the activated sludge treatment have found some use: tapered aeration and step aeration.

因此，两种改进的活性污泥处理法，逐减曝气法和分段曝气法，已得到一些应用。

(3) The two principal means of introducing sufficient oxygen into the aeration tank are by bubbling compressed air through porous diffusers or beating air in mechanically.

将充足的氧引入曝气池的两种主要的方法是通过多孔扩散器鼓入压缩空气或通过机械的方式曝气(表曝)。

(4) Of critical importance is the separation of the microorganisms in the final clarifier.

在终沉池中微生物的分离是非常重要的。本句为倒装句。

(5) Often this condition is characterized by a biomass comprised almost totally of filamentous organisms which form a kind of lattice structure with the filaments and refuse to settle.

这种状况的特征常常是生物体几乎完全由丝状的有机体构成，该有机体形成一种丝状的网状结构并不会沉降。

# 12. Tertiary Treatment

Primary and biological treatments make up the conventional wastewater treatment plant. However, secondary treatment plant effluents still contain a significant amount of various types of pollutants. Suspended solids, in addition to contributing to BOD, can settle out in streams and form unsightly mud banks. The BOD, if discharged into a stream with low flow, can still cause damage to aquatic life by depressing the DO. Neither primary nor secondary treatment is effective in removing phosphorus and other nutrients or toxic substances(1).

Suspended solids can be effectively removed by a simple device called a *pebble filter* which consists of a box full of pebbles hung over the edge of the secondary clarifier so that all water must travel through the filter bed. The actual mechanism of how the pebbles catch the floc is unknown. What is known is that the effluent through a pebble filter is clear. However, this device will never succeed in the United States because the pebble filter is entirely too simple, with no moving parts. Nobody wants to manufacture it, nor will engineers design such a "primitive" device(2).

A much more sophisticated, and hence more marketable, gadget is the microstrainer, pictured in Figure 1. The microstrainer is a large drum covered with a stainless steel sheet with small holes in it. The dirty water is pumped into the drum and the clean water filters through. As the drum rotates, the holes are cleaned with spray.

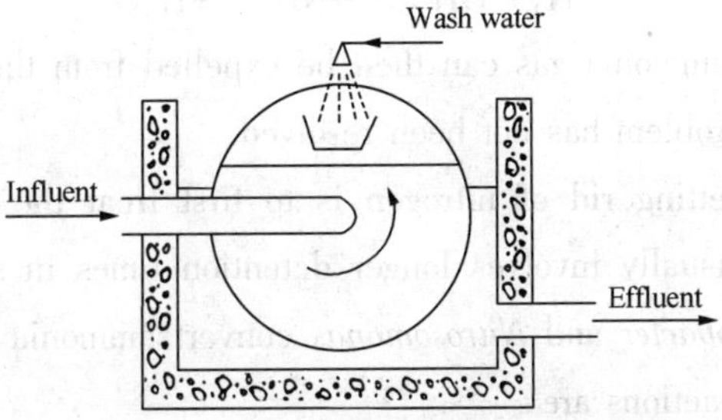


Figure 1 Microstrainer Schematic

For BOD removal, by far the most popular advanced treatment method is the polishing pond, often called the *oxidation pond*. This is essentially a hole in the ground, a large pond used to confine the plant effluent before it is discharged. Such ponds are designed to be aerobic, hence light penetration for algal growth is important, and a large surface area is needed. The reactions occurring within an oxidation pond are depicted in Figure 2. Oxidation ponds are sometimes used as the only treatment step if the waste flow is small and the pond area is large.

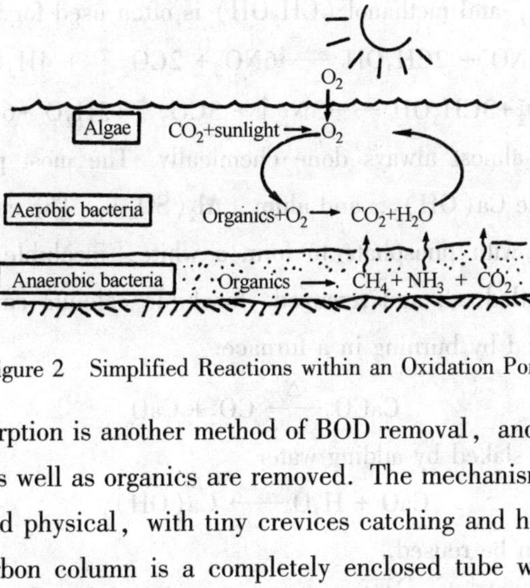


Figure 2 Simplified Reactions within an Oxidation Pond

Activated carbon adsorption is another method of BOD removal, and this process has the added advantage that inorganics as well as organics are removed. The mechanism of adsorption on activated carbon is both chemical and physical, with tiny crevices catching and holding colloidal and smaller particles. An activated carbon column is a completely enclosed tube with dirty water pumped up from the bottom and the clear water exiting at the top. As the carbon becomes saturated with various materials, it must be removed from the column and regenerated, or cleaned. Removal is often continuous, with clean carbon being added at the top of the column. The cleaning or regeneration is usually done by heating the carbon in the absence of oxygen. A slight loss in efficiency is noted with regeneration, and some virgin carbon must always be added to ensure effective performance.

*Reverse osmosis* is also finding acceptance as a treatment for various kinds of trace pollutants, organic as well as inorganic. The wastewater is forced through a semipermeable membrane which acts as a superfilter, rejecting dissolved as well as suspended solids[(3)].

Nitrogen removal can be accomplished in two ways. The first method makes use of the fact that even after secondary treatment, most of the nitrogen exists as ammonia. Increasing the pH produces the following reaction:

$$NH_4^+ + OH^- \longrightarrow NH_3 \uparrow + H_2O$$

Much of the dissolved ammonia gas can then be expelled from the water into the atmosphere. The resulting air pollution problem has not been resolved.

A second method of getting rid of nitrogen is to first treat the waste thoroughly enough to produce nitrate ions. This usually involves longer detention times in secondary treatment, during which bacteria such as *Nitrobacter* and *Nitrosomonas* convert ammonia nitrogen to $NO_2^-$, a process called *nitrification*. These reactions are

$$2NH_4^+ + 3O_2 \xrightarrow{Nitrosomonas} 2NO_2^- + 2H_2O + 4H^+$$

$$2NO_2^- + O_2 \xrightarrow{Nitrobacter} 2NO_3^-$$

These reactions are slow and thus require long retention times in the aeration tank, as well as sufficient dissolved oxygen. The kinetics constants for these reactions are low, with very low yields, so that the net sludge production is limited, making washout a constant danger.

Once the ammonia has been converted to nitrate, it can be reduced by a broad range of facultative and anaerobic bacteria such as *Pseudomonas*(4). This reduction, called *denitrification*, requires a source of carbon, and methanol ($CH_3OH$) is often used for that purpose.

$$6NO_3^- + 2CH_3OH \longrightarrow 6NO_2^- + 2CO_2 \uparrow + 4H_2O$$

$$6NO_2^- + 3CH_3OH \longrightarrow 3N_2 \uparrow + 3CO_2 \uparrow + 3H_2O + 6OH^-$$

Phosphate removal is almost always done chemically. The most popular chemicals used for phosphorus removal are lime $Ca(OH)_2$, and alum, $Al_2(SO_4)_3$. The calcium ion, in the presence of high pH, will combine with phosphate to form a white, insoluble precipitate called calcium hydroxyapatite which is settled out and removed. Insoluble calcium carbonate is also formed and removed and can be recycled by burning in a furnace.

$$CaCO_3 \xrightarrow{\Delta} CO_2 + CaO$$

Quick lime, CaO, is slaked by adding water

$$CaO + H_2O \longrightarrow Ca(OH)_2$$

thus forming lime which can be reused.

The aluminum ion from alum precipitates out as poorly soluble aluminum phosphate,

$$Al^{3+} + PO_4^{3-} \longrightarrow AlPO_4 \downarrow$$

and also forms aluminum hydroxides,

$$Al^{3+} + 3OH^- \longrightarrow Al(OH)_3 \downarrow$$

which are sticky flocs and help to settle out the phosphates. The most common point of alum dosing is in the final clarifier.

The amount of alum required to achieve a given level of phosphorus removal depends on the amount of phosphorus in the water, as well as other constituents. The sludge produced can be calculated using stoichiometric relationships.

Using such a technique as alum precipitation of phosphorus, [and a bonus of more efficient SS and BOD removal due to the formation of $Al(OH)_3$], we now have attained our effluent goal:

| | Raw Wastewater | Following Primary Treatment | Following Secondary Treatment | Following Tertiary Treatment |
|---|---|---|---|---|
| BOD/(mg/L) | 250 | 175 | 15 | 10 |
| SS/(mg/L) | 220 | 60 | 15 | 10 |
| P/(mg/L) | 8 | 7 | 6 | 0.5 |

An alternative to high-technology advanced wastewater treatment systems is to spray secondary effluent on land and allow the soil microorganisms to degrade the remaining organics. Such systems, known as land treatment, have been employed for many years in Europe, but only recently have been used in North America. They appear to represent a reasonable alternative to complex and expensive systems, especially for smaller communities.

Probably the most promising land treatment method is irrigation. Commonly, from 1000 to 2000 hectares of land are required for every 1 $m^3$/sec of wastewater flow, depending on the crop and soil. Nutrients such as N and P remaining in the secondary effluent are of course beneficial to the crops.

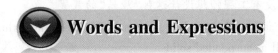

## Words and Expressions

1. unsightly mud bank 不雅观的泥岸
2. pebble filter 卵石滤池，石子滤料滤池
3. actual mechanism 实际的机理
4. gadget 小装置
5. microstrainer 微滤器
6. a large drum covered with a stainless steel sheet 覆盖了不锈钢片的大鼓状物
7. polishing pond 精处理塘，深处理塘
8. oxidation pond 氧化塘
9. confine 圈禁，禁锢，限制
10. light penetration 光线的渗入
11. algal growth 藻类生长
12. activated carbon adsorption 活性炭吸附
13. mechanism of adsorption on activated carbon活性炭吸附机理
14. crevice 缝隙，裂缝
15. enclosed tube 闭合管，封闭管
16. regenerate 再生
17. in the absence of oxygen 没有氧
18. virgin carbon 新炭，纯净炭
19. ensure effective performance 保证有效性能
20. reverse osmosis 反渗透，逆渗透
21. find acceptance 得到公认，得到认可
22. trace pollutant 微量污染物
23. superfilter 超滤器
24. reject dissolved as well as suspended solids 去除溶解和悬浮固体
25. get rid of nitrogen 去除氮
26. Nitrobacter 硝化杆菌
27. Nitrosomonas 硝化毛杆菌
28. nitrification 硝化(作用)
29. kinetics constant 动力学常数
30. make washout a constant danger 避免通常的危险
31. a broad range of 一大批
32. facultative bacteria 兼性菌
33. anaerobic bacteria 厌氧菌
34. Pseudomonas 假单孢菌属
35. denitrification 反硝化(作用)
36. methanol 甲醇
37. calcium ion 钙离子

38. calcium hydroxyapatite 羟磷灰石钙
39. insoluble calcium carbonate 不溶性碳酸钙
40. quick lime 生石灰
41. slake 消解，熟化
42. aluminum hydroxide 氢氧化铝
43. stoichiometric relationship 化学计量关系
44. bonus 红利，奖金，额外津贴
45. alternative to high-technology advanced waste- water treatment systems 对高技术的高级废水处理系统的替代方案
46. land treatment 土地处理
47. beneficial 有益的

## Notes

(1) Neither primary nor secondary treatment is effective in removing phosphorus and other nutrients or toxic substances.

一级处理和二级处理在去除磷和其他营养物或毒物方面都是无效的。

(2) Nobody wants to manufacture it, nor will engineers design such a "primitive" device.

没有人想制造，也没有工程师想设计这种"原始"的装置。

(3) The wastewater is forced through a semipermeable membrane which acts as a superfilter, rejecting dissolved as well as suspended solids.

废水被强制通过半透膜，该膜充当超滤器，能排除溶解性的和悬浮的固体。

(4) Once the ammonia has been converted to nitrate, it can be reduced by a broad range of facultative and anaerobic bacteria such as Pseudomonas.

一旦氨被转化为硝酸盐，它就能被一大批兼性和厌氧菌(如假单孢菌属)还原。

# 13. Characteristics of the Sludges

The characteristics of sludges of interest depend entirely on what is to be done with a sludge. For example, if the sludge is to be thickened by gravity, its settling and compaction characteristics are important[(1)]. On the other hand, if the sludge is to be digested anaerobically, the concentrations of volatile solids, heavy metals, etc., are important.

Another fact of immense importance in the design of sludge handling and disposal operations is the variability of the sludges. In fact this variability can be stated in terms of three "laws":

1. No two wastewater sludges are alike in all respects.
2. Sludge characteristics change with time.
3. There is no "average sludge".

The first statement reflects the fact that no two wastewaters are alike, and that if the variable of treatment is added, the sludges produced will have significantly different characteristics.

The second statement is often overlooked by designers. For example, the settling characteristics of chemical sludges from the treatment of plating wastes [e.g., $Pb(OH)_2$, $Zn(OH)_2$ or $Cr(OH)_3$] vary with time simply because of uncontrolled pH changes. Biological sludges are of course continually changing, with the greatest change occurring when the sludge changes from aerobic to anaerobic (or vice versa). It is thus quite difficult to design sludge handling equipment, since the sludge may change in some significant characteristic in only a few hours.

The third "law" is constantly violated. Tables showing "average values" for "average sludges" are useful for illustrative and comparative purposes only, and should not be used for design.

With that caveat, we now proceed to discuss some characteristics of "average sludges", tabulated in Table 1.

**Table 1 Sludge Characteristics**

| Type of Sludge | Physical | | | | Chemical | | |
|---|---|---|---|---|---|---|---|
| | Solid Concentration/ (mg/L) | Volatile Solids/% | Yield Strength/ (dyn/cm$^2$) | Plastic Viscosity/ (g/cm · s) | Nitrogen (% as N) | Phosphorus (% as $P_2O_5$) | Potassium (% as $K_2O$) |
| Water | — | — | 0 | 0.01 | — | — | — |
| Raw Primary | 60, 000 | 60 | 40 | 0.3 | 2.5 | 1.5 | 0.4 |
| Mixed Digested | 80, 000 | 40 | 15 | 0.9 | 4.0 | 1.4 | 0.2 |
| Waste Activated | 15, 000 | 70 | 0.1 | 0.06 | 4.0 | 3.0 | 0.5 |
| Alum/ppt | 20, 000 | 40 | — | — | 2.0 | 2.0 | — |
| Lime/ppt | 200, 000 | 18 | — | — | 2.0 | 3.0 | — |

The first characteristic, solids concentration, is perhaps the most important variable, defining the volume of sludge to be handled, and determining whether the sludge behaves as a liquid or a solid. The importance of volatile solids is, of course, in the disposability of the sludge. With high volatiles, a sludge would be difficult to dispose of into the environment. As the volatiles are degraded, gases and odors are produced, and thus a high volatile solids concentration would restrict the methods of disposal.

The rheological characteristics of sludge are of interest in that this is one of only a few truly basic parameters describing the physical nature of a sludge. Two-phase mixtures like sludges, however, are almost without exception non-Newtonian and thixotropic. Sludges tend to act like pseudoplastics, with an apparent yield stress and a plastic viscosity. The rheological behavior of a pseudoplastic fluid is defined by a rheogram shown as Figure 1. The term thixotropic relates to the time dependence of the rheological properties.

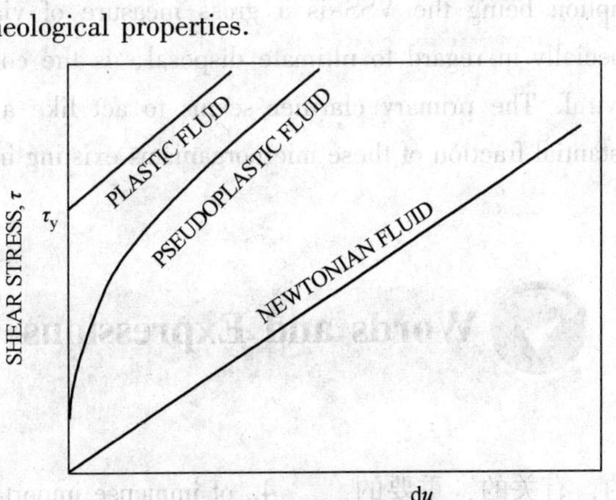


Figure 1 Typical Rheogram for Sludge

Sludges tend to act more like plastic fluids as the solids concentration increases. True plastic fluids can be described by the equation

$$\tau = \tau_y + \eta \frac{du}{dy}$$

where $\tau$ = shear stress

$\tau_y$ = yield stress

$\eta$ = plastic viscosity

$du/dy$ = rate of shear, or the slope of the velocity($u$)-depth($y$) profile.

Although sludges are seldom true plastics, the above equation can be as an approximation of the rheograrm.

The yield stress can vary from above 40 dyn/$cm^2$ for 6% raw sludge, to only 0.07 dyn/$cm^2$ for a thickened activated sludge(2). The large differences suggest that rheological parameters could well be used for scale-up purposes. Unfortunately, few researchers have bothered to measure the rheological characteristics (such analyses are not even included in *Standard Methods*) and many gaps exist in the available data.

The chemical composition is important for several reasons. First, the fertilizer value of the sludge is dependent on the availability of N, P and K, as well as trace elements. A more important measurement, however, is the concentration of heavy metals and other toxins which would make the sludge toxic to the environment. The ranges of heavy metal concentrations are very large (e. g., cadmium can range from almost zero to over 1000 mg/kg). Since a major source of such toxins is in industrial discharges, a single poorly operated industrial firm may contribute enough toxins to make the sludge worthless as a fertilizer. Most engineers agree that although it would be most practical to treat the sludge at the plant in order to achieve the removal of toxic components, there are no effective methods available for removing heavy metals, pesticides and other potential toxins from the sludge, and that the control must be over the influent (that is, tight sewer ordinances).

In addition to the physical and chemical characteristics, the biological parameters of sludge can also be important. The volatile solids parameter is in fact often interpreted as a biological characteristic, the assumption being the VSS is a gross measure of viable biomass(3). Another important parameter, especially in regard to ultimate disposal, is the concentration of pathogens, both bacteriological and viral. The primary clarifier seems to act like a viral and bacteriological concentrator, with a substantial fraction of these microorganisms existing in the sludge instead of the liquid effluent (4).

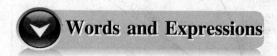

## Words and Expressions

1. of interest 值得注意的，有关的，重要的
2. compaction 压实，压缩，致密化
3. volatile solids 挥发性固体
4. of immense importance 非常重要的
5. vice versa 反之亦然
6. caveat(防止误解的)说明

7. disposability of the sludge 污泥的可处置性
8. dispose of into the environment 处置到环境中，在环境中处置
9. rheological characteristics 流变特性
10. two-phase mixture 两相混合物
11. thixotropic 触变(性)的
12. pseudoplastic 准塑性体，假塑性体
13. apparent yield stress 表观屈服应力
14. plastic viscosity 塑性黏性
15. rheological behavior 流变行为
16. rheogram 流变图
17. without exception 无例外
18. time dependence of the rheological property 流变性能对时间的依赖关系
19. plastic fluid 塑性流体
20. approximation 接近，逼近，近似
21. shear stress 剪切应力
22. chemical composition 化学成分
23. fertilizer value of the sludge 污泥的肥料价值
24. trace element 痕量元素，微量元素
25. tight sewer ordinance 严格的污水管法令
26. viable biomass 易繁殖生物量
27. in regard to 关于，相对于
28. ultimate disposal 最终处置

## Notes

(1) For example, if the sludge is to be thickened by gravity, its settling and compaction characteristics are important.

例如，如果污泥要进行重力浓缩，那么它的沉降和致密特性是重要的。

(2) The yield stress can vary from above 40 dyn/cm$^2$ for 6% raw sludge, to only 0.07 dyn/cm$^2$ for a thickened activated sludge.

屈服应力能从含量为6%的原污泥的40 dyn/cm$^2$以上变化至浓缩活性污泥的仅0.07 dyn/cm$^2$。

(3) The volatile solids parameter is in fact often interpreted as a biological characteristic, the assumption being the VSS is a gross measure of viable biomass.

挥发性固体参数实际上常被看成生物特性，因为假定VSS是易繁殖生物量的总的测定。

(4) The primary clarifier seems to act like a viral and bacteriological concentrator, with a substantial fraction of these microorganisms existing in the sludge instead of the liquid effluent.

一次澄清池似乎充当了病毒和细菌的浓缩器，这些微生物的大部分是随污泥而不是随出水排出。

# 14. Sludge Treatment

A great deal of money could be saved, and troubles averted, if sludge could be disposed of as it is drawn off the main process train. Unfortunately, the sludges have three characteristics which make such a simple solution unlikely: they are aesthetically displeasing, they are potentially harmful, and they have too much water.

The first two problems are often solved by stabilization, such as *anaerobic* or *aerobic digestion*. The third problem requires the removal of water by either thickening or dewatering. Accordingly, the next three sections cover the topics of stabilization, thickening and dewatering, followed finally by considerations of ultimate disposal.

Before we embark on a discussion on sludge treatment, however, we should be reminded that the processes should be used in wastewater treatment only when necessary. It makes little sense to spend time and money to treat sludges if their ultimate disposal does not require it. The "Get the water out!" syndrome often found in engineering offices is often associated with "Gotta digest it!" disease [(1)]. These processes all cost money, and should be considered only if direct disposal in undigested liquid form is not practical.

## Sludge Stabilization

The objective of sludge stabilization is to reduce the problems associated with two of the detrimental characteristics listed above: sludge odor and putrescibility, and the presence of pathogenic organisms.

There are three primary means of sludge stabilization

- lime
- aerobic digestion
- anaerobic digestion.

Lime stabilization is achieved by adding lime (either as hydrated lime, $Ca(OH)_2$ or as quicklime, CaO) to the sludge and thus raising the pH to about 11 or above. This significantly reduces the odor and helps in the destruction of pathogens. The major disadvantage of lime stabilization is that it is temporary. With time (days) the pH drops and the sludge once again becomes putrescible.

Aerobic stabilization is merely a logical extension of the activated sludge system. Waste activated sludge is placed in dedicated aeration tanks for a very long time, and the concentrated solids allowed to progress well into the endogenous respiration phase, in which food is obtained only by the destruction of other viable organisms. This results in a net reduction in total and volatile solids. Aerobically digested sludges are, however, more difficult to dewater than anaerobic sludges.

The third commonly employed method of sludge stabilization is anaerobic digestion. The biochemistry of anaerobic decomposition of organics is illustrated in Figure 1. Note that this is a staged process, with the solution of organics by extracellular enzymes being followed by the production of organic acids by a large and hearty group of anaerobic microorganisms known, appropriately enough, as the *acid formers*. The organic acids are in turn degraded further by a group of strict anaerobes called *methane formers*. These microorganisms are the prima donnas of wastewater treatment, getting upset at the least change in their environment. The success of anaerobic treatment thus boils down to the creation of a suitable condition for the methane formers. Since they are strict anaerobes, they are unable to function in the presence of oxygen and very sensitive to environmental conditions such as temperature, pH and toxins. If a digester goes "sour," the methane formers have been inhibited in some way. The acid formers, however, keep chugging away, making more organic acids. This has the effect of further lowering the pH and making conditions even worse for the methane formers. A sick digester is therefore difficult to cure without massive doses of lime or other antacids.

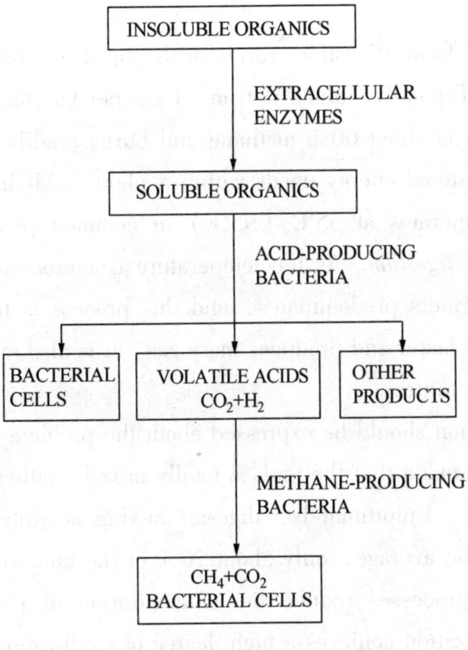


Figure 1 Generalized Biochemical Reactions to Anaerobic Digestion

Most treatment plants have two kinds of digesters-primary and secondary (Figure 2). The primary digester is covered, heated and mixed to increase the reaction rate. The temperature of the sludge is usually about 35℃ (95°F). Secondary digesters are not mixed or heated and are used for storage of gas and for concentrating the sludge by settling. As the solids settle, the liquid supernatant is pumped back to the main plant for further treatment. The cover of the secondary digester often floats up and down, depending on the amount of gas stored. The gas is high enough in methane to be used as a fuel, and is in fact usually used to heat the primary digester.

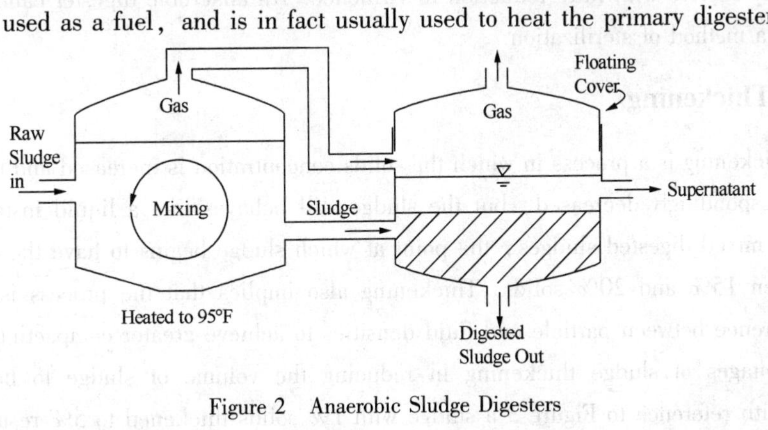


Figure 2 Anaerobic Sludge Digesters

Anaerobic digesters are commonly designed on the basis of solids loading. Experience has shown that domestic wastewaters contain about 120 g (0.27 lb) suspended solids per capita. This can be translated, knowing the population served, into total suspended solids to be handled. To this, of course, must be added the production of solids in secondary treatment(2). Once the solids production is calculated, the digester volume is estimated by assuming a reasonable hydraulic detention time for the sludge (generally 30 days) and using a loading factor such as 4 kg/($m^3$ · day) [0.27 lb/($ft^3$ · day)]. This loading factor is decreased if a higher reduction of volatile solids is desired.

The production of gas from digestion varies with the temperature, solids loading, solids volatility and other factors. Typically, about 0.6 $m^3$ of gas per kg volatile solids added (10 $ft^3$/lb) has been observed. This gas is about 60% methane and burns readily, usually being used to heat the digester and answer additional energy needs within a plant[3]. It has been found that an active group of methane formers operates at 35℃ (95°F) in common practice, and this process has become known as *mesophilic digestion*. As the temperature is increased, to about 45℃ (115°F), another group of methane formers predominates, and this process is tagged *thermophilic digestion*. Although the latter process is faster and produces more gas, it is also more difficult and expensive to maintain such elevated temperatures.

Finally, a word of caution should be expressed about the problem of mixing a primary digester. The assumption is invariably made that the tank is totally mixed, either by mechanical means or by bubbling gases through it[4]. Unfortunately, digester mixing is quite difficult, and some recent studies have shown that on the average, only about 20% of the tank volume is well mixed!

All three stabilization processes reduce the concentration of pathogenic organisms, but to varying degrees. Lime stabilization achieves a high degree of sterilization, due to the high pH. Further, if quicklime (CaO) is used, the reaction is exothermic and the elevated temperatures assist in the destruction of pathogens. Aerobic digestion at ambient temperatures is not very effective in the destruction of pathogens.

Anaerobic digesters have been well studied from the standpoint of pathogen viability since the elevated temperatures should result in substantial sterilization. As early as 1958, however, it was found that *Salmonella typhosa* organisms and many other pathogens can survive digestion. Polio viruses similarly survive with little reduction in virulence. An anaerobic digester cannot, therefore, be considered a method of sterilization.

## Sludge Thickening

Sludge thickening is a process in which the solids concentration is increased and the total sludge volume is correspondingly decreased, but the sludge still behaves like a liquid instead of a solid. Typically, for mixed digested sludges, the point at which sludge begins to have the properties of a solid is between 15% and 20% solids. Thickening also implies that the process is gravitational, using the difference between particle and fluid densities to achieve greater compacting of solids.

The advantages of sludge thickening in reducing the volume of sludge to be handled are substantial. With reference to Figure 3 a sludge with 1% solids thickened to 5% results in an 80% volume reduction. A 20% solids concentration, which might be achieved by mechanical dewatering (discussed in the next section) would result in a 95% reduction in volume. The savings in treatment, handling and disposal costs accrued can be substantial.

Two types of nonmechanical thickening operations are presently in use: the gravity thickener and the flotation thickener. These are not very good names, since the latter also uses gravity to separate the solids from the liquid. For the sake of simplicity, however, we will continue to use the two descriptive terms.

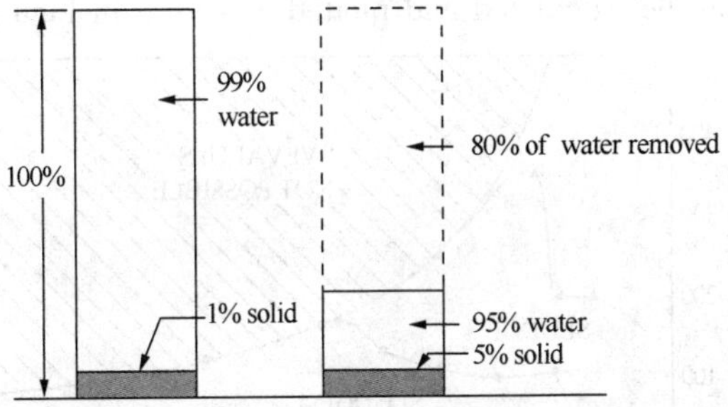


Figure 3 Volume Reduction due to Sludge Thickening

A typical gravity thickener is shown in Figure 4. The influent, or feed, enters in the middle, and the water moves to the outside, eventually leaving as the clear effluent over the weirs. The sludge solids settle as a blanket and are removed out the bottom(5).

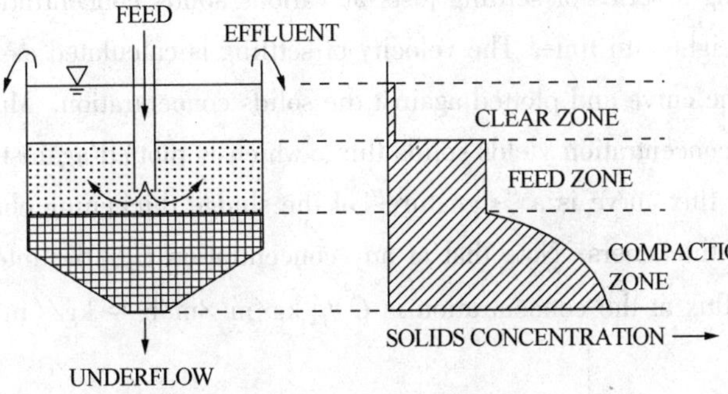


Figure 4 Gravity Thickener

The thickening characteristics of a sludge have for many years been described by the sludge volume index (SVI). The parameter is defined as the volume occupied by one gram of sludge after settling for 30 minutes(6). The sludge is commonly settled in a 1-liter cylinder and the volume of sludge (in mL) is measured after 30 minutes(7). The SVI is calculated as

$$\text{SVI} = (\text{mL of sludge after 30 min} \times 1000)/\text{sludge SS mg/L}$$

Treatment plant operators usually consider sludges with a SVI less than 100 as well settling sludges and those with a SVI greater than 200 as potential problems. These numbers are of some use in estimating the settleability of activated sludge, and the SVI is without question a valuable tool in running a secondary treatment plant. But the SVI also has some drawbacks and potential problems.

One of these problems is that the SVI is not necessarily independent of solids concentration. Substantial increases in SS can change the SVI several fold(8). Further, for high suspended solids, the SVI has a maximum above which it can never occur. Take for example a lime sludge with a suspended solids concentration of 40,000 mg/L, and suppose that it doesn't settle at all (the sludge volume after 30 min is 1000 mL)(9). Calculating the SVI gives us

$$\text{SVI} = \frac{1000 \times 1000}{40,000} = 25$$

How can a sludge that does not settle at all have a SVI = 25, when we originally noted that a SVI of less than 100 indicated a good settling sludge?

The point is that the maximum SVI for a sludge with a suspended solids of 40,000 is 25.

Similar maximum values can be calculated and plotted as shown in Figure 5.

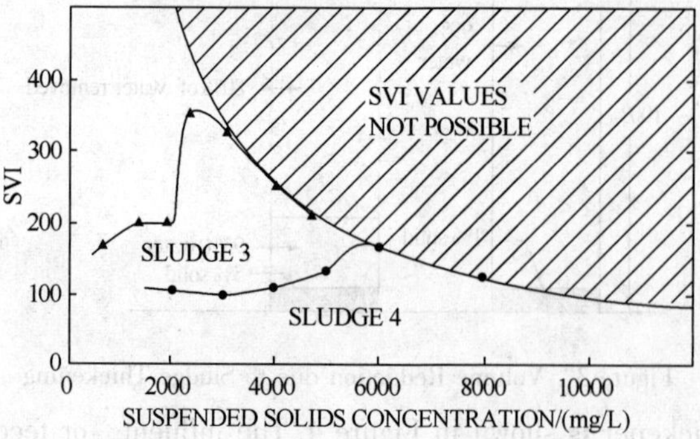


Figure 5 Maximum Values of SVI

A better way to describe the settling characteristics of a sludge is a "flux plot". This graph is developed by first running a series of settling tests at various solids concentrations and recording the sludge-water interface height with time. The velocity of settling is calculated as the slope of the initial straight line portion of the curve and plotted against the solids concentration. Multiplying a velocity by its corresponding solids concentration yields solids flux, which is plotted against the concentration (as shown in Figure 6). The flux curve is a "signature" of the sludge thickening characteristics. Further, it can be used to design thickeners. Note that at any concentration $C_i$, the interface settling velocity is $v_i$ and thus the solids flux at the concentration is $C_i v_i$ [$\text{kg/m}^3 \times \text{m/h} = \text{kg/(m}^2 \cdot \text{h)}$]

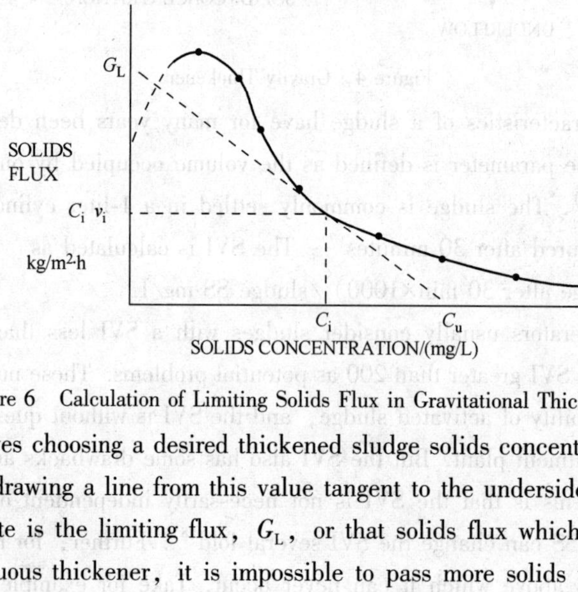


Figure 6 Calculation of Limiting Solids Flux in Gravitational Thickening

The design involves choosing a desired thickened sludge solids concentration, $C_u$ (estimated in laboratory tests) and drawing a line from this value tangent to the underside of the flux curve. The intercept at the ordinate is the limiting flux, $G_L$, or that solids flux which controls the thickening operation. In a continuous thickener, it is impossible to pass more solids than that (at the stated concentration) through a unit area.

The units of flux are kg solids/$\text{m}^2$hr (or other comparable units), and are defined as

$$G = \frac{Q_0 C_0}{A}$$

where $G$ = solids flux

$Q_0$ = inflow rate ($\text{m}^3$/h)

$C_0$ = inflow solids concentration (mg/L)

$A$ = surface area of thickener ( $m^3$ ).

If $G$ is the limiting flux, the necessary thickener area can be calculated as

$$A_L = \frac{Q_0 C_0}{G_L}$$

where the subscript L specifies the limiting area and flux.

This graphical procedure can be used to develop an optimal process. If the calculated thickener area is too big, a new (less ambitious) underflow solids concentration can be selected and a new required area calculated.

Note that the area requirement increases as the desired underflow concentration increases.

In the absence of laboratory data, thickeners are designed on the basis of solids loading, which is another way of expressing the limiting flux. Gravity thickener design loadings for some specific sludges are shown in Table 1.

**Table 1 Design Loadings for Gravity Thickeners**

| Sludge | Design Loading [kg solids/( $m^2 \cdot h$ )] | Sludge | Design Loading [kg solids/( $m^2 \cdot h$ )] |
|---|---|---|---|
| Raw Primary | 5.2 | Raw Primary + Waste Activated | 2.4 |
| Waste Activated | 1.2 | Trickling Filter Sludge | 1.8 |

A flotation thickener, shown in Figure 7, operates by forcing air under pressure to dissolve in the return flow, and releasing the pressure as the return is mixed with the feed. As the air comes out of solution, the tiny bubbles attach themselves to the solids and carry them upward, to be scraped off [10].

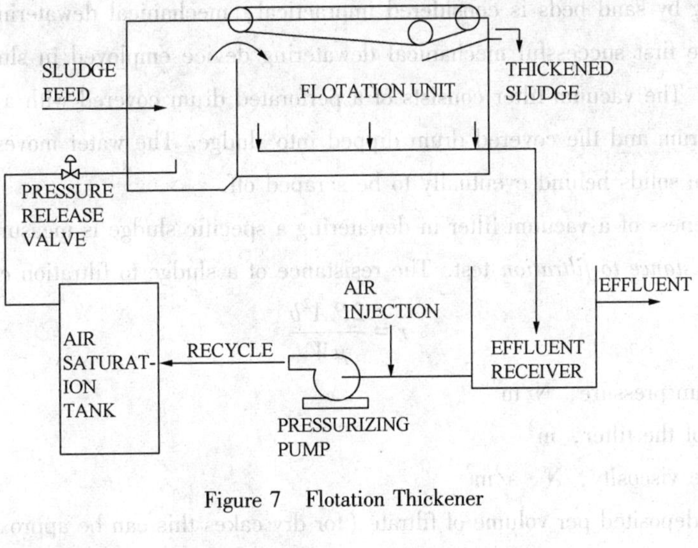


Figure 7 Flotation Thickener

## Sludge Dewatering

As defined above, dewatering differs from thickening in that the sludge should behave as a solid after it has been dewatered. Dewatering is seldom used as an intermediate process, unless the sludge is to be incinerated. Most wastewater plants use dewatering as a final method of volume reduction prior to ultimate disposal.

In the United States five dewatering techniques have been most popular: sand beds, vacuum filters, pressure filters, belt filters and centrifuges. Each of these is discussed below.

Sand beds have been in use for a great many years and are still the most cost-effective means of dewatering when land is available and labor is not exorbitant. The beds consists of tile drains in gravel, covered by about 26 cm (10 in.) of sand. The sludge to be dewatered is poured on the beds at about 15 cm (6 in.) deep. Two mechanisms combine to separate the water from the solids: seepage and evaporation. Seepage into the sand and through the tile drains, although important in the total volume of water extracted, lasts for only a few days. The sand pores are quickly clogged, and all drainage into the sand ceases. The mechanism of evaporation takes over, and this process is actually responsible for the conversion of liquid sludge to solid. In some northern areas sand beds are enclosed under greenhouses to promote evaporation as well as prevent rain from falling into the beds.

For mixed digested sludge, the usual design is to allow for 3 months drying time. Some engineers suggest that this period be extended to allow a sand bed to rest for a month after the sludge has been removed. This seems to be an effective means of increasing the drainage efficiency once the sand beds are again flooded.

Raw sludge will not drain well on sand beds and will usually have an obnoxious odor. Hence raw sludges are seldom dried on beds. Raw secondary sludges have a habit of either seeping through the sand or clogging the pores so quickly that no effective drainage takes place. Aerobically digested sludge can be dried on sand, but usually with some difficulty.

If dewatering by sand beds is considered impractical, mechanical dewatering techniques must be employed. The first successful mechanical dewatering device employed in sludge treatment was the *vacuum filter*. The vacuum filter consists of a perforated drum covered with a fabric. A vacuum is drawn in the drum and the covered drum dipped into sludge. The water moves through the filter cloth, leaving the solids behind eventually to be scraped off.

The effectiveness of a vacuum filter in dewatering a specific sludge is measured most frequently by the *specific resistance to filtration* test. The resistance of a sludge to filtration can be stated as

$$r = \frac{2PA^2b}{\mu W}$$

where $P$ = vacuum pressure, $N/m^2$

$A$ = area of the filter, $m^2$

$\mu$ = filtrate viscosity, $N \cdot s/m^2$

$W$ = cake deposited per volume of filtrate (for dry cakes this can be approximated as the feed solids concentration), $kg/m^3$

$b$ = slope of the filtrate volume vs time/filtrate volume curve.

Using the units above, specific resistance is in terms of m/kg.

The factor b can be determined from a simple test with a Buchner funnel, as shown in Figure 8. The sludge is poured onto the filter, the vacuum exerted, and the volume of the filtrate recorded against time[11]. The plot of these data yields a straight line with a slope b.

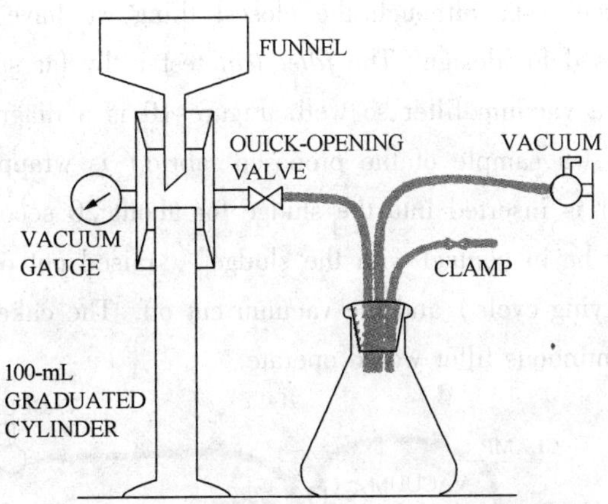


Figure 8 Buchner Funnel Test for Determining Specific Resistance to Filtration

Again, although there are no "average sludges", it may be instructive to list some data from experience and compare the filterability of different sludges. Such a listing is shown in Table 2.

**Table 2 Specific Resistance of Typical sludges**

| | Specific Resistance①/(m/kg) | | Specific Resistance①/(m/kg) |
|---|---|---|---|
| Raw Primary | $(10\sim30)\times10^{14}$ | Lime and Biological Sludge | $(1\sim5)\times10^{14}$ |
| Mixed Digested | $(3\sim30)\times10^{14}$ | Lime Slurry | $(5\sim10)\times10^{13}$ |
| Waste Activated | $(5\sim20)\times10^{14}$ | Alum | $(2\sim10)\times10^{13}$ |

① As a rule of thumb, a sludge will not filter well if the specific resistance exceeds $1\times10^{12}$ m/kg.

The Buchner funnel test for specific resistance is rather clumsy and time consuming. An indirect method for estimating how well a sludge will filter uses the *capillary suction time* (CST) apparatus, developed in Great Britain. The CST device, as illustrated in Figure 9, allows the water to seep out of a sludge (in effect a filtration process) and onto a blotter. The speed at which the water is taken up by the blotter is measured by timing of the travel over a set distance. This time, in seconds, can be correlated to specific resistance. A short time would indicate a highly filterable sludge, whereas a long time would portend problems in filtration. Some typical CST values are shown in Table 3.

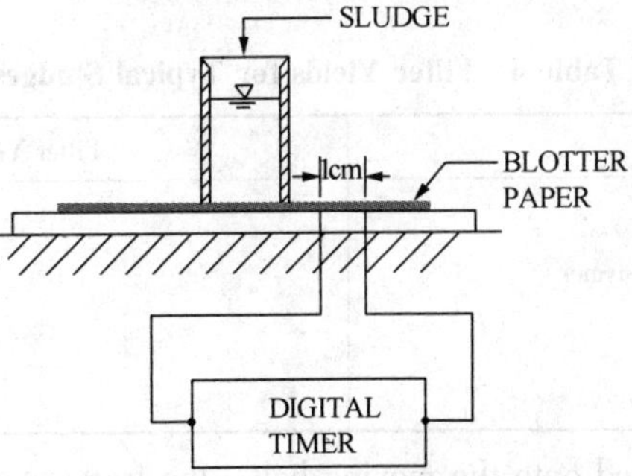


Figure 9 Capillary Suction Time Apparatus

**Table 3 Typical CST Values**

| Sludge | CST/s | Sludge | CST/s |
|---|---|---|---|
| Raw Primary Sludge | 80 | Mixed Digested Sludge | 50 |
| Activated Sludge | 10 | Alum Sludge | 30 |

The specific resistance test, although the closest thing we have to a basic parameter for filtration, should not be used for design. The *filter leaf* test is by far superior, simply because it simulates the workings of a vacuum filter so well. Figure 10 is a diagram showing the filter leaf apparatus. The filter cloth (a sample of the proposed fabric) is wrapped around a hollowed-out disc, the disc (filter leaf) is inserted into the sludge for about 20 seconds (the length of time a typical vacuum filter might be in contact with the sludge), raised out of the sludge for maybe 40 seconds (simulating the drying cycle) and the vacuum cut off. The cake developed is a very good approximation of how a continuous filter would operate.

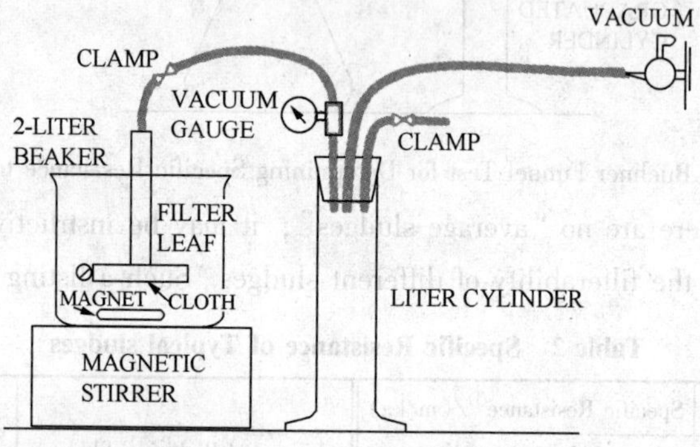


Figure 10 Filter Leaf Apparatus for Vacuum Filter Design

The design parameter calculated from the filter leaf test is the *filter yield*, in terms of kg of dry solids/$m^2$ of filter area/hr. If this value is known, the required filter size can be determined by selecting the hours of operation and establishing how much sludge needs to be removed from the treatment plant.

Typical filter yields for some representative sludges with polymer conditioning are shown in Table 4.

Typically, the pressure filters are built as plate-and-frame filters, where the sludge solids are captured in the spaces between the plates and frames, which are then pulled apart to allow for sludge cleanout.

**Table 4 Filter Yields for Typical Sludges**

| | Filter Yield/[kg/($m^2$ · h)] |
|---|---|
| Raw Primary (with polymer) | 40 |
| Mixed anaerobically Digested (with polymer) | 20 |
| Waste Activated | (will not filter) |
| Lime and Biological Sludge | 15 |

As the sludge is introduced onto the moving belt, the free water drips through the belt but the solids are retained. The belt then moves into the dewatering zone where the sludge is squeezed between two belts. These machines are quite effective in dewatering many different kinds of sludges and are being widely installed in small wastewater treatment plants.

*Centrifugation* became popular in wastewater treatment only after organic polymers were available for sludge conditioning. Although the centrifuge will work on any sludge (unlike the vacuum filter which will not pick up some sludges, resulting in zero filter yield), without good conditioning, most sludges can not be centrifuged with greater than 60% or 70% solids recovery.

The centrifuge most widely used is the solid bowl decanter, which consists of a bullet-shaped body rotating on its long axis. The sludge is placed into the bowl, the solids settle out under about 500 to 1000 gravities (centrifugally applied) and are scraped out of the bowl by a screw conveyor.

Although laboratory tests are of some value in estimating centrifuge applicability, tests using continuous models are considerably better and highly recommended whenever possible(12).

A centrifuge must be able to first settle the solids, and then to move them out of the bowl. Accordingly, two parameters have been suggested for scale-up between two geometrically similar machines.

Settling characteristics are measured by the *Sigma equation*. Without going into the derivation of this parameter, it is simply assumed that if two machines (1 and 2) are to have equal effects on the settling within the bowl, the relationship

$$\frac{Q_1}{\sum_1} = \frac{Q_2}{\sum_2}$$

must hold. The symbol $Q$ represents the liquid flow rate into the machine, and $\Sigma$ is a parameter comprised of machine (not sludge!) characteristics. For solid bowl centrifuges, Sigma is calculated as

$$\sum = \frac{\omega}{g\ln(r_2/r_1)}$$

where $\omega$ = rotational velocity of bowl, rad/s

$g$ = gravitational constant, m/sec$^2$

$r_2$ = radius from centerline to inside bowl wall, m

$r_1$ = radius from centerline to surface of sludge, m

Thus if machine 1, at a certain $\sum_1$ as calculated from the machine parameters, at the flow rate $Q_1$, produced a satisfactory result, it is expected that a second (geometrically similar) machine with a (larger) $\sum_2$ will achieve equal dewatering performance at a flow rate of $Q_2$.

This analysis does not, however, consider the movement of the solids out of the bowl; a very necessary component of centrifugation. The solids movement can be calculated based on the *Beta equation*, where again, for two machines,

$$\frac{Q_1}{\beta_1} = \frac{Q_2}{\beta_2}$$

where Q now is the terms of solids (lb/hr or other such dimension) and where $\beta = (\Delta\omega) SN\pi Dz$. These terms are defined as

$\Delta\omega$ = difference in rotational velocity between the bowl and conveyor, rad/sec

$S$ = scroll pitch (distance between blades), m

$N$ = number of leads

$D$ = bowl diameter, m

$z$ = depth of sludge in the bowl, m

The scale-up procedure involves the calculation of $Q_2$ for liquid as well as solids throughput, and the lowest value would govern the centrifuge capacity.

## Words and Expressions

1. save 节约，节省，营救
2. avert 避免，转移，防止
3. main process train 主要工艺序列，主要工艺系统
4. embark on 着手，开始做，从事
5. anaerobic digestion 厌氧消化
6. aerobic digestion 好氧消化
7. makes little sense 几乎没有意义
8. putrescibility 易腐败性
9. logical extension of the activated sludge system 活性污泥系统的合理延伸
10. dedicated 专用的
11. endogenous respiration phase 内源呼吸阶段
12. viable organism 能生存的生物，活有机体，活生物
13. extracellular enzyme 胞外酶
14. acid former 酸形成菌
15. methane former 甲烷形成菌
16. prima donna 主角
17. boils down to 归根到底是
18. antacid 抗酸剂
19. digester 消化池，消化器
20. supernatant 上层清液；(浮在)表层的
21. hydraulic detention time 水力停留时间
22. loading factor 负荷因子
23. mesophilic digestion 中温消化
24. thermophilic digestion 高温消化
25. maintain elevated temperature 维持高温，保持高温
26. tag 贴标签于，把……称为；标签
27. sterilization 灭菌，消毒，绝育
28. at ambient temperature 在环境温度下
29. Polio virus 脊髓灰质炎病毒，小儿麻痹症病毒
30. virulence 毒性
31. difference between particle and fluid densities 颗粒密度和流体密度之差
32. accrue 产生，出现，增长，增殖
33. gravity thickener 重力浓缩装置
34. flotation thickener 浮选浓缩装置
35. for the sake of simplicity 为简化起见
36. weir 堰
37. sludge volume index (SVI) 污泥体积指数
38. drawback 缺陷，缺点
39. be independent of 与……无关
40. not necessarily 未必
41. flux plot 通量曲线(图)
42. initial straight line portion of the curve 曲线的初始直线段
43. plot $A$ against $B$ 将 $A$ 对 $B$ 作图
44. multiply a velocity by its corresponding solids concentration 将流速与相应的固体浓度相乘
45. tangent to the underside of the flux curve 正切于通量曲线的底边
46. intercept at the ordinate 纵坐标上的截距
47. unit area 单位面积
48. subscript 下标
49. graphical procedure 图解法
50. underflow solids concentration 底流固体浓度
51. underflow concentration 底流浓度
52. gravity thickener design loadings for some

specific sludges 某些特定污泥的重力浓缩池设计负荷
53. in the absence of laboratory data 没有实验室数据
54. intermediate process 中间过程
55. incinerate 焚烧，煅烧
56. sand bed 砂床
57. vacuum filter 真空过滤器
58. pressure filter 压滤机
59. belt filter 带式过滤机
60. centrifuge 离心机
61. the most cost-effective means of dewatering 脱水的最合算的手段，脱水的最经济有效的手段
62. exorbitant 过度的，过高的，非法的
63. greenhouse 温室，暖房
64. seepage 渗透，渗出，渗漏
65. drainage efficiency 排水效率
66. obnoxious odor 不良气味，讨厌的气味
67. perforated drum covered with a fabric 覆盖了布料的穿孔鼓状物
68. leave the solids behind eventually to be scraped off 留下固体最终被刮去
69. specific resistance to filtration test 过滤试验的比阻力
70. Buchner funnel 布氏漏斗，平底瓷漏斗
71. instructive 指导性的，有教益的
72. filterability 可滤性
73. as a rule of thumb 作为经验法则
74. clumsy 笨拙的
75. capillary suction time（CST）毛细管空吸时间
76. blotter 吸墨纸，吸墨材料，吸油(集料)
77. portend 预示
78. polymer conditioning 聚合物调理
79. positive pressure 正压
80. plate-and-frame filter 板框压滤机
81. cleanout 清扫，清理，清除口，清理孔
82. squeeze 压榨，挤干
83. solid bowl decanter 无孔转鼓倾析器
84. screw conveyor 螺旋式输送装置
85. geometrically similar machines 几何相似机械
86. flow rate 流量
87. rotational velocity 旋转速度
88. govern 控制，支配

## Notes

（1）The “Get the water out!” syndrome often found in engineering offices is often associated with “Gotta digest it!” disease.

在工程办公室常有的“将水排出去”综合症常常与“对污泥进行消化”的病症相联系。

（2）To this, of course, must be added the production of solids in secondary treatment.

当然，在二级处理中产生的固体量必须加入到该量上。本句为倒装句。

（3）This gas is about 60% methane and burns readily, usually being used to heat the digester and answer additional energy needs within a plant.

该气体含有约60%的甲烷，易燃，通常被用来加热消化池和弥补厂内的额外能量需求。

（4）The assumption is invariably made that the tank is totally mixed, either by mechanical means or by bubbling gases through it.

总是假定池子是完全混合的，或是通过机械方式，或是通过鼓入气体的方式。

(5) The sludge solids settle as a blanket and are removed out the bottom.

污泥固体成层沉降，并在底部被去除。

(6) The parameter is defined as the volume occupied by one gram of sludge after settling for 30 minutes.

该参数被定义为1g污泥在沉降30 min后所占据的体积。

(7) The sludge is commonly settled in a 1-liter cylinder and the volume of sludge (in mL) is measured after 30 minutes.

污泥通常在1L的量筒中沉降，污泥的体积(以mL计)在(沉降)30 min后测定。

(8) Substantial increases in SS can change the SVI several fold.

SS的显著增加能增加SVI几倍。

(9) Take for example a lime sludge with a suspended solids concentration of 40,000 mg/L, and suppose that it doesn't settle at all (the sludge volume after 30 min is 1000 mL)

以SS浓度为40,000 mg/L的石灰污泥为例，假设它一点都不会沉降(30 min后污泥体积为1000 mL)。

(10) As the air comes out of solution, the tiny bubbles attach themselves to the solids and carry them upward, to be scraped off.

当空气从水中逸出，微气泡就会依附在固体上，并将固体往上携带，最终被刮去。

(11) The sludge is poured onto the filter, the vacuum exerted, and the volume of the filtrate recorded against time.

污泥被倒在滤池内，施以真空，然后记录滤液体积随时间的变化。

(12) Although laboratory tests are of some value in estimating centrifuge applicability, tests using continuous models are considerably better and highly recommended whenever possible.

尽管实验室试验在估计离心可应用性方面有一些价值，但采用连续模式的试验还是相当好的，只要有可能，是非常可取的。

# Unit 3 Air

## 15. Air Pollution Control Equipment

### Introduction

In solving an air pollution control equipment problem an engineer must first carefully evaluate the system or process in order to select the most appropriate type(s) of collector(s). After making preliminary equipment selection, suitable vendors can be contacted for help in arriving at a final answer. An early and complete definition of the problem can help reduce a poor decision that can lead to wasted pilot trials or costly inadequate installations.

Selecting an air pollution control device for cleaning a process gas stream can be a challenge. Some engineers, after trying to find shortcuts, employ quick estimates for both gas flow and collection efficiency that may be the entire extent of the collector specification. The end result can be an ineffective installation that has to be replaced. Treating a gas stream, especially to control pollution, is usually not a moneymaker, but costs—both capital and operating—can be minimized, not by buying the cheapest collector but by thoroughly engineering the whole system as is normally done in process design areas(1).

Controlling the emission of pollutants from industrial and domestic sources is important in protecting the quality of air. Air pollutants can exist in the form of particulate matter or gases. Air-cleaning devices have been reducing pollutant emissions from various sources for many years. Originally, air-cleaning equipment was used only if the contaminant was highly toxic or had some recovery value. Now with recent legislation, control technologies have been upgraded and more sources are regulated in order to meet the National Ambient Air Quality Standards (NAAQS)(2). In addition, state and local air pollution agencies have adopted regulations that are in some cases more stringent than the federal emission standards.

Equipment used to control particulate emissions are gravity settlers (often referred to as settling chambers), mechanical collectors (cyclones), electrostatic precipitators (ESPs), scrubbers (venturi scrubbers), and fabric filters (baghouses). Techniques used to control gaseous emissions are absorption, adsorption, combustion, and condensation. The applicability of a given technique depends on the physical and chemical properties of the pollutant and the exhaust stream. More than one technique may be capable of controlling emissions from a given source. For example, vapors generated from loading gasoline into tank trucks at large bulk terminals are controlled by using any of the above four gaseous control techniques. Most often, however, one control technique is used more frequently than others for a given source-pollutant combination. For example, absorption is commonly used to remove sulfur dioxide ($SO_2$) from boiler flue gas.

The material presented in this chapter regarding air pollution control equipment contains, at best, an overview of each control device. Equipment diagrams and figures, operation and maintenance procedures, and so on, have not been included in this development. More details, including predictive and design calculational procedures are available in the literature.

## Air Pollution Control Equipment for Particulates

As described above, the five major types of particulate air pollution control equipment are:

1. Gravity Settlers.
2. Cyclones.
3. Electrostatic Precipitators.
4. Venturi Scrubbers.
5. Baghouses.

Each of these devices is briefly described below.

## Gravity Settlers

Gravity settlers, or gravity settling chambers, have long been utilized industrially for the removal of solid and liquid waste materials from gaseous streams. Advantages accounting for their use are simple construction, low initial cost and maintenance, low pressure losses, and simple disposal of waste materials[(3)]. Gravity settlers are usually constructed in the form of a long, horizontal parallel-epipeds with suitable inlet and outlet ports. In its simplest form the settler is an enlargement (large box) in the duct carrying the particle-laden gases: the contaminated gas stream enters at one end, while the cleaned gas exits from the other end. The particles settle toward the collection surface at the bottom of the unit with a velocity at or near their settling velocity. One advantage of this device is that the external force leading to separation is provided free by nature. Its use in industry is generally limited to the removal of large particles, i. e., those larger than 40 microns (or micrometers).

## Cyclones

Centrifugal separators, commonly referred to as cyclones, are widely used in industry for the removal of solid and liquid particles (or particulates) from gas streams. Typical applications are found in mining and metallurgical operations, the cement and plastics industries, pulp and paper mill operations, chemical and pharmaceutical processes, petroleum production (cat-cracking cyclones) and combustion operations (fly ash collection)[(4)].

Particulates suspended in a moving gas stream possess inertia and momentum and are acted upon by gravity. Should the gas stream be forced to change direction, these properties can be utilized to promote centrifugal forces to act on the particles. In the conventional unit the entire mass of the gas stream with the entrained particles enter the unit tangentially and is forced into a constrained vortex in the cylindrical portion of the cyclone. Upon entering the unit, a particle develops an angular velocity. Because of its greater inertia, it tends to move across the gas streamlines in a tangential rather than rotary direction; thus, it attains a net outward radial velocity. By virtue of its rotation with the carrier gas around the axis of the tube (main vortex) and its high density with respect to the gas, the entrained particles are forced toward the wall of the unit. Eventually the particle may reach the outer wall, where they are carried by gravity and assisted by the downward movement of the outer vortex and/or secondary eddies toward the dust collector at the bottom of the unit. The flow vortex is reversed in the lower (conical) portion of the unit, leaving most of the entrained particles behind. The cleaned gas then passes up through the center of the unit (inner vortex) and out of the collector.

Multiple-cyclone collectors (multicones) are high efficiency devices that consist of a number of small-diameter cyclones operating in parallel with a common gas inlet and outlet. The flow pattern differs from a conventional cyclone in that instead of bringing the gas in at the side to initiate the swirling action, the gas is brought in at the top of the collecting tube and the swirling action is then imparted by a stationary vane positioned in the path of the incoming gas. The diameters of the collecting tubes usually range from 6 to 24 inches. Properly designed units can be constructed and operated with a collection efficiency as high as 90 percent for particulates in the 5 to 10 micron range. The most serious problems encountered with these systems involve plugging and flow equalization.

## Electrostatic Precipitators

Electrostatic precipitators (ESPs) are satisfactory devices for removing small particles from moving gas streams at high collection efficiencies. They have been used almost universally in power plants for removing fly ash from the gases prior to discharge.

Two major types of high-voltage ESP configuration currently used are tubular and plate. Tubular precipitators consist of cylindrical collection tubes with discharge electrodes located along the axis of the cylinder(5). However, the vast majority of ESPs installed are the plate type. Particles are collected on a flat parallel collection surface spaced 8 to 12 inches apart, with a series of discharge electrodes located along the centerline of the adjacent plates. The gas to be cleaned passes horizontally between the plates (horizontal flow type) or vertically up through the plates (vertical flow type). Collected particles are usually removed by rapping.

Depending on the operating conditions and the required collection efficiency, the gas velocity in an industrial ESP is usually between 2.5 and 8.0 ft/sec. A uniform gas distribution is of prime importance for precipitators, and it should be achieved with a minimum expenditure of pressure drop. This is not always easy, since gas velocities in the duct ahead of the precipitator may be 30 to 100 ft/sec in order to prevent dust buildup. It should be clear that the best operating condition for a precipitator will occur when the velocity distribution is uniform. When significant maldistribution occurs, the higher velocity in one collecting plate area will decrease efficiency more than a lower velocity at another plate area will increase the efficiency of that area.

The maximum voltage at which a given field can be maintained depends on the properties of the gas and the dust being collected. These parameters may vary from one point to another within the precipitator, as well as with time. In order to keep each section working at high efficiency, a high degree of sectionalization is recommended. This means that the many separate power supplies and controls will produce better performance on a precipitator of a given size than if there were only one or two independently controlled sections. This is particularly true if high efficiencies are required.

## Venturi Scrubbers

Wet scrubbers have found widespread use in cleaning contaminated gas streams because of their ability to effectively remove both particulate and gaseous pollutants(6). Specifically, wet scrubbing involves a technique of bringing a contaminated gas stream into intimate contact with a liquid. Wet scrubbers include all the various types of gas absorption equipment (to be discussed later). The term "scrubber" will be restricted to those systems that utilize a liquid, usually water, to achieve or assist in the removal of particulate matter from a gas stream. The use of wet scrubbers to remove gaseous pollutants from contaminated streams is considered in the next section.

Another important design consideration for the venturi scrubber (as well as absorbers) is concerned with suppressing the steam plume. Water-scrubber systems removing pollutants from high-temperature processes (i. e., combustion) can generate a supersaturated water vapor that becomes a visible white plume as it leaves the stack. Although not strictly an air pollution problem, such a plume may be objectionable for aesthetic reasons. Regardless, there are several ways to avoid or eliminate the steam plume. The most obvious way is to specify control equipment that does not use water in contact with the high-temperature gas stream, (i. e., ESP, cyclones or fabric filters). Should this not be possible or practical, a number of suppression methods are available:

1. Mixing with heated and relatively dry air
2. Condensation of moisture by direct contact with water, then mixing with heated ambient air
3. Condensation of moisture by direct contact with water, then reheating the scrubber exhaust gas

## Baghouses

The basic filtration process may be conducted in many different types of fabric filters in which the physical arrangement of hardware and the method of removing collected material from the filter media will vary. The essential differences may be related, in general, to:

1. Type of fabric
2. Cleaning mechanism
3. Equipment
4. Mode of operation

Gases to be cleaned can be either pushed or pulled through the baghouse. In the pressure system (push through), the gases may enter through the cleanout hopper in the bottom or through the top of the bags. In the suction type (pull through), the dirty gases are forced through the inside of the bag and exit through the outside.

Baghouse collectors are available for either intermittent or continuous operation. Intermittent operation is employed where the operational schedule of the dust-generating source permits halting the gas cleaning function at periodic intervals (regularly defined by time or by pressure differential) for removal of collected material from the filter media (cleaning). Collectors of this type are primarily utilized for the control of small-volume operations such as grinding and polishing, and for aerosols of a very coarse nature. For most air pollution control installations and major particulate control problems, however, it is desirable to use collectors that allow for continuous operation. This is accomplished by arranging several filter areas in a parallel flow system and cleaning one area at a time according to some preset mode of operation.

Baghouses may also be characterized and identified according to the method used to remove collected material from the bags[7]. Particle removal can be accomplished in a variety of ways, including shaking the bags, blowing a jet of air on the bags, or rapidly expanding the bags by a pulse of compressed air. In general, the various types of bag cleaning methods can be divided into those involving fabric flexing and those involving a reverse flow of clean air. In pressure-jet or pulse-jet cleaning, a momentary burst of compressed air is introduced through a tube or nozzle attached at the top of the bag. A bubble of air flows down the bag, causing the bag walls to collapse behind it.

A wide variety of woven and felted fabrics are used in fabric filters. Clean felted fabrics are more efficient dust collectors than are woven fabrics, but woven materials are capable of giving equal filtration efficiency after a dust layer accumulates on the surface. When a new woven fabric is placed in service, visible penetration of dust within the fabric may occur. This normally takes from a few hours to a few days for industrial applications, depending on the dust loadings and the nature of the particles.

Baghouses are constructed as single units or compartmental units. The single unit is generally used on small processes that are not in continuous operation, such as grinding and paint-spraying processes. Compartmental units consist of more than one baghouse compartment and are used in continuous operating processes with large exhaust volumes such as electric melt steel furnaces and industrial boilers. In both cases, the bags are housed in a shell made of rigid metal material.

## AIR POLLUTION CONTROL EQUIPMENT FOR GASEOUS POLLUTANTS

As described in the Introduction, the four generic types of gaseous control equipment include:

1. Absorption
2. Adsorption
3. Combustion
4. Condensation

### Absorption

Absorption is a mass transfer operation in which a gas is dissolved in a liquid. A contaminant (pollutant exhaust stream) contacts a liquid and the contaminant diffuses (is transported) from the gas phase into the liquid phase. The absorption rate is enhanced by (1) high diffusion rates, (2) high solubility of the contaminant, (3) large liquid-gas contact area, and (4) good mixing between liquid and gas phases (turbulence).

The liquid most often used for absorption is water because it is inexpensive, is readily available, and can dissolve a number of contaminants. Reagents can be added to the absorbing water to increase the removal efficiency of the system. Certain reagents merely increase the solubility of the contaminant in the water. Other reagents chemically react with the contaminant after it is absorbed. In reactive scrubbing the absorption rate is much higher, so in some cases a smaller, economical system can be used. However, the reactions can form precipitates that could cause plugging problems in the absorber or in associated equipment.

If a gaseous contaminant is very soluble, almost any of the wet scrubbers will adequately remove this contaminant. However, if the contaminant is of low solubility, the packed tower or the plate tower is more effective. Both of these devices provide long contact time between phases and have relatively low pressure drops. The packed tower, the most common gas absorption device, consists of an empty shell filled with packing. The liquid flows down over the packing, exposing a large film area to the gas flowing up the packing. Plate towers consist of horizontal plates placed inside the tower. Gas passes up through the orifices in these plates while the liquid flows down across the plate, thereby providing desired contact.

### Adsorption

Adsorption is a mass transfer process that involves removing a gaseous contaminant by adhering to the surface of a solid. Adsorption can be classified as physical or chemical. In physical adsorption, a gas molecule adheres to the surface of the solid due to an imbalance of natural forces (electron distribution). In chemisorption, once the gas molecule adheres to the surface, it reacts chemically with it. The major distinction is that physical adsorption is readily reversible whereas chemisorption is not.

All solids physically adsorb gases to some extent. Certain solids, called adsorbents, have a high attraction for specific gases; they also have a large surface area that provides a high capacity for gas capture. By far the most important adsorbent for air pollution control is activated carbon. Because of its unique surface properties, activated carbon will preferentially adsorb hydrocarbon vapors and odorous organic compounds from an airstream. Most other adsorbents (molecular sieves, silica gel, and activated aluminas) will preferentially adsorb water vapor, which may render them useless to remove other contaminants.

For activated carbon, the amount of hydrocarbon vapors that can be adsorbed depends on the physical and chemical characteristics of the vapors, their concentration in the gas stream, system temperature, system pressure, humidity of the gas stream, and the molecular weight of the vapor. Physical adsorption is a reversible process; the adsorbed vapors can be released (desorbed) by increasing the temperature, decreasing the pressure or using a combination of both. Vapors are normally desorbed by heating the adsorber with steam.

Adsorption can be a very useful removal technique, since it is capable of removing very small quantities (a few parts per million) of vapor from an airstream. The vapors are not destroyed; instead, they are stored on the adsorbent surface until they can be removed by desorption. The desorbed vapor stream is normally highly concentrated. It can be condensed and recycled, or burned as an ultimate disposal technique.

The most common adsorption system is the fixed bed adsorber. These systems consist of two or more adsorber beds operating on a timed adsorbing/desorbing cycle. One or more beds are adsorbing vapors, while the other bed(s) is being regenerated. If particulate matter or liquid droplets are present in the vapor-laden airstream, this stream is sent to pretreatment to remove them. If the temperature of the inlet vapor stream is high (much above 120 °F), cooling may also be required. Since all adsorption processes are exothermic, cooling coils in the carbon bed itself may also be needed to prevent excessive heat buildup. Carbon bed depth is usually limited to a maximum of 4 ft, and the vapor velocity through the adsorber is held below 100 ft/min to prevent an excessive pressure drop.

## Combustion

Combustion is defined as a rapid, high-temperature gas-phase oxidation. Simply, the contaminant (a carbon-hydrogen substance) is burned with air and converted to carbon dioxide and water vapor. The operation of any combustion source is governed by the three T's of combustion: temperature, turbulence, and time. For complete combustion to occur, each contaminant molecule must come in contact (turbulence) with oxygen at a sufficient temperature, while being maintained at this temperature for an adequate time. These three variables are dependent on each other. For example, if a higher temperature is used, less mixing of the contaminant and combustion air or shorter residence time may be required. If adequate turbulence cannot be provided, a higher temperature or longer residence time may be employed for complete combustion.

Combustion devices can be categorized as flares, thermal incinerators, or catalytic incinerators[(8)]. Flares are direct combustion devices used to dispose of small quantities or emergency releases of combustible gases. Flares are normally elevated (from 100 to 400 ft) to protect the surroundings from the heat and flames. Flares are often designed for steam injection at the flare tip. The steam provides sufficient turbulence to ensure complete combustion; this prevents smoking. Flares are also very noisy, which can cause problems for adjacent neighborhoods.

Thermal incinerators are also called afterburners, direct flame incinerators, or thermal oxidizers[(9)]. These are devices in which the contaminant airstream passes around or through a burner and into a refractory-line residence chamber where oxidation occurs. To ensure complete combustion of the contaminant, thermal incinerators are designed to operate at a temperature of 700 to 800℃ (1300 to 1500 °F) and a residence time of 0.3 to 0.5 sec. Ideally, as much fuel value as possible is supplied by the waste contaminant stream; this reduces the amount of auxiliary fuel needed to maintain the proper temperature.

In catalytic incineration the contaminant-laden stream is heated and passed through a catalyst bed that promotes the oxidation reaction at a lower temperature. Catalytic incinerators normally operate at 370 to 480℃ (700 to 900 °F). This reduced temperature represents a continuous fuel savings. However, this may be offset by the cost of the catalyst. The catalyst, which is usually platinum, is coated on a cheaper metal or ceramic support base. The support can be arranged to expose a high surface area, which provides sufficient active sites on which the reaction(s) occur. Catalysts are subject to both physical and chemical deterioration. Halogens and sulfur-containing compounds act as catalyst suppressants and decrease the catalyst usefulness. Certain heavy metals such as mercury, arsenic, phosphorous, lead, and zinc are particularly poisonous.

### Condensation

Condensation is a process in which the volatile gases are removed from the contaminant stream and changed into a liquid. Condensation is usually achieved by reducing the temperature of a vapor mixture until the partial pressure of the condensable component equals its vapor pressure. Condensation requires low temperatures to liquefy most pure contaminant vapors. Condensation is affected by the composition of the contaminant gas stream. The presence of additional gases that do not condense at the same conditions-such as air-hinders condensation.

Condensers are normally used in combination with primary control devices. Condensers can be located upstream of (before) an incinerator, adsorber, or absorber. These condensers reduce the volume of vapors that the more expensive equipment must handle. Therefore, the size and the cost of the primary control device can be reduced. Similarly, condensers can be used to remove water vapors from a process stream with a high moisture content upstream of a control system. A prime example is the use of condensers in rendering plants to remove moisture from the cooker exhaust gas. When used alone, refrigeration is required to achieve the low temperatures required for condensation. Refrigeration units are used to successfully control gasoline vapors at large gasoline dispensing terminals.

Condensers are classified as being either contact condensers or surface condensers. Contact condensers cool the vapor by spraying liquid directly on the vapor stream. These devices resemble a simple spray scrubber. Surface condensers are normally shell-and-tube heat exchangers. Coolant flows through the tubes, while vapor is passed over and condenses on the outside of the tubes. In general, contact condensers are more flexible, simpler, and less expensive than surface condensers. However, surface condensers require much less water and produce nearly twenty times less wastewater that must be treated than do contact condensers. Surface condensers also have an advantage in that they can directly recover valuable contaminant vapors.

### Hybrid System

Hybrid systems are defined as those types of control devices that involve combinations of control mechanisms—for example, fabric filtration combined with electrostatic precipitation. Unfortunately, the term hybrid system has come to mean different things to different people. The two most prevalent definitions employed today for hybrid systems are:

1. Two or more different air pollution control equipment connected in series, e. g., a baghouse followed by an absorber.

2. An air pollution control system that utilizes two or more collection mechanisms simultaneously to enhance pollution capture, for example, an ionizing wet scrubber (IWS), that will be discussed shortly.

The two major hybrid systems found in practice today include ionizing wet scrubbers and dry scrubbers. These are briefly described below.

### Ionizing Wet Scrubbers

The ionizing wet scrubber (IWS) is a relatively new development in the technology of the removal of particulate matter from a gas stream. These devices have been incorporated in commercial incineration facilities. In the IWS, high-voltage ionization in the charge section places a static electric charge on the particles in the gas stream, which then passes through a crossflow packed-bed scrubber. The packing is normally polypropylene in the form of circular-wound spirals and gearlike wheel configurations, providing a large surface area. Particles with sizes of 3 microns or larger are trapped by inertial impaction within the bed. Smaller charged particles pass close to the surface of either the packing material or a scrubbing water droplet. An opposite charge on that surface is induced by the charged particle, which is then attracted to an ion attached to the surface. All collected particles are eventually washed out of the scrubber. The scrubbing water also can function to absorb gaseous pollutants.

According to Celicote (the IWS vendor), the collection efficiency of the two-stage IWS is greater than that of a baghouse or a conventional ESP for particles in the 0.2 to 0.6 micron range. For 0.8 microns and above, the ESP is as effective as the IWS. Scrubbing water can include caustic soda or soda ash when needed for efficient adsorption of acid gases. Corrosion resistance of the IWS is achieved by fabricating its shell and most internal parts with fiberglass-reinforced plastic (FRP) and thermoplastic materials. Pressure drop through a single-stage IWS is approximately 5 in $H_2O$ (primarily through the wet scrubber section). All internal areas of the ionizer section are periodically deluge-flushed with recycled liquid from the scrubber system[10].

### Dry Scrubbers

The success of fabric filters in removing fine particles from flue gas streams has encouraged the use of combined dry scrubbing/fabric filter systems for the dual purpose of removing both particulates and acid gases simultaneously. Dry scrubbers offer potential advantages over their wet counterparts, especially in the areas of energy savings and capital costs. Furthermore, the dry-scrubbing process design is relatively simple, and the product is a dry waste rather than a wet sludge.

There are two major types of so-called dry scrubber systems: spray drying and dry injection. The first process is often referred to as a wet-dry system. When compared to the conventional wet scrubber, it uses significantly less liquid. The second process has been referred to as a dry-dry system because no liquid scrubbing is involved. The spray-drying system is predominately used in utility and industrial applications.

The method of operation of the spray dryer is relatively simple, requiring only two major items: a spray dryer similar to those used in the chemical food-processing and mineral-preparation industries, and a baghouse or ESP to collect the fly ash and entrained solids. In the spray dryer, the sorbent solution, or slurry, is atomized into the incoming flue gas stream to increase the liquid-gas interface and to promote the mass transfer of the $SO_2$ from the gas to the slurry droplets where it is absorbed. Simultaneously, the thermal energy of the gas evaporates the water in the droplets to produce a dry powdered mixture of sulfite-sulfate and some unreacted alkali. Because the flue gas is not saturated and contains no liquid carryover, potentially troublesome mist eliminators are not required. After leaving the spray dryer, the solids-bearings gas passes through a fabric filter (or ESP), where the dry product is collected and where a percentage of unreacted alkali reacts with the $SO_2$ for further removal. The cleaned gas is then discharged through the fabric-filter plenum to an induced draft (ID) fan and to the stack.

Among the inherent advantages that the spray dryer enjoys over the wet scrubbers are:

1. Lower capital cost
2. Lower draft losses
3. Reduced auxiliary power
4. Reduced water consumption
5. Continuous, two-stage operation, from liquid feed to dry product

The sorbent of choice for most spray-dryer systems is a lime slurry.

Dry-injection processes generally involve pneumatic introduction of a dry, powdery alkaline material, usually a sodium-base sorbent, into the flue gas stream with subsequent fabric filter collection. The injection point in such processes can vary from the boiler-furnace area all the way to the flue gas entrance to the baghouse, depending on operating conditions and design criteria.

## Factors in Control Equipment Selection

There are a number of factors to be considered prior to selecting a particular piece of air pollution control hardware (ETS, 1989). In general, they can be grouped into three categories: environmental, engineering, and economic. These are detailed below.

### Environmental

1. Equipment location
2. Available space
3. Ambient conditions
4. Availability of adequate utilities (i. e., power, water, etc.) and ancillary system facilities (i. e., waste treatment and disposal, etc.)
5. Maximum allowable emissions (air pollution regulations)
6. Aesthetic considerations (i. e., visible steam or water vapor plume, impact on scenic vistas, etc.)
7. Contribution of air pollution control system to wastewater and solid waste
8. Contribution of air pollution control system to plant noise levels

### Engineering

1. Contaminant characteristics (i. e., physical and chemical properties, concentration, particulate shape and size distribution, etc.; in the case of particulates, chemical reactivity, corrosivity, abrasiveness, toxicity, etc.)
2. Gas stream characteristics (i. e., volume flow rate, temperature, pressure, humidity, composition, viscosity, density, reactivity, combustibility, corrosivity, toxicity, etc.)
3. Design and performance characteristics of the particular control system (i. e., size and weight, fractional efficiency curves, mass transfer and/or contaminant destruction capability, pressure drop, reliability and dependability, turndown capability, power requirements, utility requirements, temperature limitations, maintenance requirements, flexibility toward complying with more stringent air pollution regulations, etc.)

### Economic

1. Capital cost (equipment, installation, engineering, etc.)
2. Operating cost (utilities, maintenance, etc.)
3. Expected equipment lifetime and salvage value

Proper selection of a particular system for a specific application can be extremely difficult and complicated. In view of the multitude of complex and often ambiguous pollution regulations, it is in the best interest of the prospective user to work closely with regulatory officials as early in the process as possible. Finally, previous experience on a similar application cannot be over-emphasized.

## Comparing Control Equipment Alternatives

The final choice in equipment selection is usually dictated by that equipment capable of achieving compliance with the regulatory codes at the lowest uniform annual cost (amortized capital investment plus operation and maintenance costs). In order to compare specific control equipment alternatives, knowledge of the particular application and site is essential. A preliminary screening, however, may be performed by reviewing the advantages and disadvantages of each type of air pollution control equipment. For example, if water or a waste stream treatment is not available at the site, this may preclude use of a wet scrubber system and instead focus on particulate removal by dry systems, such as cyclones or baghouses and/or ESP. If auxiliary fuel is unavailable on a continuous basis, it may not be possible to combust organic pollutant vapors in an incineration system. If the particle-size distribution in the gas stream is relatively fine, cyclone collectors would probably not be considered. If the pollutant vapors can be reused in the process, control efforts may be directed to adsorption systems. There are many more situations where the knowledge of the capabilities of the various control options, combined with common sense will simplify the selection procedure. General advantages and disadvantages of the most popular types of air pollution control equipment for gases and particulates are too detailed to present here but are available in literature.

## Future Trends

The basic design of air pollution control equipment has remained relatively unchanged since first used in the early part of the twentieth century. Some modest equipment changes and new types of devices have appeared in the last twenty years, but all have essentially employed the same capture mechanisms used in the past. One area that has recently received some attention is hybrid systems (see earlier section)—equipment that can in some cases operate at higher efficiency more economically than conventional devices. Tighter regulations and a greater concern for environmental control by society has placed increased emphasis on the development and application of these systems. The future will unquestionably see more activity in this area.

Recent advances in this field have been primarily involved in the treatment of metals. A dry scrubber followed by a wet scrubber has been employed in the United States to improve the collection of fine particulate metals in hazardous-waste incinerators; the dry scrubber captures metals that condense at the operating temperature of the unit and the wet scrubber captures residue metals (particularly mercury) and dioxin/furan compounds. Another recent application in Europe involves the injection of powdered activated carbon into a flue gas stream from an hazardous waste incinerator at a location between the spray dryer (the dry scrubber) and the baghouse (or electrostatic precipitator). The carbon mixing with the lime particulates from the dry scrubbing system and the gas stream itself adsorb the mercury vapors and residual dioxin/furan compounds and are separated from the gas stream by a particulate control device. More widespread use of these types of systems is anticipated in the future.

## Words and Expressions

1. vendor 卖主，卖方，自动售货机
2. arrive at a final answer 得出最终的答案
3. pilot trial 小批试生产，小规模试车
4. process gas stream 工艺气流
5. shortcut 捷径，近路
6. collection efficiency 集尘效率
7. capital and operating costs 基建和运行成本［费用］
8. engineer 设计
9. particulate matter 颗粒物
10. air-cleaning device 空气净化装置
11. upgrade 改进，提高
12. stringent 严格的，严厉的
13. gravity settler 重力沉降器
14. settling chamber 沉降室
15. mechanical collector 机械集尘装置
16. cyclone 旋风除尘器，旋流，旋风
17. electrostatic precipitator（ESP）静电除尘器
18. scrubber 涤气器，洗涤器
19. Venturi scrubber 文丘里涤气器
20. fabric filter 布袋除尘器，布袋［纤维］过滤器
21. baghouse 布袋，袋滤室
22. gaseous emission 气体排放物，排气
23. absorption 吸收
24. adsorption 吸附
25. combustion 燃烧
26. condensation 冷凝
27. exhaust stream 废气
28. tank truck 槽车
29. large bulk terminal 大型散装场
30. sulfur dioxide 二氧化硫
31. boiler flue gas 锅炉烟气
32. operation and maintenance procedure 操作和维护方式
33. particle-laden gas 含颗粒物气体
34. centrifugal separator 离心分离机
35. mining and metallurgical operation 采矿和冶金操作［作业］
36. cat-cracking cyclone 催化裂解气旋器
37. fly ash 飞灰
38. inertia 惯性
39. momentum 动量
40. entrained particles 夹带的颗粒物
41. tangentially 成切线
42. vortex 涡流，旋涡
43. cylindrical 圆筒形的，圆柱体的
44. angular velocity 角速度
45. rotary direction 旋转方向
46. radial velocity 径向速度
47. by virtue of 借助于，由于
48. eddy 旋涡（运动），涡（流）；旋转，（使）起旋涡
49. conical portion of the unit 装置的锥形部分
50. leave... behind 留下，遗留
51. flow pattern 流型
52. swirling action 旋流作用，打旋作用
53. impart 给予，传递
54. stationary vane 固定的风扇
55. rap 轻拍，轻轻敲打
56. of prime importance 非常重要的
57. discharge electrode 放电电极
58. minimum expenditure of pressure drop 压降的最少消耗［支出］
59. buildup 集结，积累，聚结，增长，增加
60. maldistribution 分布不良，分布不均
61. sectionalization 分区，分段，分组

62. intimate contact with a liquid 与液体的密切接触
63. restrict 限制
64. suppress the steam plume 抑制蒸汽羽，消除蒸汽羽
65. suppression method 抑制方法，消除方式
66. supersaturated water vapor 过饱和水蒸气
67. stack 烟囱
68. objectionable 讨厌的，不适合的，不能采用的，不好的，有害的
69. filter media 过滤装置介质，滤池滤料[介质]
70. cleanout hopper 清除漏斗，出灰漏斗
71. intermittent operation 间歇式操作
72. halt 阻止，停止
73. periodic intervals 周期性间断
74. aerosol 气溶胶，烟雾剂，按钮式喷雾器
75. preset mode of operation 预置操作模式
76. a pulse of compressed air 一阵压缩空气
77. a momentary burst of compressed air 一股压缩空气
78. woven fabrics 纺织品
79. felted fabrics 黏结织物
80. compartmental units 空间分隔装置
81. paint-spraying process 喷漆工艺
82. rigid 刚性的
83. gaseous pollutants 气态污染物
84. generic 一般的，普通的
85. packed tower 填料塔
86. packing 填料
87. plate tower 板式塔
88. orifice 孔，洞，口，通气口
89. chemisorption 化学吸附，化学吸着
90. distinction 差异
91. reversible 可逆的
92. molecular sieve 分子筛
93. silica gel 硅胶
94. activated alumina 活性铝
95. preferentially 优先地，优惠地
96. render 给……初涂，打底，粉刷，提炼
97. humidity 湿度
98. molecular weight 分子量
99. fixed bed adsorber 固定床吸附装置
100. cooling coils 冷却盘管，冷却蛇管
101. residence time 停留时间
102. auxiliary fuel 辅助燃料
103. contaminant-laden stream 含污染物气流
104. offset 补偿，弥补，抵消
105. platinum 铂
106. deterioration 恶化
107. halogen 卤(素)
108. catalyst suppressants 催化剂抑制剂
109. partial pressure 分压
110. liquefy 液化
111. rendering plant 炼油厂
112. refrigeration units 冷冻装置
113. gasoline dispensing terminals 汽油配给站
114. hybrid system 混合系统
115. prevalent 盛行的
116. ionizing wet scrubber（IWS）电离湿式涤气器
117. static electric charge 静电荷
118. crossflow packed-bed scrubber 错流填充床涤气器
119. polypropylene 聚丙烯
120. circular-wound spirals and gearlike wheel configurations 绕成圆形的螺旋管和齿轮状轮子的构形
121. inertial impaction 惯性碰撞
122. induce 诱导，感应，导致
123. caustic soda 苛性钠
124. soda ash 苏打灰
125. fibreglass-reinforced plastics（FRP）玻璃纤维增强塑料，玻璃钢
126. thermoplastic materials 热塑性材料
127. utility and industrial applications 公用事业和工业应用
128. slurry 淤泥，泥浆
129. atomize 雾化，原子化
130. mist 雾

131. solids-bearings gas 含固体气体
132. induced draft（ID）fan 引风机，吸风机，排烟机
133. pneumatic 气动的，(有)空气的，
134. ancillary system facilities 辅助系统设施
135. scenic vista 风景优美的景色
136. noise level 噪声级
137. abrasiveness 磨耗，磨损性，耐磨性
138. volume flow rate 体积流量
139. comply with more stringent air pollution regulations 遵守更严格的空气污染规章，符合更严格的空气污染条例
140. expected equipment lifetime and salvage value 预期设备寿命和综合利用价值
141. ambiguous 模糊的，意义不明确的
142. prospective user 可能的用户，预期的用户
143. regulatory code 规章
144. preclude 预防，防止
145. common sense 常识
146. dioxin 二噁英
147. furan 呋喃

## Notes

(1) Treating a gas stream, especially to control pollution, is usually not a moneymaker, but costs—both capital and operating—can be minimized, not by buying the cheapest collector but by thoroughly engineering the whole system as is normally done in process design areas.

处理气流，特别是控制污染，通常不是一件赚钱的事情，但是成本—包括基建和运行成本，能够最小化，不要购买最廉价的集尘器而要购买充分设计的完整系统(正如在工艺设计领域所做的那样)。

(2) Now with recent legislation, control technologies have been upgraded and more sources are regulated in order to meet the National Ambient Air Quality Standards (NAAQS).

现在有了新的立法，为了符合国家环境空气质量标准(NAAQS)，控制技术已得到改进，更多的污染源受到管理。

(3) Advantages accounting for their use are simple construction, low initial cost and maintenance, low pressure losses, and simple disposal of waste materials.

说明它们应用的优点有结构简单，初期成本和维护费低，压力损失小和废物处置简单。

(4) Typical applications are found in mining and metallurgical operations, the cement and plastics industries, pulp and paper mill operations, chemical and pharmaceutical processes, petroleum production (cat-cracking cyclones) and combustion operations (fly ash collection).

典型的应用在采矿和冶金作业、水泥和塑料工业、制浆和造纸工业、化学和制药过程、石油生产(催化裂解气旋器)和燃烧作业(飞灰收集)中可见。

(5) Tubular precipitators consist of cylindrical collection tubes with discharge electrodes located along the axis of the cylinder.

管式除尘器包含筒形收集管，筒中轴向布置放电电极。

(6) Wet scrubbers have found widespread use in cleaning contaminated gas streams because of their ability to effectively remove both particulate and gaseous pollutants.

湿式涤气器由于能有效去除颗粒状和气态污染物而在净化污染气体中得到广泛应用。

(7) Baghouses may also be characterized and identified according to the method used to

remove collected material from the bags.

布袋除尘器也可根据从布袋中去除集尘的方法表征和识别。

(8) Combustion devices can be categorized as flares, thermal incinerators, or catalytic incinerators.

燃烧装置可分为明火燃烧炉、热焚烧炉或催化焚烧炉。

(9) Thermal incinerators are also called afterburners, direct flame incinerators, or thermal oxidizers.

热焚烧炉也称为后燃烧炉、直接火焰燃烧炉或热氧化装置。

(10) All internal areas of the ionizer section are periodically deluge-flushed with recycled liquid from the scrubber system.

电离装置部分的所有内表面都用来自于涤气器的循环水定期大水流冲洗。

# 16. Indoor Air Quality

## Introduction

Indoor air pollution is rapidly becoming a major worldwide health issue. Although research efforts are still underway to better define the nature and extent of the health implications for the general population, recent studies have shown significant amounts of harmful pollutants in the indoor environment(1). The serious concern over pollutants in indoor air is due largely to the fact that indoor pollutants are not easily dispersed or diluted as are pollutants outdoors. Thus, indoor pollutant levels are frequently higher than outdoors, particularly where buildings are tightly constructed to save energy. In some cases, these indoor levels exceed the Environmental Protection Agency (EPA) standards already established for outdoors. Research by the EPA in this area, called the Total Exposure Assessment Methodology (TEAM) studies, has documented the fact that levels indoors for some pollutants may exceed outdoor levels by 200 to 500 percent (EPA, 1988)

Since most people spend 90 percent of their time indoors, many may be exposed to unhealthy concentrations of pollutants(2). People most susceptible to the risks of pollution—the aged, the ill, and the very young—spend nearly all of their time indoors. These indoor environments include such places as homes, offices, hotels, stores, restaurants, warehouses, factories, government buildings, and even vehicles. In these environments, people are exposed to pollutants emanating from a wide array of sources.

Some common indoor air contaminants are:

1. Radon
2. Formaldehyde
3. Volatile Organic Compounds (VOCs)
4. Combustion gases
5. Particulates
6. Biological contaminants

In addition to air contaminants, other factors need to be observed in Indoor Air Quality (IAQ) monitoring programs to fully understand the significance of contaminant measurements. Important factors to be considered in IAQ studies include:

1. Air exchange rates
2. Building design and ventilation characteristics
3. Indoor contaminant sources and sinks
4. Air movement and mixing
5. Temperature
6. Relative humidity
7. Outdoor contaminant concentrations and meteorological conditions

Designers, builders, and homeowners must make crucial decisions about the kinds and potential levels of existing indoor air pollutants at proposed house sites. Building structure design, construction, operation, and household furnishings, all rely on specific design parameters being set down to handle the reduction of these pollutants at their sources.

The health effects associated with IAQ can be either short- or long-term. Immediate effects experienced after a single exposure or repeated exposures include irritation of the eyes, nose, and throat; headaches; dizziness; and fatigue. These short-term effects are usually treatable by some means, oftentimes by eliminating the person's exposure to the source of pollution.

The likelihood of an individual developing immediate reactions to indoor air pollutants depends on several factors, including age and pre-existing medical conditions. Also, individual sensitivity to a reactant varies tremendously. Some people can become sensitized to biological pollutants after repeated exposures, and it appears that some people can become sensitized to chemical pollutants as well. Other health effects may show up either years after exposure has occurred, or only after long or repeated periods of exposure. These effects range from impairment of the nervous system to cancer; emphysema and other respiratory diseases; and heart disease, which can be severely debilitating or fatal. Certain symptoms are similar to those of other viral diseases and difficult to determine if it is a result of IA pollution. Therefore, special attention should be paid to the time and place symptoms occur(3).

Further research is needed to better understand which health effects can arise after exposure to the average pollutant concentrations found in homes. These can arise from the higher concentrations that occur for short periods of time. Yet, both the amount of pollutant, called the dose, and the length of time of exposure are important in assessing health effects. The effects of simultaneous exposure to several pollutants are even more uncertain. Indoor air quality can be severely debilitating or even fatal. Indoor air pollutants of special concern are described below in separate sections.

It is not possible to provide estimates of typical mixtures of pollutants found in residences. This is because the levels of pollutants found in homes vary significantly depending on location, use of combustion devices, existing building materials, and use of certain household products. Also, emissions of pollutants into the indoor air may be sporadic, as in the case of aerosols or organic vapors that are released during specific household activities or when woodstoves or fireplaces are in use. Another important consideration regarding indoor pollutant concentrations is the interaction among pollutants. Pollutants often tend to attach themselves to airborne particles that get caught more easily in the lungs. In addition, certain organic compounds released indoors could react with each other to form highly toxic substances.

The data provided in this chapter consists of approximate ranges of indoor pollutants based on studies conducted around the United States. These provide an overview of several major pollutants that have been measured in residences at levels that may cause health problems ranging from minor irritations or allergies to potentially debilitating diseases.

## Radon

Radon is a unique environmental problem because it occurs naturally. Radon results from the radioactive decay sequence of uranium-238, a long-lived precursor to radon. The isotope of most concern, radon-222, has a half-life (time for half to disappear) of 3.8 days. Radon itself decays and produces a series of short-lived decay products called radon progeny or daughters. Polonium-218 and polonium-214 are the most harmful because they emit charged alpha particles more dangerous than x-rays or gamma rays. They also tend to adhere to other particles (attachment) or surfaces (plate out). These larger particles are more susceptible to becoming lodged in the lungs when inhaled and cause irreparable damage to surrounding lung tissue (which may lead to lung cancer).

Radon is a colorless, odorless gas that is found everywhere at very low levels. Radon becomes a cause for concern when it is trapped in buildings and concentrations build up. In contrast, indoor air has approximately two to ten times higher concentrations of radon than outdoor air. Primary sources of radon are from soil, well water supplies, and building materials.

Most indoor radon comes from the rock and soil around a building and enters structures through cracks or openings in the foundation or basement. High concentrations of radon are also found in wells, where storage, or hold-up time, is too short to allow time for radon decay. Building materials, such as phosphate slag (a component of concrete used in an estimated 74,000 U.S. homes) has been found to be high in radium content. Studies have shown concrete to have the highest radon content when compared to all other building materials, with wood having the least.

It is becoming increasingly apparent that local geological factors play a dominant role in determining the distribution of indoor radon concentrations in a given area. To date, no indoor radon standard has been promulgated for all residential housing in the United States. However, various organizations have proposed ranges of guidelines and standards.

Data taken from various states suggest an average indoor radon-222 concentration of 1.5 pCi/L (picoCuries per liter, a concentration of radiation term), and approximately 1 million homes with concentrations exceeding 8 pCi/L. One curie is equal to a quantity of a material with 37 billion radioactive decays per second. One trillionth of a curie is a pCi. Assuming residents in these homes spend close to 80 percent of their time indoors, their radon exposure would come close to the level for recommended remedial action set by the U.S. National Council on Radiation Protection and Measurements. The EPA believes that up to 8 million homes may have radon levels exceeding 4 pCi/L air, the level at which the EPA recommends corrective action. In comparison, the maximum level of radon set for miners by the U.S. Mine Safety and Health Administration is as high as

16 pCi/L.

Radon may be the leading cause of lung cancer among nonsmokers. Several radiation protection groups have approximated the number of annual lung cancer deaths attributable to indoor radon. The EPA estimates that radon may be responsible for 5,000 to 20,000 lung cancer deaths among nonsmokers[(4)]. Also, scientific evidence indicates that smoking, coupled with the effects of exposure to radon, increases the risk of cancer by ten times that of nonsmokers (EPA, 1988).

A variety of measures can be employed to help control indoor concentrations of radon and/or radon progeny. Mitigation methods for existing homes include placing barriers between the source material and living space itself using several techniques, such as:

1. Covering exposed soil inside a structure with cement
2. Eliminating and sealing any cracks in the floors or walls
3. Adding traps to underfloor drains
4. Filling concreteblock walls

Soil ventilation prevents radon from entering the home by drawing the gas away before it can enter the home. Pipes are inserted into the stone aggregate under basement floors or onto the hollow portion of concrete walls to ventilate radon gas accumulating in these locations. Pipes can also be attached to underground drain tile systems drawing the radon gas away from the house. Fans are often attached to the system to improve ventilation. Crawl space ventilation is also generally regarded as an effective and cheap method of source reduction. This allows for exchange of outdoor air by placing a number of openings in the crawl space walls.

Home ventilation involves increasing a home's air exchange rate-the rate at which incoming outdoor air completely replaces indoor air—either naturally (by opening windows or vents) or mechanically (through the use of fans). This method works best when applied to houses with low initial exchange rates. However, when indoor air pressure is reduced, pressure-driven radon entry is induced, increasing levels in the home instead of decreasing them. The benefits of increased ventilation can be achieved without raising radon exposure by opening windows evenly on all sides of the home.

Mechanical devices can also be used to help rid indoor air of radon progeny. Air cleaning systems use high efficiency filters or electronic devices to collect dust and other airborne particles, some with radon products attached to them. These devices decrease the concentration of airborne particles, but do not decrease the concentration of smaller unattached radon decay products, which can result in a higher radiation dose when inhaled.

## Formaldehyde

Formaldehyde is a colorless, water-soluble gas that has a pungent, irritating odor noticeable at less than 1ppm. It is an inexpensive chemical with excellent bonding characteristics that is produced in high volume throughout the world. A major use is in the fabrication of urea-formaldehyde (UF) resins used primarily as adhesives when making plywood, particleboard, and fiberboard. Formaldehyde is also a component of UF foam insulation, injected into sidewalls primarily during the 1970s. Many common household cleaning agents contain formaldehyde. Other minor sources in the residential environment include cigarette smoke and other combustion sources such as gas stoves, woodstoves, and unvented gas space heaters. Formaldehyde can also be found in paper products such as facial tissues, paper towels, and grocery bags, as well as stiffeners and wrinkle resisters.

Although information regarding emission rates is limited, in general, the rate of formaldehyde release has been shown to increase with temperature, wood moisture content, humidity, and with decreased formaldehyde concentration in the air.

UF foam was used as a thermal insulation in the sidewalls of many buildings. It was injected directly into wall cavities through small holes that were then sealed. When improperly installed, UF foam emits significant amounts of formaldehyde. The Consumer Product Safety Commission (CPSC) measured values as high as 4 ppm and imposed a nationwide ban on UF foam, but it was later overturned.

The superior bonding properties and low cost of formaldehyde polymers make them the resins of choice for the production of building materials. Plywood is composed of several thin sheets of wood glued together with UF resin. Particleboard (compressed wood shavings mixed with UF resin at high temperatures), can emit formaldehyde continuously for a long time, from several months to several years. Medium density fiberboard was found to be the highest emitter of formaldehyde.

Indoor monitoring data on formaldehyde concentrations are variable because of the wide range of products that may be present in the home. However, elevated levels are more likely to be found in mobile homes and new homes with pressed-wood construction materials. Indoor concentrations also vary with home age since emissions decrease as products containing formaldehyde age and cure. In general, indoor formaldehyde concentration exceed levels found outdoors.

Although individual sensitivity to formaldehyde varies, about 10 to 20 percent of the population appears to be highly sensitive to even low concentrations. Its principal effect is irritation of the eyes, nose, and throat, as well as asthma-like symptoms. Allergic dermatitis may possibly occur from skin contact. Exposure to higher concentrations may cause nausea, headache, coughing, constriction of the chest, and rapid heartbeat (EPA, 1988).

One of the most promising techniques for reducing indoor formaldehyde concentrations is to modify the source materials to reduce emission rates. This can be accomplished by measures performed during manufacture or after installation. A variety of production changes, that is, changes in raw materials, processing times, and temperatures, are promising methods for reducing emission rates. Applying vinyl wallpaper or nonpermeable paint to interior walls, venting exterior walls, and increased ventilation are other methods employed after installation.

## Volatile Organic Compounds (VOCs)

In addition to formaldehyde, many other organic compounds may be present in the indoor environment. More than 800 different compounds can be attributed to volatile vapors alone. Common sources in the home are building materials, furnishings, pesticides, gas or wood burning devices, and consumer products (cleaners, aerosols, deodorizers). In addition, occupant activities such as smoking, cooking, or arts and crafts activities can contribute to indoor pollutant levels.

Organic contaminants in the home are usually present as complex mixtures of many compounds at low concentrations. Thus, it is very difficult to provide estimates of typical indoor concentrations or associated health risks. It is likely, however, that organic compounds may be responsible for health-related complaints registered by residents where formaldehyde and other indoor pollutants are found to be low or undetectable. The sources of three major types of organic contaminants include solvents, polymer components, and pesticides.

Volatile organic solvents commonly pollute air. Exposure occurs when occupants use spot removers, paint removers, cleaning products, paint adhesives, aerosols, fuels, lacquers and varnishes, glues, cosmetics, and numerous other household products. Halogenated hydrocarbons such as methyl chloroform and methylene chloride are widely used in a variety of home products. Aromatic hydrocarbons such as toluene have been found to be present in more than 50 percent of samples taken on indoor air. Alcohols, ketones, ethers, and esters are also present in organic solvents. Some of them, especially esters, emit pleasant odors and are used in flavors and perfumes, yet are still potentially harmful.

Polymer components are found in clothes, furniture, packages, and cookware. Many are used for medical purposes—for example, in blood transfusion bags and disposable syringes. Fortunately, most polymers are relatively nontoxic. However, polymers contain unreacted monomers, plasticizers, stabilizers, fillers, colorants, and antistatic agents, some of which are toxic. These chemicals diffuse from the polymers into air. Certain monomers (acrylic acid esters, toluenediisocynate, and epichlorohydrin) used to produce plastics, polyurethane, and epoxy resins in tile floors, are all toxic.

Most American households use pesticides in the home, garden, or lawn, and many people become ill after using these chemicals. According to an EPA survey, nine out of ten U.S. households use pesticides and another study suggests that 80 to 90 percent of most people's exposure to pesticides has been found in the air inside homes. Pesticides used in and around the home include products to control insects, termites, rodents, and fungi. Chlordane, one of the most harmful active ingredients in pesticides, has been found in structures up to twenty years after its application. In addition to the active ingredient, pesticides are also made up of inerts that are used to carry the active agent. These inerts may not be toxic to the targeted post, but are capable of causing health problems. Methylene chloride, discussed earlier as an organic pollutant, is used as an inert (EPA, 1988).

Human beings can also be significant sources of organic emissions. Human breath contains trace amounts of acetone and ethanol at 20℃ and 1 atmosphere. Measurements taken in schoolrooms while people were present averaged almost twice the amount of acetone and ethanol present in unoccupied rooms. At least part of this increase for ethanol was presumed to be due to perfume and deodorant, in addition to breath emissions.

As mentioned earlier, large numbers of organic compounds have been identified in residences. Studies have shown that of the forty most common organics, nearly all were found at much higher concentrations indoors than outdoors. Another EPA study identified eleven chemicals present in more than half of all samples taken nationwide. Although individual compounds are usually present at low concentrations, which are well below outdoor air quality standards, the average total hydrocarbon concentration can exceed both outdoor concentrations and ambient air quality standards.

Little is known of the short- and long-term health effects of many organic compounds at the low levels of exposure occurring in nonindustrial environments[5]. Yet cumulative effects of various compounds found indoors have been associated with a number of symptoms, such as headache, drowsiness, irritation of the eyes and mucous membranes, irritation of the respiratory system, and general malaise. In general, volatile organic compounds are lipid soluble and easily absorbed through the lungs. Their ability to cross the blood-brain barrier may induce depression of the central nervous system and cardiac functions. Some known and suspected human and animal carcinogens found indoors are benzene, trichloroethane, tetrachloroethylene, vinyl chloride, and dioxane.

One of the best methods to reduce health risks from exposure to organic compounds is for residents or consumers to increase their awareness of the types of toxic chemicals present in household products[6]. Attention to warnings and instructions for storage and use are important, especially regarding ventilation conditions. In some instances, substitution of less hazardous products is possible, as in use of a liquid or dry form of a product rather than an aerosol spray. Consumers should also be wary of the simultaneous use of various products containing organic compounds, since chemical reactions may occur if products are mixed, and adverse health effects may result from the synergism between/among components[7].

## Combustion Gases

Combustion gases, such as carbon monoxide, nitrogen oxides, and sulfur dioxide, can be introduced into the indoor environment by a variety of sources. These sources frequently depend on occupant activities or lifestyles and include the use of gas stoves, kerosene and unvented gas space heaters, woodstoves, and fireplaces. In addition, tobacco smoke is a combustion product that contributes to the contamination of indoor air. More than 2, 000 gaseous compounds have been identified in cigarette smoke, and carbon monoxide and nitrogen oxide are among them (EPA, 1988).

This section focuses on nitrogen oxides (primarily nitrogen dioxide) and carbon monoxide because they are frequently occurring products of combustion often found at higher indoor concentrations than outdoors. Other combustion products such as sulfur dioxide, hydrocarbons, formaldehyde, and carbon dioxide are produced by combustion sources to a lesser degree or only under unusual or infrequent circumstances.

Unvented kerosene and gas space heaters can provide an additional source of heat for homes in cold climates or can serve as a primary heating source when needed for homes in warm climates. There are several basic types of unvented kerosene and gas space heaters which can be classified by the type of burner and type of fuel. Unvented gas space heaters can be convective or infrared and can be fueled by natural gas or propane. Kerosene heaters can be convective, radiant, two-stage, and wickless. A recent study found that emission rates from the various types of heaters fall into three distinct groups. The two-stage kerosene heaters emitted the least CO and the least $NO_2$. The radiant/infrared heater group emitted the most CO under well-tuned conditions; and the convective group emitted the most $NO_2$. Many studies have also noted that some heaters have significantly higher emission rates than heaters of other brands or than models of the same type. Older or improperly used heaters will also increase emission rates.

The kitchen stove is one of the few modern gas appliances that emit combustion products directly into the home. It is estimated that natural gas is used in over 45 percent of all U. S. homes, and studies show that most of these homes do not vent the combustion-produced emissions to the outside. Combustion gas emissions vary considerably and are dependent upon factors such as the fuel consumption rate, combustion efficiency, age of burner, and burner design, as well as the usage pattern of the appliance. An improperly adjusted gas stove is likely to have a yellow-tipped flame rather than a blue-tipped flame, which can result in increased pollutant emissions (mostly $NO_2$ and CO).

Increasing energy costs, consumer concerns about fuel availability, and desire for self-reliance, are some of the factors that have brought about an upswing in the use of solid fuels for residential heating. These devices include woodburning stoves, furnaces, and fireplaces. Although woodstoves and fireplaces are vented to the outdoors, a number of circumstances can cause combustion products to be emitted to the indoor air: improper installation (such as insufficient stack height), cracks or leaks in stovepipes, negative air pressure indoors, downdrafts, refueling, and accidents (as when a log rolls out of a fireplace). The type and amount of wood burned also influences pollutant emissions, which vary from home to home. Although elevated levels of CO and $NO_2$ have been reported, the major impact of woodburning appears to be on indoor respirable suspended particles.

The term nitrogen oxides ($NO_x$) refers to a number of compounds, all of which have the potential to affect humans. $NO_2$ and NO have been studied extensively as outdoor pollutants, yet cannot be ignored in the indoor environment. There is evidence that suggests these oxides may be harmful at levels of exposure that can occur indoors. Both NO and $NO_2$ combine with hemoglobin in the blood, forming methemoglobin, which reduces the oxygen-carrying capacity of the blood. It is about four times more effective than CO in reducing the oxygen-carrying capacity of the blood. $NO_2$ produces respiratory illnesses that range from slight burning and pain in the throat and chest to shortness of breath and violent coughing. It places stress on the cardiovascular system and causes short-term and long-term damage to the lungs. Concentrations typically found in kitchens with gas stoves do not appear to cause chronic respiratory diseases, but may affect sensory perception and produce eye irritation.

Carbon monoxide (CO) is a poisonous gas that causes tissue hypoxia (oxygen starvation) by binding with blood hemoglobin and blocking its ability to transport oxygen. CO has in excess of 200 times more binding affinity for hemoglobin than oxygen does. The product, carboxyhemoglobin, is an indicator of reduction in oxygen-carrying capacity. A small amount of CO is even produced naturally in the body, producing a concentration in unexposed persons of about 0. 5 percent

CO-bound hemoglobin. Under chronic exposure (for example, cigarette smoking), the body compensates somewhat by increasing the concentration of red blood cells and the total amount of hemoglobin available for oxygen transport. The central nervous system, cardiovascular system, and liver are most sensitive to CO-induced hypoxia. Hypoxia of the central nervous system causes a wide range of effects in the exposure range of 5 to 15 percent carboxyhemoglobin. These include loss of alertness and impaired perception, loss of normal dexterity, reduced learning ability, sleep disruption, drowsiness, confusion, and at very high concentrations, coma and death. Health effects related to hypoxia of the cardiovascular system include decrease in exercise time required to produce angina pectoris (chest pain); increase in incidences of myocardosis (degeneration of heart muscle); and a general increase in the probability of heart failure among susceptible individuals.

Population groups at special risk of detrimental effects of CO exposure include fetuses, persons with existing health impairments (especially heart disease), persons under the influence of drugs, and those not adapted to high altitudes who are exposed to both CO and high altitudes.

Proper installation, operation, and maintenance of combustion devices can significantly reduce the health risks associated with these appliances. Manufacturers' instructions regarding the proper size space heater in relation to room size, ventilation conditions, and tuning should be observed. This includes using vented range hoods when operating gas stoves. Studies have indicated reductions in CO, $CO_2$, and $NO_2$ levels as high as 60 to 87 percent with the use of range hoods during gas stove operation. Unvented forced draft and unvented range hoods with charcoal filters can be effective for removing grease, odors, and other molecules, but cannot be considered a reliable control for CO and other small molecules. Fireplace flues and chimneys should be inspected and cleaned frequently, and opened completely when in use.

## Particulates

Environmental tobacco smoke, ETS (smoke that nonsmokers are exposed to from smokers), has been judged by the Surgeon General, the National Research Council, and the International Agency for Research on Cancer to pose a risk of lung cancer to nonsmokers. Nonsmokers' exposure to environmental tobacco smoke is called "passive smoking", "second-hand smoking", and "involuntary smoking". Tobacco smoke contains a number of pollutants, including inorganic gases, heavy metals, particulates, VOCs, and products of incomplete burning, such as polynuclear aromatic hydrocarbons. Smoke can also yield a number of organic compounds. Including both gases and particles, tobacco smoke is a complex mixture of over 4700 compounds (EPA, 1988).

There are two components of tobacco smoke: (1) mainstream smoke, which is the smoke drawn through the tobacco during inhalation, (2) sidestream smoke, which arises from the smoldering tobacco. Sidestream smoke accounts for 96% of gases and particles produced.

Studies indicate that exposure to tobacco smoke may increase the risk of lung cancer by an average of 30 percent in the nonsmoking spouses of smokers. Published risk estimates of lung cancer deaths among nonsmokers exposed to tobacco smoke conclude that ETS is responsible for 3,000 deaths each year. It also seriously affects the respiratory health of hundreds of thousands of children. Very young children exposed to smoking at home are more likely to be hospitalized for bronchitis and pneumonia. Recent studies suggest that environmental tobacco smoke can also cause other diseases, including other cancers and heart disease in healthy nonsmokers (EPA, 1988).

The best way to reduce exposure to cigarette smoke in the house is to quit smoking and discourage smoking indoors. Ventilation is the most common method of reducing exposure to these pollutants, but it will not eliminate it altogether. Smoking produces such large amounts of pollutants that neither natural nor mechanical methods can remove them from the air as quickly as they build up. In addition, ventilation practices sometimes lead to increased energy costs.

Respirable suspended particles (RSP) are particles or fibers in the air that are small enough to be inhaled. Particles can exist in either solid or liquid phase or in a combination. Where these particles are deposited and how long they are retained depends on their size, chemical composition, and density. Respirable suspended particles (generally less than 10 micrometers in diameter), can settle on the tissues of the upper respiratory tract, with the smallest particles (those less than 2.5 micrometers) penetrating the alveoli, the small air sacs in the lungs.

Particulate matter is a broad class of chemically and physically diverse substances that present risks to health. These effects can be attributed to either the intrinsic toxic chemical or physical characteristics, as in the case of lead and asbestos, or to the particles acting as a carrier of adsorbed toxic substances, as in the case of attachment of radon daughters. Carbon particles, such as those created by combustion processes, are efficient adsorbers of many organic compounds and are able to carry toxic gases such as sulfur dioxide into the lungs.

Asbestos is a mineral fiber used mostly before the mid-seventies in a variety of construction materials. Home exposure to asbestos is usually due to aging, cracking, or physical disruption of insulated pipes or asbestos-containing ceiling tiles and spackling compounds. Apartments and school buildings may have an asbestos compound spayed on certain structural components as a fire retardant. Exposure occurs when asbestos materials are disturbed and the fibers are released into the air and inhaled. Consumer exposure to asbestos has been reduced considerably since the mid-seventies, when use of asbestos was either prohibited or stopped voluntarily in sprayed-on insulation, fire protection, soundproofing, artificial logs, patching compounds, and handheld hair dryers. Today, asbestos is most commonly found in older homes in pipe and furnace insulation materials, asbestos shingles, millboard, textured paints and other coating materials, and floor tiles. Elevated concentrations of airborne asbestos can occur after asbestos- containing materials are disturbed by cutting, sanding, or other remodeling activities. Improper attempts to remove these materials can release asbestos fibers into the air in homes, thereby increasing asbestos levels and endangering the people living in those homes. The most dangerous asbestos fibers are too small to be visible. After they are inhaled, they can remain and accumulate in the lungs. Asbestos can cause lung cancer, mesothelioma (a cancer of the chest and abdominal linings), and asbestosis (irreversible lung scarring that can be fatal). Symptoms of these diseases do not show up until many years after exposure began. A more detailed presentation on asbestos can be found in Chapter 46.

Lead has long been recognized as a harmful environmental pollutant. There are many ways in which humans are exposed to lead, including air, drinking water, food, and contaminated soil and dust. Airborne lead enters the body when an individual breathes lead particles or swallows lead dust once it has settled. Until recently, the most important airborne source of lead was automobile exhaust. Lead-based paint has long been recognized as a hazard to children who eat lead-contained paint chips. A 1988 National Institute of Building Sciences Task Force report found that harmful exposures to lead can be created when lead-based paint is removed from surfaces by sanding or open-flame burning. High concentrations of airborne lead particles in homes can also result from the lead dust from outdoor sources, contaminated soil tracked inside, and use of lead in activities such as soldering, electronics repair, and stained-glass artwork. Lead is toxic to many organs within the body at both low and high concentrations. Lead is capable of causing serious damage to the brain, kidneys, peripheral nervous system (the sense organs and nerves controlling the body), and red blood cells. Even low levels of lead may increase high blood pressure in adults. Fetuses, infants, and children are more vulnerable to lead exposure than are adults because lead is more easily absorbed into growing bodies, and the tissues of small children are more sensitive to the damaging effects of lead. The effects of lead exposure on fetuses and young children include delays in physical and mental development, lower IQ levels, shortened attention spans, and increased behavioral problems.

Particles present a risk to health out of proportion to their concentration in the atmosphere because they deliver a high-concentration package of potentially harmful substances(8). So, while few cells may be affected at any one time, those few that are can be badly damaged. Whereas larger particles deposited in the upper respiratory portion of the respiratory system are continuously cleared away, smaller particles deposited deep in the lung may cause adverse health effects. Particle sizes vary over a broad range, depending on source characteristics.

Major effects of concern attributed to particle exposure are impairment of respiratory mechanics, aggravation of existing respiratory and cardiovascular disease, and reduction in particle clearance and other host defense mechanisms. Respiratory effects can range from mild transient changes of little direct health significance to incapacitating impairment of breathing.

One method of reducing RSP concentrations is to properly design, install, and operate combustion sources. One should make sure there are no existing leaks or cracks in stovepipes, and that these appliances are always vented to the outdoors.

Also available are particulate air cleaners, which can be separated into mechanical filters and electrostatic filters(9). Mechanical filtration is generally accomplished by passing the air through a fibrous media (wire, hemp, glass, etc.). These filters are capable of removing almost any sized particles. Electrostatic filtration operates on the principle of attraction between opposite electrical charges. Ion generators, electrostatic precipitators, and electric filters use this principle for removing particles from the air.

The ability of these various types of air-cleaning devices to remove respirable particles varies widely. High efficiency particulate air (HEPA) filters can capture over 99 percent of particles, and are advantageous in that filters only need changing every three to five years, but costs can reach \$ 500 to \$ 800. It is also important to note the location of air-cleaning device inlets in relation to the contaminant sources as an important factor influencing removal efficiencies.

## Biological Contaminants

Heating, ventilation, and air conditioning systems and humidifiers can be breeding grounds for biological contaminants when they are not properly cleaned and maintained. They can also bring biological contaminants indoors and circulate them. Biological contaminants include bacteria, mold and mildew, viruses, animal dander and cat saliva, mites, cockroaches, and pollen. There are many sources for these pollutants. For example, pollens originate from plants; viruses are transmitted by people and animals; bacteria are carried by people, animals, and soil and plant debris; and household pets are sources of saliva, hair, and dead skin (known as dander).

Available evidence indicates that a number of viruses that infect humans can be transmitted via the air. Among them are the most common infections of mankind. Airborne contagion is the mechanism of transmission of most acute respiratory infections, and these are the greatest of all causes of morbidity.

The primary source of bacteria indoors is the human body. Although the major source is the respiratory tract, it has been shown that 7 million skin scales are shed per minute per person, with an average of 4 viable bacteria per scale. Airborne transmission of bacteria is facilitated by the prompt dispersion of particles. Infectious contact requires proximity in time and space between host and contact, and is also related to air filtration and air exchange rate.

Although many important allergens—such as pollen, fungi, insects, and algae—enter buildings from outdoors, several airborne allergens originate predominately in homes and office buildings. House dust mites, one of the most powerful biologicals in triggering allergic reactions, can grow in any damp, warm environment. Allergic reactions can occur on the skin, nose, airways, and alveoli.

The most common respiratory diseases attributable to these allergens are rhinitis, affecting about 15 percent of the population, and asthma, affecting about 3 to 5 percent. These diseases are most common among children and young adults, but can occur at any age. Research has shown that asthma occurs four times more often among poor, inner-city families than in other families. Among the suspected causes are mouse urine antigens, cockroach feces antigens, and a type of fungus called Alternia.

Hypersensitivity pneumonitis (HP), characterized by shortness of breathe, fever, and cough, is a much less common disease, but is dangerous if not diagnosed and treated early. HP is most commonly caused by contaminated forced-air heating systems, humidifiers, and flooding disasters. It can also be caused by inhalation of microbial aerosols from saunas, home tap water, and even automobile air conditioners. Humidifiers with reservoirs containing stagnant water may be important sources of allergens in both residential and public buildings.

Some biological contaminants trigger allergic reactions, while others transmit infectious illnesses, such as influenza, measles, and chicken pox. Certain molds and mildews release disease-causing toxins. Symptoms of health problems caused by biologicals include sneezing, watery eyes, coughing, shortness of breathe, dizziness, lethargy, fever, and digestive problems.

Attempts to control airborne viral disease have included quarantine, vaccination, and inactivation or removal of the viral aerosol. Infiltration and ventilation play a large role in the routes of transmission. Because many contaminants originate outdoors, attempts to reduce the ventilation rate might lower indoor pollutant concentrations. However, any reduction in fresh air exchange should be supplemented by a carefully filtered air source.

Central electrostatic filtration (as part of a home's forced-air system) has proven effective in reducing indoor mold problems. Careful cleaning, vacuuming, and air filtration are effective ways to reduce dust levels in a home. Ventilation of attic and crawl spaces help prevent moisture buildup, keeping humidity levels between 30 to 50 percent. Also, when using cool mist or ultrasonic humidifiers, one should remember to clean and refill water trays often, since these areas often become breeding grounds for biological contaminants.

## Monitoring Methods

Methods and instrumentation for measuring indoor air quality vary in their levels of sensitivity (what levels of pollutant they can detect) and accuracy (how close they can come to measuring the true concentration). Instruments that can measure low levels of pollutant very accurately are likely to be expensive and require special expertise to use. Some level of sensitivity and accuracy is required, however, to ensure that data collected are useful in assessing levels of exposure and risk.

In choosing methods for monitoring indoor air quality, a tradeoff must be made between cost and the levels of sensitivity, accuracy, and precision achieved in a monitoring program. Required levels for each pollutant are based on ranges found in residential buildings. In providing detailed information concerning specific methods or instruments, emphasis is placed on those that are readily available, easy to use, reasonably priced, and that provide the required levels of sensitivity and accuracy[10].

Methods to monitor indoor air fall into several broad categories. Sampling instruments may be fixed-location, portable, or small personal monitors designed to be carried by an individual. These samplers may act in an active or passive mode. Active samplers require a pump to draw in air. Passive samplers rely on diffusion or permeation.

Monitors may be either analytical instruments that provide a direct reading of pollutant concentration, or collectors that must be sent to a laboratory for analysis. Instruments may also be categorized according to the time period over which they sample. These include grab samplers, continuous samplers, and time-integrated samplers, each of which is briefly described below.

1. Grab sampler: Collects samples of air in a bag, tube, or bottle, providing a short-term average.

2. Continuous sampling: Allows sampling of real-time concentration of pollutants, providing data on peak short-term concentrations and average concentrations over the sampling period.

3. Time-integrated sampling: Measures an average air concentration over some period of time (active or passive), using collector monitors that must be sent out for analysis; cannot determine peak concentrations.

More details regarding monitoring methods for specific indoor air pollutants can be found in the IAQ Handbook.

CosaTron is just one example of a company that produces mechanical air cleaning devices. The patented CosaTron system has been handling IAQ successfully in thousands of installations for over twenty-five years. CosaTron is not a filter that ionizes air. It cleans the air electronically, causing the submicron particles of smoke, odor, dirt and gases to collide and adhere to each other until they become larger and airborne and are easily carried out of the conditioned space by the system air flow to be exhausted or captured in the filter. Mechanical air devices such as this one improve IAQ so much that outside air requirements can be reduced significantly.

## Future Trends

In recent years, the EPA has increased efforts to address IAQ problems through a building systems approach. EPA hopes to bolster awareness of the importance of prevention and encourage a whole systems perspective to resolve indoor air problems. The EPA Office of Research and Development is also conducting a multidisciplinary IAQ research program that encompasses studies of the health effects associated with indoor air pollution exposure, assessments of indoor air pollution sources and control approaches, building studies and investigation methods, risk assessments of indoor air pollutants, and a recently initiated program on biocontaminants.

Federal research on air quality issues is driven in part by the increasing attention that IAQ has attracted from journalists as well as scientists and engineers. EPA has performed comparative studies that have consistently ranked indoor air pollution among the top five environmental risks to public health. In analyzing over 500 IAQ investigations conducted through the end of 1988, the National Institute for Occupational Safety and Health (NIOSH) categorized its findings into seven broad sources of poor indoor air quality: inadequate ventilation (53%), inside contamination (15%), outside contamination (13%), microbiological contamination (5%), building materials contamination (4%), and unknown sources (13%). Since then, ventilation has been the primary focus of most EPA programs.

Requirements for clean air are still changing rapidly and most buildings will need to be refitted with different filters to meet with these new standards and guidelines. EPA's research will continue in these and other areas to try to ensure comfortable and clean air conditions for the indoor environment.

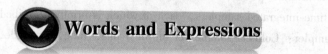

## Words and Expressions

1. implication 牵连，含意，本质，含蓄
2. the Environmental Protection Agency 美国环保局
3. susceptible 易受影响的，敏感的
4. warehouse 仓库
5. radon 氡
6. formaldehyde 甲醛
7. Volatile Organic Compounds(VOCs) 挥发性有机化合物
8. sources and sinks 源和汇
9. crucial 关键的
10. household furnishings 家用家具，家用器具
11. irritation 发炎，疼痛，刺激(物)

12. dizziness 眩晕
13. sensitize 使敏感，敏化，活化
14. impairment 损害，损伤，削弱
15. nervous system 神经系统
16. cancer 癌症
17. emphysema 气肿，肺气肿
18. respiratory disease 呼吸道疾病
19. heart disease 心脏疾病
20. debilitate 使衰弱
21. sporadic 间歇性的，偶尔发生的
22. woodstove 烧柴炉
23. fireplace 壁炉，火炉
24. allergy 过敏症，变态反应，反感
25. radioactive decay sequence 放射性衰变序列
26. precursor 前驱物，先驱物，预兆，先锋
27. progeny 子孙，后代，结果
28. lodge 停留，住宿，寄宿，投宿
29. inhale 吸入，吸气
30. irreparable 不可弥补的，不能恢复的
31. play a dominant role in... 在……方面起主要作用
32. promulgate 颁布，公布，传播
33. mitigation 缓和，减轻，调节
34. aggregate 使聚集，合计为
35. basement floor 地下室地面
36. underground drain tile 地下排水瓦管
37. crawl space ventilation(屋顶、地板等下面)供电线或水管通过的狭小空隙的通风
38. pungent 刺激性的，刺鼻的，尖锐的
39. bonding characteristics 黏结特性
40. urea-formaldehyde(UF) resin 尿醛树脂
41. adhesive 胶黏的，带黏性的；胶黏剂
42. plywood 胶合板，层压板
43. fireboard 防火板
44. cleaning agent 清洁剂
45. facial tissue 面巾纸，擦面用的纸
46. paper towel 擦手纸，擦脸纸
47. stiffener 硬化剂，增稠剂，加强杆
48. wrinkle resister 抗皱物，抗皱剂
49. medium density fireboard 中密度防火板
50. age 使变老，陈化，熟化，硬化
51. cure 硫化，熟化，固化，治愈，处理
52. allergic dermatitis 过敏性皮炎
53. deodorizer 除臭剂，脱臭机
54. spot remover 去污剂
55. paint remover 油漆去除剂
56. paint adhesive 油漆黏结剂，涂料黏结剂
57. lacquer 漆，硝基漆，真漆
58. varnish 清漆
59. cosmetic 化妆品；化妆用的，装饰性的
60. halogenated hydrocarbon 卤代烃
61. methyl chloroform 甲基氯仿，甲基三氯甲烷
62. methylene chloride 二氯甲烷
63. aromatic hydrocarbon 芳香烃
64. toluene 甲苯
65. ketone 酮
66. ether 醚，乙醚
67. blood transfusion bag 输血袋
68. disposable syringe 一次性注射器
69. plasticizer 增塑剂，塑化剂
70. stabilizer 稳定器，稳定剂
71. filler 填充料
72. colorant 着色剂，颜料，染料
73. antistatic agent 抗静电剂
74. polyurethane 聚氨酯
75. epoxy resin 环氧树脂
76. termite 白蚁
77. rodent 啮齿动物(如兔、鼠等)
78. chlordane 氯丹
79. ethanol 乙醇，酒精
80. cardiac function 心脏功能
81. trichloroethane 三氯乙烷
82. tetrachloroethylene 四氯乙烯
83. vinyl chloride 氯乙烯
84. dioxane 二噁烷，二氧六环
85. synergism 协同作用
86. unvented kerosene heater 不通风煤油加热器

87. gas space heater 燃气空间加热器
88. convective 传递性的，对流的
89. infrared 红外线的，产生红外辐射的；红外线
90. propane 丙烷
91. upswing 提高，进步，改进，改善，上升
92. downdraft 下向通风，下降气流
93. respirable suspended particles 可吸入悬浮颗粒
94. hemoglobin 血红蛋白，血红素
95. methemoglobin 高铁血红蛋白
96. cardiovascular system 心血管系统
97. chronic respiratory disease 慢性呼吸疾病
98. sensory perception 感觉
99. carboxyhemoglobin 碳氧血红蛋白
100. dexterity 灵巧，灵活，熟练，技巧
101. coma 昏迷
102. angina pectoris 心绞痛
103. vented range hood 通风炉灶罩
104. unvented forced draft 无孔强制通风
105. charcoal filter 木炭过滤器
106. smolder 用文火闷烧；闷烧，闷火
107. bronchitis 支气管炎
108. pneumonia 肺炎
109. alveoli 小窝，气泡，腔区
110. sac 囊，液囊
111. ceiling tile 天花板瓦
112. fire retardant 阻燃剂
113. patching compounds 修补化合物
114. automobile exhaust 汽车排气，汽车尾气
115. lead-based paint 铅基涂料，铅基油漆
116. paint chip 涂料碎片，油漆碎屑
117. solder 焊接；焊料，焊锡
118. stained-glass artwork 着色玻璃工艺品
119. peripheral nervous system 周边的神经系统
120. fetus 胎，胎儿
121. out of proportion to 不成比例，与……不相称
122. incapacitate 使无能力，使残疾
123. humidifier 增湿器，湿润器
124. breeding ground 繁殖地，繁殖场
125. mold 霉，霉菌；发霉
126. mildew 霉，霉病；使发霉，发霉
127. dander 头皮屑
128. saliva 唾液
129. mite 螨，蛆
130. cockroach 蟑螂
131. pollen 花粉；传花粉给
132. debris 碎片，碎屑，废石，尾矿，瓦砾堆
133. morbidity 发病率
134. allergen 能引起过敏症的东西(如花粉、食物、药物)
135. trigger 触发，启动，发射，引起，激发起
136. rhinitis 鼻炎
137. asthma 气喘病，哮喘病
138. antigen 抗原
139. cockroach feces 蟑螂排泄物，蟑螂粪便
140. hypersensitivity pneumonitis 高敏感性局部急性肺炎
141. influenza 流行性感冒，(马、猪等的)流行性热病
142. measles 麻疹，家畜囊虫病
143. sneeze 打喷嚏；喷嚏，喷嚏声
144. lethargy 嗜眠症，冷淡
145. quarantine 检疫，检疫处，(因传染病流行对人、畜等的)隔离；隔离，对……进行检疫
146. vaccination 种痘，接种
147. tradeoff 权衡，换位，比较评定，交替，放弃
148. permeation 渗透(作用)，贯穿
149. grab sampler 临时样品采样仪，单一样品采样仪
150. continuous sampler 连续采样仪
151. time-integrated sampler 时间积分采样仪
152. bolster 支持，支撑，垫；支持物
153. multidisciplinary 多学科的

## Notes

(1) Although research efforts are still underway to better define the nature and extent of the health implications for the general population, recent studies have shown significant amounts of harmful pollutants in the indoor environment.

为了更好地确定对普通人群健康影响的性质和程度，研究仍在进行中，新的研究已表明在室内环境中存在大量的有害污染物。

(2) Since most people spend 90 percent of their time indoors, many may be exposed to unhealthy concentrations of pollutants.

由于大多数人在室内度过90%的时间，许多人可能处在不健康的污染物浓度中。

(3) Therefore, special attention should be paid to the time and place symptoms occur.

因此，应特别关注症状出现的时间和地点。

(4) Several radiation protection groups have approximated the number of annual lung cancer deaths attributable to indoor radon. The EPA estimates that radon may be responsible for 5,000 to 20, 000 lung cancer deaths among nonsmokers. Also, scientific evidence indicates that smoking, coupled with the effects of exposure to radon, increases the risk of cancer by ten times that of non-smokers .

几个辐射防护小组已近似估算了每年因室内氡引起的肺癌死亡数。美国环保局估计氡在不吸烟人群中可能引起5000~20000人患肺癌死亡。科学证据还表明吸烟并接触氡的人患癌症的风险比不吸烟的要增加10倍。

(5) Little is known of the short- and long-term health effects of many organic compounds at the low levels of exposure occurring in nonindustrial environments.

对在非工业区接触许多低浓度有机化合物的短期和长期健康影响还了解得很少。

(6) One of the best methods to reduce health risks from exposure to organic compounds is for residents or consumers to increase their awareness of the types of toxic chemicals present in household products.

减少由于接触有机化合物带来的健康风险的一种最好方法是居民或消费者增强他们对家用产品中有毒化学品的意识。

(7) Consumers should also be wary of the simultaneous use of various products containing organic compounds, since chemical reactions may occur if products are mixed, and adverse health effects may result from the synergism between/among components.

消费者也应谨慎同时使用含有机物的各种产品，因为产品混合时会发生化学反应，各组分间的协同作用可能产生不良的健康影响。

(8) Particles present a risk to health out of proportion to their concentration in the atmosphere because they deliver a high-concentration package of potentially harmful substances.

超过其在大气中浓度的颗粒物会对健康产生危险，因为它们传送着高浓度的可能有害的物质。

(9) Also available are particulate air cleaners, which can be separated into mechanical filters

and electrostatic filters.

还有空气颗粒物净化器，可分为机械过滤器和静电过滤器。本句为倒装句。

(10) In providing detailed information concerning specific methods or instruments, emphasis is placed on those that are readily available, easy to use, reasonably priced, and that provide the required levels of sensitivity and accuracy.

在提供有关具体方法或仪器的详细信息时，重点是那些容易得到、容易使用、合理报价及具有所要求的灵敏度和准确性的方法或仪器。

# Unit 4 Solid Waste

## 17. Resource Recovery

Recovery of resources from solid waste, commonly known as recycling, is theoretically very appealing. Unfortunately, our present economic system makes efficient (money-making) recycling difficult, although the picture is changing rapidly and resource recovery may in the near future become the most desirable means of solid waste management.

In the heat of environmental concern, little attention has been paid to the total process necessary for resource recovery. "Recycling" has, in fact, been confused with "collection", but collection is only one step in the process. After the material has been collected from consumers, it must be cleaned, sold to an industry, transported, remanufactured and (most importantly) sold once again to consumers. This last step in fact controls the entire operation. If the remanufactured material cannot be sold, there is little sense in doing anything else. We must therefore not think of the collection of newspapers as "recycling", since the cycle is complete only when the paper is reused by consumers.

The two basic reasons for recycling are (a) conservation of resources, and (b) reduction in volume of refuse to be disposed. Some of the common materials which have been suggested as recyclable are paper, metals, glass and organics.

Paper is one product which is in plentiful supply and fairly clean. But only about 20 % of our present supply is recycled since, unfortunately, virgin paper is much too cheap to make. The realization that each ton of paper recycled saves about 17 trees from the ax often prompts community paper drives(1). This type of free labor is necessary in order to keep most repulping operations solvent.

Unfortunately, the ton of waste paper so lovingly collected seldom saves the 17 trees. Only a small fraction of recycled paper ends up as paper, and most of that is shipped overseas. Recycling paper to paper is simply more expensive than making virgin paper.

The only way paper can be recycled so as to truly save our forests is to create a market for recycled paper, and make virgin paper artificially expensive(2). This can be accomplished either through public support, legislation, or taxes.

Metals can be easily recycled from industrial scrap, and this is the largest source of "secondary metals". The second largest source of waste metal is wrecked automobiles. Unfortunately, the present methods of making steel can tolerate only limited scrap input, and thus scrap steel has low market value.

Some aluminum companies have conducted successful drives to collect aluminum cans. These cans are especially obnoxious as litter since they do not rust when discarded and remain as visible trash almost indefinitely.

Glass is the perfect product for recycling. It is clean, plentiful, easy to reprocess, and can be used in many ways. Unfortunately, it is also cheap. It is about as expensive to make a new glass bottle as to recycle or refill an old one. In addition, the raw materials for glass are in such plentiful supply that there does not seem to be anything gained through recycling.

Organics can be converted into several useful products. The most common process, used extensively in many countries, is composting.

Composting is, in contrast to the landfill, an aerobic method of decomposing solid waste. Typically, a composting operation involves (a) the segregation of refuse into organic and inorganic components (either by the household or at the plant); (b) grinding of the organic portion; and (c) stabilizing in either open piles or in mechanical "digesters".

The segregation operation is the weak link in the process. Most communities have not had much luck in asking the citizens to separate paper and garbage from glass and tin cans; hand sorting at the compost plant, an expensive and often unreliable operation, has been necessary.

Considerable research is directed toward finding better separation methods, because it is felt that if this problem were solved, composting would be a much more attractive method of resource recovery.

After grinding and usually after the separation of metals, the organic material is commonly placed in long piles called windrows, 8 to 10 feet wide and 4 to 6 feet high. Under these conditions, sufficient moisture and oxygen are available to support aerobic life. The piles must be turned periodically to allow sufficient oxygen to penetrate to all parts of the pile.

Temperatures within a windrow approach 140°F, due entirely to biological activity. The pH will approach neutrality after an initial drop. With most wastes, additional nutrients are not needed. The composting of bark and other materials, however, is successful only with the addition of nitrogen and phosphorus nutrients.

Moisture must usually be controlled. Excessive moisture makes it difficult to maintain aerobic conditions while a dearth inhibits biological life. A 40% ~ 60% moisture content is considered desirable.

There has been some controversy over the use of inoculants, freeze-dried cultures, used to speed up the process. Once the composting pile is established, which requires about two weeks, the inoculants have not proven to be of any significant value[3]. Most municipal refuse contains all the organisms required for successful composting, and "mystery cultures" are thus not needed.

The end point of a composting operation can be measured by noting a drop in temperature. The compost should have an earthy smell, similar to peat moss, and should have a dark brown color.

Several composting plants are in operation in the U. S., most with some type of "automated windrows". These units, often referred to as "digesters", aerate the mixture and help maintain an optimum moisture concentration. The detention time can thus be reduced to a few days instead of a few weeks.

Although compost is an excellent soil conditioner, it is not widely used by U. S. farmers.

Inorganic fertilizers are cheap and easy to apply and most farms are located where soil conditions are good. The plentiful food supply in most developed countries does not dictate the use of marginal lands where compost would be of real value.

## Words and Expressions

1. recycle 再循环，重复利用
2. appealing 动人的，吸引人的
3. virgin 新的，原始的，未用过的
4. free labor 免费劳动力，无偿劳动
5. repulp 重新制浆，重新化成纸浆
6. solvent 有偿付能力的，(有)溶解(力)的
7. scrap 废金属，切屑，废料；废弃，废置
8. wrecked 遭严重破坏的
9. trash 废物，垃圾；废弃
10. refill 再填充，再装填
11. hand sorting 手工分拣，手工分类
12. aerobic life 好氧生物
13. windrow 料堆；堆成条形长堆
14. inoculant 变质剂，孕育剂
15. peat moss 泥炭沼泽
16. marginal 边缘的，勉强够格的，页边的

## Notes

(1) The realization that each ton of paper recycled saves about 17 trees from the ax often prompts community paper drives. This type of free labor is necessary in order to keep most repulping operations solvent.

认识到每回用一吨纸等于少砍17颗树会推动社会回用纸张。为了使大部分的再制浆作业有利可图，这种无偿劳动是必需的。"The realization that…"中 that 引出同位语从句。

(2) The only way paper can be recycled so as to truly save our forests is to create a market for recycled paper, and make virgin paper artificially expensive.

回用纸张以拯救森林的唯一办法是为回用纸创建一个市场，人为地使原始纸张昂贵。

(3) Once the composting pile is established, which requires about two weeks, the inoculants have not proven to be of any significant value.

已证明一旦堆肥料堆建立起来，这约需2周时间，孕育剂就没有多大价值了。

# 18. Solid Wastes

When one thinks of solid waste, often the problems and processes which immediately come to mind are those associated with municipal disposal[(1)]. These substances generally are, of course, the products of various industries, but their disposal is not directly the responsibility of the industry which created them. The industries have their own problems: their own types of solid wastes which must be disposed of.

The ideal solution, economically, energetically, and environmentally, would be to recover and reuse many of the solid wastes[2]. Many industries have been attempting to recover their wastes, with varying degrees of success. As with most industry-related issues, pollution control and economics can't be separated. The primary responsibility of an industry official is to protect his company's financial position; if he doesn't, the company will soon be out of business, and the shareholders will suffer. The financial incentive may be to avoid fines, court cases, or costly enforcement squabbles, or it may be byproduct recovery, but unless the incentive is there, little progress will be made.

There are several types of solid wastes an industry may have to handle. There are, of course, the sludges which result from water treatment. There are also the process solids, such as collected particulates and slags. Many of these are composed of various minerals, though their form and actual chemical composition may vary significantly, depending upon the source.

More than one billion tons of solid wastes are produced annually by the minerals processing industries alone. These ores usually contain only small percentages of the desired substances (such as copper, iron, gold, or silver); thus the spent ores, or tailings, accumulate very rapidly. Tailings are typically composed of silica (sand), and various silicates and carbonates of calcium, magnesium, and possibly aluminum. These tailings, often consisting of very fine particles, are piled near the processing plant, creating a nuisance because of their size and physical instability; plant growth often must be encouraged to stabilize the piles. Few recovery methods have been found to be economically feasible for many of these wastes (Table 1).

**Table 1 Composition of Typical Fly Ash**

| Compound | Percent | Compound | Percent |
|---|---|---|---|
| $SiO_2$ | 31.6~39.7 | MgO | 3.0~3.6 |
| $Al_2O_3$ | 16.9~19.0 | $K_2O$ | 1.0~1.6 |
| $Fe_2O_3$ | 10.6~18.8 | $Na_2O$ | 0.6~0.7 |
| TiO | 0.5~0.7 | Carbon | ~2 |
| CeO | 15.3~18.8 | | |

Many industries also generate fly ash, the coal ash which results from, among other things, power generation[3]. Fly ash is one substance on which much research has been done, looking for more and better ways for its usage. The composition of a typical fly ash is given in Table 1. In 1968, about 18% of the fly ash was recovered for byproduct reuse. Hopes are that this percentage will increase (although, at the same time the amount generated is also increasing).

Fly ash has been used fairly widely as an additive for cement. The fine fly ash can be added to the ground cement clinker, increasing for some purposes the desirable cement characteristics. For example, the U.S. Corps of Engineers uses fly ash in much of its concrete. Unfortunately, this is not the solution for fly ash disposal. Many cement companies have found the cement not to be marketable, primarily due to its dark color.

Fly ash is also used as one of the raw materials for the production of sintered lightweight aggregate, such as used in concrete blocks and other precast forms. It can be used as a filler in asphalt pavings, as a soil stabilizer for embankments, as raw material for bricks, and in the bases for road beds. Recent studies have shown its feasibility as a plastics filler. Some of the fly ash generated by power companies is bought by companies such as American Admixtures of Chicago for reuse. The remainder of the fly ash must be landfilled or piled near the plant.

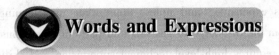

## Words and Expressions

1. shareholder 股票持有人，股东
2. incentive 刺激，鼓励；刺激的
3. fine 罚款
4. court case 法院事例，法院案例
5. slag 矿渣，熔渣；(使)成渣
6. tailing(矿石)尾砂，尾渣，渣滓
7. nuisance 公害，损害，有害的东西
8. fly ash 飞灰
9. clinker 熔渣，烧结块；烧结，从……清除熔渣
10. sinter 烧结，矿渣，烧结物；烧结
11. aggregate 聚集，共计为；集料，聚集体
12. filler 填充物，填充料
13. asphalt 沥青，(铺路用)沥青混合料；铺沥青于
14. embankment 筑堤(工程)，路堤，(河、海的)堤岸

## Notes

(1) When one thinks of solid waste, often the problems and processes which immediately come to mind are those associated with municipal disposal.

当人们想到固体废物时常常会立刻想到与城市垃圾处置有关的问题和过程。

(2) The ideal solution, economically, energetically, and environmentally, would be to recover and reuse many of the solid wastes.

从经济、能量和环境的角度来看，理想的解决办法是回收和回用许多固体废物。

(3) Many industries also generate fly ash, the coal ash which results from, among other things, power generation.

许多工业部门也产生飞灰，即煤灰，它连同其他一些废物一起来源于发电过程。

# 19. Solid Waste Disposal

Solid and hazardous waste may be treated or processed prior to final disposal. Waste treatment or processing offers several advantages. First, it can serve to reduce the total volume and weight of material that requires disposal. It can also change the form of the waste and improve it handing characteristics. Garbage and other organic wastes, for example, can be rendered inoffensive and even useful by a process called composting. Finally, processing can serve to recover natural resources and energy in the waste material, for recycling or reuse. Much of the "waste" material can actually be used as raw material for productive purposes. However, a basic disadvantage inherent in any waste processing or recovery system is the additional cost of constructing and operating the facility[1].

One of the most effective methods to reduce the volume and weight of solid waste is to burn it in a properly designed furnace, under suitable temperature and operating conditions. This process is called incineration. It is expensive, and unless appropriate air cleaning devices are provided, atmospheric pollution from the discharge of gaseous and particulate combustion products can occur. It also is a process that requires high-level technical supervision and skilled employees for proper operation and maintenance.

The advantages of incineration, however, often overweight these disadvantages[2]. Incineration can reduce the total volume of ordinary refuse by more than 80 percent. In densely populated urban areas, where large sites suitable for landfilling are not available within reasonable hauling distances, incineration may be the only economical option for solid waste management[3]. In some cases it is feasible to design and operate the incinerator so that heat from combustion can be recovered and used to produce steam or electricity. Incineration may also be used to destroy certain types of hazardous waste material.

Heat recovery and reuse from MSW (municipal solid waste) incineration is an attractive waste management option from an environmental and ecological perspective. But the problems just mentioned, along with the very high costs for equipment and controls, the need for skilled technical personnel, and the need for auxiliary fuel system, can make it a less attractive option.

Incineration without heat recovery is simpler to manage and can be less than one third as costly as recovery system. Nevertheless, plain incineration does result in a total loss of recoverable energy[4]. Because of public and political interest in "recycling", there will probably be an increasing emphasis on the design and construction of MSW heat recovery systems in the coming years. Incineration, with or without heat recovery, is becoming more "attractive" for solid waste management than burial of MSW in a landfill since suitable sites for burial of solid waste are becoming increasingly difficult to find.

Another relatively new development in MSW treatment by thermal-chemical conversion is a process known as pyrolysis, which is also called destructive distillation. It differs from conventional incineration in that it is an endothermic process, that is, it requires continuous input of heat energy to occur. (Incineration, on the other hand, is an exothermic process which gives off heat as oxidation occurs.)

Pyrolysis is a high-temperature process (1100 ℃ or 2000 ℉) which takes place in a low-oxygen or oxygen-free environment. Combustion of natural gas is used to start the process, but if about 70 percent of the gaseous pyrolysis by-products are recycled back to the gas burners, the process can become self-sustaining[5].

Instead of combustion, pyrolysis involves a complex series of chemical reactions. These reaction decompose or convert the organic carbon components of the solid waste into potentially useful by-products. Pyrolysis also substantially reduces the volume of the solid waste. The gaseous, liquid, and solid by-products of pyrolysis include methane, methanol, tar and char-coal. They are combustible and can be used as fuels, or they can serve as raw materials for other synthetic chemical products.

The actual composition of pyrolysis end-products may vary, and it is very dependent on the nature of the solid waste as well as on the temperature and pressure under which the process operates. The quality of the by-products can be significantly improved if glass, metal, and other inorganic material is first removed or separated from the solid waste that is fed into the pyrolysis furnace. Although waste separation adds to the expense, the pyrolysis process still has great potential as an effective solid waste management method. In addition to reducing waste volume and producing useful by-products, it poses less of a threat to air quality than does incineration.

## Words and Expressions

1. handling characteristics 处理特性，运转特性，输送特性
2. haul 搬运；运输，用力拖或拉
3. from an environmental and ecological perspective 从环境和生态的观点
4. auxiliary 辅助的
5. plain incineration 简单[单纯]焚烧
6. pyrolysis 热解(作用)
7. destructive distillation 破坏性蒸馏，分解蒸馏
8. endothermic 吸热的
9. exothermic 放热的

## Notes

(1) However, a basic disadvantage inherent in any waste processing or recovery system is the additional cost of constructing and operating the facility.

然而，所有废物处理或回收系统的根本缺点是建造和运行相关设施的附加费用。

(2) The advantages of incineration, however, often overweight these disadvantages.

然而，焚烧的优点常常超过缺点。

(3) In densely populated urban areas, where large sites suitable for landfilling are not available within reasonable hauling distances, incineration may be the only economical option for solid waste management.

在人口稠密的城市地区，在合理运输距离内适宜填埋作业的大型场地是缺乏的，焚烧可能是固体废物管理唯一经济的选择。

(4) Nevertheless, plain incineration does result in a total loss of recoverable energy.

然而，简单焚烧不会导致可回收能量的完全丧失。

(5) Combustion of natural gas is used to start the process, but if about 70 percent of the gaseous pyrolysis by-products are recycled back to the gas burners, the process can become self-sustaining.

天然气的燃烧是用来引发该过程的，但如果有约70%的气态热解副产物返回到气体燃烧炉，则该过程可以自我维持。

## 20. Disposal of Solid Waste—Landfilling

The dump is the most inexpensive and thus most popular means of solid waste disposal. The dump does, however, have drawbacks, including rats, insects, odor, and fires. More acceptable means of solid waste disposal are not being used by most communities, usually at the insistence of federal or state governments.

Sanitary landfills differ markedly from open dumps. The latter are simply places to dump wastes, while sanitary landfills are engineered operations, designed and operated according to acceptable standards(1).

The basic principle of a landfill operation is to deposit the refuse, compact it with bulldozers, and cover the material with at least 6 inches of dirt at the conclusion of each day's operation and a final cover of 2 feet when the area is full. The 2 feet is necessary to prevent rodents from burrowing into the refuse.

The selection of a landfill site is a sticky problem. The engineering aspects include (a) drainage—rapid runoff will lessen mosquito problems, but proximity to streams or well supplies might result in water pollution; (b) wind—it is preferable that the landfill be downwind from the community; (c) distance from collection; (d) size—a small site with limited capacity is generally not acceptable since the trouble of finding a new site is considerable; (e) ultimate use— can the area be utilized for public or private use after the operation is complete?

Perhaps even more important than the engineering problems are the social and psychological problems. No one in his right mind will be happy about having a sanitary landfill in his back yard. Right?

Surprisingly enough, however, there have been many cases where property values have actually been enhanced by a landfill or, more correctly, by what was done with the landfill site after the operation was complete(2). Golf courses, playgrounds and tennis courts can be rewards for tolerating a landfill operation for a few years(3). If the operation is conducted according to accepted practice, there should be little adverse environmental impact from landfills. This is, as you might suspect, a difficult thing to explain to the community, especially since most "sanitary landfills" have in the past been glorified dumps. The landfill operation is actually a biological method of waste treatment. Municipal refuse deposited as a fill is anything but insert. In the absence of oxygen, an aerobic decomposition steadily degrades the organic material to more stable forms. But this process is very slow. After 25 years the decomposition can still be going strong.

The end products of this decomposition are mostly gases: $CO_2$, $CH_4$, $NH_3$, and a little $H_2S$. These must obviously find some means of escape, and it is good practice to install vents in landfills to prevent the build-up of these gases. The decomposition reactions are self-sustaining, and temperatures often attain 135 ℉ or higher.

The biological aspects of landfills as well as the structural properties of compacted refuse dictate the ultimate use of landfill sites. Uneven settling is often a problem, and it is generally suggested that nothing be constructed on a landfill for at least two years after completion. With poor initial compaction, it is not unreasonable to expect 50% settling within the first five years.

Landfills should never be disturbed. Not only will this cause additional structural problems, but trapped gases can be a hazard(4). Buildings constructed on landfill sites should have spread footings (large concrete slabs) as foundations, although some have been constructed on pilings which extend through the fill and onto rock or other adequately strong material.

The cost of operating a landfill varies from about \$ 1 to \$ 2 per ton of refuse and represents the least-cost method of acceptable disposal for many communities.

Ocean disposal is another alternative for getting rid of solid wastes(5). Fortunately, this uncivilized practice is strongly discouraged by the governments of many countries and hopefully will eventually cease to be an acceptable alternative.

## Words and Expressions

1. dump 垃圾堆；倾倒(垃圾)
2. sanitary landfill 卫生填埋，垃圾填坑
3. bulldozer 推土机，压路机
4. piling 打桩工程，桩基
5. ocean disposal 海洋处置
6. alternative(供选择的)方案，(供选择的)对象

## Notes

(1) The latter are simply places to dump wastes, while sanitary landfills are engineered operations, designed and operated according to acceptable standards.

后者仅仅是倾倒废物的场地，而卫生填埋是经设计的作业，是根据认可的标准设计和操作的。

(2) Surprisingly enough, however, there have been many cases where property values have actually been enhanced by a landfill or, more correctly, by what was done with the landfill site after the operation was complete.

然而，令人惊奇的是有许多案例表明，通过垃圾填埋，或更确切地说，在垃圾填埋工作结束后，通过对垃圾填埋场所做的一切，地产实际上增值了。

(3) Golf courses, playgrounds and tennis courts can be rewards for tolerating a landfill operation for a few years.

几年忍受垃圾填埋作业可能得到的是高尔夫球场、运动场和网球场的回报。

(4) Not only will this cause additional structural problems, but trapped gases can be a hazard.

这不仅会引起额外的结构问题，而且捕集的气体可能是一种危险。

(5) Ocean disposal is another alternative for getting rid of solid wastes.

去除固体废物的另一种办法是海洋处置。

# Unit 5 Noise

## 21. Noise Pollution

### Introduction

By definition, noise is a sound that is annoying and has a long-term physiological effect on an individual. Noise is a subtle pollutant. Although it can be hazardous to a person's health and wellbeing, noise usually leaves no visible evidence. Noise pollution has grown to be a major environmental problem today. An estimated 14.7 million Americans are exposed to noises that pose a threat to hearing on their jobs. Another 13.5 million Americans are exposed to dangerous noise levels, such as from trucks, airplanes, motorcycles, and stereos without knowing it. Moreover, noise can cause temporary stress reactions like increasing the heart-rate and blood pressure, and produce negative effects on the digestive and respiratory system.

Sound is a disturbance that propagates through a medium having the properties of inertia (mass) and elasticity(1). The medium by which audible sound is transmitted is air. The higher the wave, the greater its power; the greater the number of waves a sound has, the larger is its frequency or pitch(2). The frequency can be described as the rate of vibration that is measured in Hertz (Hz, cycles per second). The human ear does not hear all of the frequencies. The normal hearing range for humans is from 20 Hz to 20,000 Hz. And also, the human ear cannot define all sounds equally. Very low and very high notes sound more faint to the ear than do 1000 Hz sounds of equal strength; that is how the ear functions. The human voice in conversation covers a median range of 300 to 4000 Hz; and, the musical scale ranges from 30 to 4000 Hz. Hearing also varies widely between individuals.

The unit of the strength of sound is measured in decibels (dB). Although the degree of loudness depends on personal judgments, precise measurement of sound is made possible by the use of the decibel scale (see Table 1). The decibel scale ranges from 0 (minimum) to 194 (maximum). Because the decibel scale is in logarithm form, at high levels, even a small reduction in level values can make a significant difference in noise intensity. This decibel scale measures sound pressure or energy according to international standards. By comparing some common sounds, the scale shows how they rank in potential harm. Recent scientific evidence showed that relatively continuous exposures to sound exceeding 70 dB can be harmful to hearing. Noise begins to harm hearing at 70 dB; and, each 10-dB increase seems twice as loud.

**Table 1 Sound Levels and Human Response**

| Common Sounds | Noise Level/dB | Effect |
|---|---|---|
| Carrier deck jet operation | 140 | Painfully loud |
| Air raid siren | 130 | |
| Jet takeoff (200 ft) | | Thunderclap |
| Discotheque | 120 | Maximum vocal effort |
| Auto horn (3 ft) | | |
| Pile drivers | 110 | |
| Garbage truck | 100 | |
| Heavy truck (50 ft) | 90 | Very annoying |
| City traffic | | Hearing damage (8 hr) |
| Alarm clock (2 ft) | 80 | Annoying |
| Hair dryer | | |
| Noisy restaurant | 70 | Phone use difficult |
| Freeway traffic | | |
| Man's voice (3 ft) | | |
| Air-conditioning unit (20 ft) | 60 | Intrusive |
| Light auto traffic (100 ft) | 50 | Quiet |
| Living room | | |
| Bedroom | 40 | |
| Quiet office | | |
| Library | 30 | Very quiet |
| Soft whisper (15 ft) | | |
| Broadcasting studio | 20 | |
| | 10 | Just audible |
| | 0 | Hearing begins |

## Noise Legislation

Because noise pollution has become such a threat to the health of so many lives, many regulations have been established to monitor and control the level of unwanted harmful sounds. The Occupational Safety and Health Act (OSHA) was signed on December 29, 1970 and went into effect April 28, 1971. The purpose of this Act is "to assure so far as possible every working man and woman in the nation safe and healthful working conditions and to preserve our human resources". The OSHA does not apply to working conditions that are protected by other federal occupational safety and health laws such as the Federal Coal Mine Health and Safety Act, the Atomic Energy Act, the Metal and Nonmetallic Mines Safety and Health Standards, and the Open Pit and Quarries Safety and Health Standards. This Act puts all state and federal occupational safety and health enforcement programs under federal control with the goal of establishing more uniform standards, regulations, and codes with stricter enforcement. Several of the major aspects of the Act will maintain federal supervision of state programs to obtain more uniform state inspection under federal standards. The OSHA will also make it mandatory for employers to keep accurate records of employee exposures to harmful agents that are required by safety and health standards(3). The law provides procedures in investigating violations by delivering citations and monetary penalties upon the request of an employee. The OSHA establishes a National Institute of Occupational Safety and

Health (NIOSH) whose members have the same powers of inspection as members of the OSHA. The Act also delegates to the Secretary of Labor the power to issue safety and health regulations and standards enforceable by law. This last provision is implemented by the Occupational Safety and Health Administration.

The OSHA enforces two basic duties which must be carried out by employers. First, it provides each employee with a working environment free of recognized hazards that cause or have the potential to cause physical harm or death. Second, it fully complies with the Occupational Safety and Health Standards under the Act. To carry out the first duty, employers must have proper instrumentation for the evaluation of test data provided by an expert in the area of industrial hygiene. This instrumentation must be obtained because the presence of health hazards cannot be evaluated by visual inspection. This duty can be used by the employees to allege a hazardous working situation without any requirement of expert judgment. It also provides the employer with substantial evidence to disprove invalid complaints. This law also gives employers the right to take full disciplinary action against those employees who violate safe practices in working methods[(4)].

Section 50-204. 10 of the Act establishes acceptable noise levels and exposures for safe working conditions, and gives various means of actions which must be taken if these levels are exceeded. A 90 dBA level of exposure to sound energy absorbed is taken as the limit of exposure that will not cause any type of hearing loss in more than 20 percent of those exposed. Workers in any industry must not be exposed to sound levels greater than 115 dBA for any amount of time. Noise levels must be measured on the A scale of a standard sound level meter at slow response. The sound level meter is a measuring device that indicates sound intensity. "Slow response" is a particular setting on the meter, and when the meter is at this setting it will average out high-level noise of short-lived duration. For impact noise, a higher level of 140 dB is acceptable because the noise impulse due to impacts is over before the human ear has time to fully react to it.

Regulations of variable noise levels are covered under paragraph (c) of Section 50-204. 10 in the OSHA. This paragraph states, "if the variations in noise levels involve maxima at intervals of 1 second or less, it is considered to be continuous". Therefore, when the level on the meter goes from a relatively steady reading to a higher reading, at intervals of one second or less, the higher reading is taken as the continuous sound level. Sounds of short duration occurring at intervals greater than 1 second should be measured in intensity and duration over the total work day. Such sounds may be analyzed using a sound level meter and should not be treated as impact sound.

The Federal Walsh-Healey Public Contracts Act took effect on May 20, 1969. To comply with the Walsh-Healey regulations on industrial noise exposure, industry must measure the noise level of its working environment. It provides valuable data with which an inspector can evaluate working conditions. The data may be obtained by sound survey meters with A-, B-, and C-weighted filters; and, all measurements weigh all the frequencies equally in that range.

On April 30, 1976, the Administrator of the U.S. Environmental Protection Agency (EPA) established the Noise Enforcement Division under the Deputy Assistant Administrator for Mobile Source and Noise Enforcement, Office of Enforcement. The division originally had a staff of twentyone individuals whose responsibilities were divided into the following four general enforcement areas:

1. General products noise regulations
2. Surface transportation noise regulations
3. Noise enforcement testing
4. Regional (EPA), state, and local assistance

On December 31, 1975, the EPA set forth noise standards and regulations for the control of noise from portable air compressors. These regulations became effective on January 1, 1978. Additional regulations are currently being developed to control noise from truck-mounted solid waste compactors, truck transport, refrigeration units, wheel and track loaders, and dozers, which have been identified pursuant to Section 5 of the NCA (Noise Control Act). The surface transportation group will have similar responsibilities with respect to transportation-related products. The first such products to be regulated under Section 6 for the control of noise are new medium- and heavy-duty trucks (in excess of 10, 000 lb GVWR). Regulations for trucks were set forth on April 13, 1976 and became effective on January 1, 1978. Motorcycles and buses are additional major noise sources that have been identified and for which regulations are presently being developed.

Noise enforcement testing is conducted by the EPA Noise Enforcement Facility located in Sandusky, Ohio. The facility is used to conduct enforcement testing; to monitor and correlate manufacturers' compliance testing; and, to train regional, state, and local personnel for noise enforcement. This program defines and develops the EPA enforcement responsibilities under the NCA. It also provides assistance to state and local agencies regarding enforcement of the federal noise control standards and regulations, and enforcement aspects of additional state and local noise control regulations.

To assist state and local governments in drafting noise control ordinances, EPA has published a Model Community Noise Ordinance, which is available in EPA regional offices and in the EPA headquarters in Washington, DC.

## Effects of Noise

It is estimated that between 8.7 and 11.1 million Americans suffer a permanent hearing disability (EPA, 1982). This section will examine the overall effects of noise on an industrial worker, not only in terms of hearing loss, but also in work quality.

The ear has its own defense mechanism against noise—the acoustic reflex. However, this reflex has vital weak points in its defenses. First of all, the muscles within the middle ear can become fatigued and slow if overused. A person who works in an environment with high noise levels gradually loses the strength in these muscles and thus more noise will reach the inner ear. Secondly, these muscles can be affected by chemicals within the working environment. Finally, the acoustic reflex is an ear-to-brain-to-ear circuit that takes at least nine-thousandths (0.009) of a second to perform. Individuals with poor acoustic reflex are usually subjected to temporary hearing loss when they come in contact with a loud noise. Most of the hearing loss caused by noise occurs during the first hour of exposure. Recovery of hearing can be complete several hours after the noise stops. The period of recovery depends upon individual variation and the level of noise that caused the deafness.

Noises that pose the greatest threat to the human body are those that are the highest pitched, loudest, poorest in tone, and longest lasting. Another dangerous type of sound is the sound of an explosion. Deafness due to noise usually occurs in conjunction with a fairly common hearing disorder known as recruitment of loudness. The person who has this disorder will have a smaller range of zone of hearing. However, the recruitment ear will retain its sensitivity for loud sound levels. Another problem that a person with recruited ears faces is the discomfort of using hearing aids. The hearing aid is a microphone that transmits sounds from the surrounding environment to an amplifier connected to a small loudspeaker built into an earplug and aimed at the eardrum. The major problem is that the sounds entering the hearing aid have to be amplified enough to be heard loudly, and at that level the sound may produce discomfort.

Researchers have analyzed noise and its effects on the human ear and have come up with several properties of noise that contribute to the loss of hearing. They include the "overall sound level of the noise spectrum", "the shape of the noise spectrum", and "total exposure duration". A final characteristic of noise that should be mentioned is the temporal distribution of noise. However, energy in noise is distributed across time and its final effect on the threshold shift is a function of total energy. It has been determined that partial noise exposures are related closely to the continuous A-weighted noise level (a means of correlating speech-interference level and NC (Noise Criteria) or PNC (Preferred Noise Criteria) level, and the unit of this scale is dBA) by equal energy amounts. The relation between energy and the amount of exposure is: twice the energy is acceptable for every halving of exposure time, without any increase in danger.

Noise affects the mind and changes emotions and behavior in many ways. Most of the time, individuals are unaware that noise is directly affecting their minds. It interferes with communication, disturbs sleep, and arouses a sense of fear. Psychologically, noise stimulates individuals to a nervous peak. Too much arousal makes a person overly anxious and as a result, tends to cause the person to make more mistakes. The effects of noise increase the frequency of momentary lapses in efficiency.

Noise has its effects on manual workers. From a case history from Dr. Jansen, the employees who worked in the quieter surroundings were easier to interview than the employees who worked in the noisier surroundings. Noise also affects a worker's behavior at home. This study revealed that the workers exposed to higher noise levels had more than twice as many family problems. Since noise affects a worker's attitude and personality, it also affects his or her output. It can interfere with communication greatly. Noise also can cause a decrease in the quality of work output when the background noise exceeds 90 dB. The effects of noise on work output depend largely on the type of work. High noise levels tend to cause a higher rate of mistakes and accidents rather than a direct slowdown of production. Results show that a worker's attention to the job at hand will tend to drift as noise levels increase.

Dr. G. Lehmann, Director of the Max Planck Institute, had determined that noise has an explicit effect on the blood vessels, especially the smaller ones known as precapillaries. Overall, noise makes these blood vessels narrow. It was also found that noise causes significant reductions in the blood supply to various parts of the body. Tests were also conducted employing a ballistocardiogram, which is used to measure the heart with each beat. When the test was conducted on a patient in noisy surroundings, the findings led to one conclusion: noise at all levels causes the peripheral blood vessels in the toes, fingers, skin, and abdominal organs to constrict, thereby decreasing the amount of blood normally supplied to these areas. The vaso-constriction is triggered by various body chemicals, predominantly adrenaline, which is produced when the body is under stress. Finally noise affects the nervous system. Noise wears down the nervous system, breaks down the human's natural resistance to disease and natural recovery, thus lowering the quality of general health.

## Sources of Noise

As more and more noise-generating products become available to consumers, the sources of noise pollution are extremely diverse and are constantly increasing. Commonly encountered motor vehicle noise comes from cars, trucks, buses, motorcycles, and emergency vehicles with sirens. Noise levels near major airports have become so intolerable that residents sometimes are forced to relocate, and property values sometimes depreciate because of noise pollution. Airport noise is the most common source of noise pollution that will produce an immediate effect ranging from temporary deafness to a prolonged irritation(5).

The noise levels a source produces can be separated into four categories. Machines, such as refrigerators and clothes dryers, are in the first group, usually produce sound levels lower than 60 dB. The second group includes clothes washers and food mixers that produce noise from 65 to 75 dB. The third group includes vacuum cleaners and noisy dishwashers, which produce a noise range from 85-95 dB. This group also includes yard-care and shop tools. The fourth group include pneumatic chippers and jet engines, which produce noise levels above 100 dB. Any amount of exposure to such equipment will probably interfere with activities, disrupt a neighbor's sleep, cause annoyance and stress, and may contribute to hearing loss.

## Noise Abatement

Noise abatement measures are under the jurisdiction of local government, except for occupational noise abatement efforts. It is impossible for an active person to avoid exposure to potentially harmful sound levels in today's mechanized world(6). Therefore, hearing specialists now recommend that individuals get into the habit of wearing protectors to reduce the annoying effects of noise.

Muffs worn over the ears and inserts worn in the ears are two basic types of hearing protectors. Since ear canals are rarely the same size, inserts should be separately fitted for each ear. Protective muffs should be adjustable to provide a good seal around the ear, proper tension of the cups against the head, and comfort. Both types of protectors are well worth the small inconvenience they cause for the wearer and they are available at most sport stores and drugstores. Hearing protectors are recommended at work and during recreational and home activities such as target shooting and hunting, power tool use, lawnmowing, and snowmobile riding.

One should be aware of major noise sources near any residence, for example, airport flight paths, heavy truck routes, and high-speed freeways, when choosing a new house or apartment. When buying a house, check the area zoning master plan for projected changes. In some places, one cannot obtain Federal Housing Administration (FHA) loans for housing in noisy locations. Use the Department of Housing and Urban Development (HUD) "walkaway test". By means of this method, potential buyers can assess background noise around a house. Simply have one person stand with some reading material at chest level and begin reading in a normal voice while the other slowly backs away. If the listener cannot understand the words within 7 feet, the noise level is clearly unacceptable. At 7 to 25 feet, it is normally unacceptable; at 26 to 70 feet, normally acceptable; and over 70 feet, clearly acceptable.

Furthermore, look for wall-to-wall carpeting. Find out about the wall construction. Staggered-stud interior walls provide better noise control. Studs are vertical wooden supports located behind walls. Staggering them breaks up the pattern of sound transmission. Check the electrical outlet boxes because noise will pass through the wall if the boxes are back-to-back. Also, check the door construction; solid or core-filled doors with gaskets or weather stripping provide better noise control. Make sure sleeping areas are displaced from rooms with noise-producing equipment. Finally, insulating the heating and air-conditioning ducts help control noise.

There are some helpful hints to make a quieter home, including the use of carpeting to absorb noise. Hang heavy drapes over windows closest to outside noise sources. Put rubber or plastic treads on uncarpeted stairs. Use upholstered rather than hard-surfaced furniture to deaden noise. Use insulation and vibration mounts when installing dishwashers. When listening to a stereo, keep the volume down. Place window air conditioners where their hum can help mask objectionable noises. Use caution in buying children's toys that make intensive or explosive sounds. Also, compare the noise outputs of different makes of an appliance before making a selection.

Housing developments often are located near high-speed highways. Poor housing placement is on the increase in many communities across the country. To cope with the problem of light-weight construction and poor planning, HUD has developed "Noise Assessment Guidelines" to aid in community planning, construction, modernization, and rehabilitation of existing buildings. In addition, the Veterans Administration (VA) requires disclosure of information to prospective buyers about the exposure of existing VA-financed houses to noise from nearby airports.

The EPA is preparing a model building code for various building types. The code will spell out extensive acoustical requirements and will make it possible for cities and towns to regulate construction in a comprehensive manner to produce a quieter local environment [(7)]

The Noise Control Act of 1972 provides the EPA with the authority to require labels on all products that generate noise capable of adversely affecting public health or welfare and on those products sold wholly or in part for their effectiveness in reducing noise. The EPA also initiated a study to rate home appliances and other consumer products by the noise generated and the impact of the noise on users and other persons normally exposed to it. Results will be used to determine whether noise labeling or noise emission standards are necessary.

## Future Trends

Since more and more noise generators have been developed in recent years, the chances of noise affecting individuals in this and next century will certainly increase. Many scientists and engineers are working on different plans or projects to reduce noise in the future. Two of the examples that have been developed are the electric trains and electric automobiles. In Japan and many eastern countries, electric trains are one of the most popular modes of transportation because they do not cause air pollution and produce minimum noise. The electric automobiles also reduce the noise, because they do not need to burn gasoline to run their engines. Although they are not very common today, this development is a good starting point for reducing noise. In addition, the development of new equipment and tools in the future, is certain to reduce noise. Individuals should also minimize noises surrounding or caused by them; for example, try not to use noise generators, such as vacuum cleaners, dishwashers, and high-watt stereos. If every individual does his or her level best to help reduce noise, the noise pollution will be lowered in the future and humans will hopefully live in a better environment.

## Words and Expressions

1. physiological effect 生理(学)的影响
2. subtle 微妙的，细微的，难以捉摸的
3. wellbeing 幸福，福利
4. digestive and respiratory system 消化和呼吸系统
5. disturbance 扰动，干扰，破坏，妨碍，故障，失调
6. propagate 传播，宣传，普及，繁殖，增殖
7. medium 介质，媒介，方法，手段；中间的，平均的，普通的
8. inertia 惯性，惯量
9. elasticity 弹性，弹力，伸缩性，灵活性
10. pitch 音调，程度，强度，高度；(为……)定(音)调，调节
11. vibration 振动，振荡
12. note 拍，节，律音，音调，声调
13. note frequency 声频
14. decibel 分贝(测量音强的单位)
15. loudness 响度，音量，高声，大声
16. decibel scale 分贝计，分贝标尺
17. monetary penalty 金钱上的惩罚，罚款
18. instrumentation 仪器，工具，装置，方法，手段，实行，实现
19. visual inspection(用)目检(验)
20. allege 提出，断言，宣称，辩解
21. standard sound level 标准声级
22. impact noise 碰撞噪声，振动噪声
23. noise impulse 噪声冲击，噪声脉冲
24. enforcement 实施，执行，强制，强迫
25. portable air compressor 移动式[便携式]空气压缩机
26. heavy-duty truck 重型卡车，重负载卡车
27. draft noise control ordinance 起草噪声控制条例[法令]
28. headquarters 本部，总部，司令部，指挥部
29. hearing disability 听力丧失，听力障碍
30. overall effects of noise on an industrial worker 噪声对工人的综合[全面]影响
31. acoustic reflex 声反射
32. deafness 耳聋，聋度
33. recruitment of loudness 响度复原，音量恢复
34. hearing aids 助听器
35. amplifier 放大器，扩音器，增强剂
36. loudspeaker 扬声器，扩音器，喇叭
37. earplug 耳塞
38. eardrum 耳膜，中耳
39. threshold shift 阈值[临界值]变化[转化]
40. momentary lapses in efficiency 效率的瞬时下降

41. explicit effect 明显的影响
42. blood vessel 血管
43. ballistocardiogram 投影心搏图
44. abdominal organs 腹部器官
45. constrict 使收缩，压缩，使变小
46. trigger 触发，起动，引起
47. adrenaline 肾上腺素
48. wear down 磨损，损耗
49. break down 分解，击穿，破坏，发生故障
50. vacuum cleaner 真空吸尘器
51. dishwasher 洗碗[碟]机
52. pneumatic chipper 气錾
53. jet engine 喷气式发动机
54. noise abatement measure 噪声消除措施，降噪措施
55. under the jurisdiction of local government 在地方政府的管辖下
56. muff 套，套筒
57. master plan 总(平面，布置)图，总计划，总体规划
58. projected change 预计的变化
59. background noise 本底噪声，背景噪声
60. carpet 毡层，地毯；铺毡，铺地毯
61. staggering 惊人的，交错的，摇摆的
62. gasket 衬圈，衬垫；装衬垫，密封
63. heavy drape 重帘，重幕
64. tread 踩，踏
65. upholster 装潢，布置，为……装垫子，为……装套子
66. insulation and vibration mount 隔声和振动固定架
67. objectionable 讨厌的，不适合的，不能采用的，不好的，有害的
68. disclosure of information 信息的泄露
69. prospective buyer 预期的买家，可能的买家
70. code 法规，规范，标准；编码，制定法规
71. noise emission standards 噪声排放标准
72. high-watt stereos 高功率立体声系统

## Notes

(1) Sound is a disturbance that propagates through a medium having the properties of inertia (mass) and elasticity.

声音是一种扰动，该扰动通过具有惯性和弹性的介质传播。

(2) The higher the wave, the greater its power; the greater the number of waves a sound has, the larger is its frequency or pitch.

声波越高，它的功率越大，声音的波数越大，它的频率或音调也就越高。

(3) The OSHA will also make it mandatory for employers to keep accurate records of employee exposures to harmful agents that are required by safety and health standards.

美国职业安全和卫生局规定雇主必须有雇员接触安全与卫生标准要求的有害物质的准确记录。

(4) This law also gives employers the right to take full disciplinary action against those employees who violate safe practices in working methods.

该法律也授权雇主采取充分的惩罚性的行动来反对雇员违反安全作业法。

(5) Airport noise is the most common source of noise pollution that will produce an immediate effect ranging from temporary deafness to a prolonged irritation.

飞机场噪声是最常见的噪声污染源，它会产生(对人的)直接影响，从短暂的耳聋到长期的损伤。

(6) Noise abatement measures are under the jurisdiction of local government, except for occupational noise abatement efforts.

除职业噪声降噪措施外，降噪措施都在地方政府的管辖下。

(7) The EPA is preparing a model building code for various building types. The code will spell out extensive acoustical requirements and will make it possible for cities and towns to regulate construction in a comprehensive manner to produce a quieter local environment.

美国环保局正在为各种建筑制定一部样板建筑规范。该规范将说明广泛的声学要求并使城镇能以综合的方式管理建筑以产生一个较安静的地方环境。

# Unit 6 Environmental Assessment

## 22. Assessment of Environmental Impacts

The primary goal of the EIA procedure is to predict any adverse (or beneficial) effects of a proposed project on the natural and urban environment. This is done so that measures can be taken to minimize or eliminate the harmful impacts when the project is implemented. The prediction or assessment of environmental impacts is not an easy task. It must be conducted by an interdisciplinary team, including civil engineers and technicians, urban planners, and biologists or ecologists. For large and complex projects, and particularly for sensitive environmental settings, the team may also include geologists, archaeologists, architects, and social scientists.

Certain environmental impacts can be evaluated directly and objectively. They are not really subject to conflicting subjective or personal opinions. For example, the expected increase in stormwater runoff due to the project can be computed and compared to the existing runoff rates and volumes. The effect of the increase on the site and on downstream properties may then be predicted. These effects might include flooding, soil erosion, and water pollution.

Air quality impacts can also be assessed using sophisticated mathematical models. Usually, emission of carbon monoxide from cars is of particular significance in land development' projects; the increase in automobile traffic would contribute directly to this effect. Basic traffic engineering principles can be applied in order to estimate the increase in traffic, as a function of population density and land use. Using this information, along with data on existing air quality and prevailing weather conditions, the impact of the project on ambient air quality can be anticipated.

Impacts upon vegetation and wildlife are more difficult to evaluate objectively. Although it is relatively easy to estimate how many hectares or acres of woodland will be destroyed by the project, it is much more difficult to agree upon the value or importance of this impact. Of course, if the project site is the last remaining woodland in an urban community, or if some of the trees are among an endangered species, the impacts would be considered more severe than otherwise.

It is important to distinguish between short-term impacts and long-term impacts. For example, the impacts of construction activities might include a temporary increase in neighborhood noise levels from the heavy machinery. But once the project is completed, these impacts cease; they are therefore considered short term. But the effect of the project on, say, runoff patterns and local aquifer recharge rates would not cease when construction is finished; these would be long-term impacts.

Many procedures for conducting an environmental assessment have been developed over the years. They share the basic goal of providing a comprehensive and systematic environmental evaluation of the project, with the greatest degree of objectivity. These procedures range in complexity from simple checklists to more complex "matrix" methods [1].

In the checklist method, all potential environmental impacts for the various project alternatives are listed, and the anticipated magnitude of each impact is described qualitatively. For example, negative impacts can be indicated with minus signs. A small or moderate impact could be shown with, say, two minus signs (- -), whereas a relatively more severe impact could be shown with three or four minus signs (- - - -). Beneficial or positive impacts can be shown with plus (+) sign. If the environmental impact is not applicable for a particular project alternative, a zero (0) would be shown. Such a list would serve to present a visual overview of the assessment.

In the so-called matrix methods, an attempt is made to quantify or "grade" the relative impacts of the project alternatives and to provide a numerical basis for comparison and evaluation [2]. The anticipated magnitude of each potential impact may be rated on a scale of, say, 0 to 10; the higher numbers may represent severe adverse impacts, whereas the lower numbers represent minor or negligible effects. Zero (0) would indicate no expected impact for a particular activity or environmental component.

Numerical weighting factors are also used in the matrix method, to indicate the relative importance of a particular impact. These weighting factors would be agreed upon by the assessment team, and would be site and project specific. For example, the impacts on groundwater quality may be considered more important in a particular area than impacts on air quality, particularly if the groundwater is a sole source of potable water. Groundwater quality could be assigned a relative importance or weight of, say, 0.5, compared to 0.2 for air quality.

Weighting factors can be multiplied by the respective impact magnitudes, to put each impact in perspective. For example, consider that impact on groundwater quality has a magnitude of 4 and the impact on air quality has a higher magnitude of 6. But after weighting the impacts (multiplying by the weighting factors), we see that the overall significance of the impact on water quality, $0.5 \times 4 = 2$, is more important or severe than the impact on air quality, $0.2 \times 6 = 1.2$. If the weighted impacts for all the listed items are added together, a composite score or environmental quality index can be obtained for each project alternative. The alternative with the lowest index would be the one that would cause the least harmful environmental impact, overall.

## Environmental Impact

On January 1, 1970, then-President Nixon signed into law the National Environmental Policy Act, which declared a national policy to encourage productive and enjoyable harmony between people and their environment. This law established the Council on Environmental Quality (CEQ), which monitors the environmental effects of federal activities and assists the President in evaluating environmental problems and determining the best solutions to these problems. But few people realized the National Environmental Policy Act (NEPA) contained a real sleeper article: Section 102 (2) (C), which requires federal agencies to evaluate the consequences of any proposed action on the environment.

The Congress authorizes and directs that, to the fullest extent possible: (1) the policies, regulations, and public laws of the United States shall be interpreted and administered in accordance with the policies set forth in this chapter, and (2) all agencies of the Federal Government shall include in every recommendation or report on proposals for legislation and other major Federal actions significantly affecting the quality of the human environment, a detailed statement by the responsible official on —

1. the environmental impact of the proposed action,

2. any adverse environmental effects which cannot be avoided should the proposal be implemented,

3. alternatives to the proposed action,

4. the relationship between local short-term uses of man's environment and the maintenance and enhancement of long-term productivity, and

5. any irreversible and irretrievable commitments of resources which would be involved in the proposed action should it be implemented.

In other words, each project funded by the federal government must be accompanied with an "Environmental Impact Statement" (EIS). Such a published statement must assess in detail the potential environmental impacts of a proposed action and alternative actions. All federal agencies are required to prepare statements for projects and programs (a programmatic EIS) under their jurisdiction. Additionally, the agencies must generally follow a detailed and often lengthy public review of each EIS before proceeding with their projects and programs.

The original idea of the EIS is to introduce environmental factors into the decision-making machinery. The purpose of the EIS is not to provide justification for a construction project, but rather to introduce environmental concerns and have them discussed in public before the decision on a project is made(3). However, this objective is difficult to apply in practice. Historically, interest groups in and out of government articulated plans to their liking, the sum of which provided the engineer with a set of alternatives to be evaluated(4). In many other instances, the engineer is left to create his own alternatives, or participate in a group decision-making process like the Delphi technique. In either case there are normally one or two plans which, from the outset, seem eminently more feasible and reasonable, and these can be legitimatized by juggling time scales or standards of enforcement patterns just slightly and calling them alternatives (as they are in a limited sense). As a result, non-decisions are made, i. e. wholly different ways of perceiving the problem and conceiving of solutions have been overlooked, and the primary objective of an EIS has been circumvented. Over the past few years, court decisions and guidelines by various agencies have, in fact, helped to mold this procedure for the development of environmental impact statements.

Ideally, an EIS must be thorough, interdisciplinary, and as quantitative as possible. The writing of an EIS involves three distinct phases: Inventory, assessment and evaluation. The first is a cataloging of environmentally susceptible areas, the second is the process of estimating the impact of the alternatives, and the last is the interpretation of these findings.

## Environmental Assessment

The process of calculating projected effects that a proposed action or construction project will have on environmental quality is called environmental assessment(5). It is necessary to develop a methodical, reproducible and reasonable method of evaluating both the effect of the proposed project and the effects of alternatives which may achieve the same ends but which may have different environmental impacts. A number of semiquantitative approaches have been used, among them the checklist, the interaction matrix, and the checklist with weighted rankings(6).

Checklists are listings of potential environmental impacts, both primary and secondary. Primary effects occur as a direct result of the proposed project, such as the effect of a dam on aquatic life. Secondary effects occur as an indirect result of the action. For example, an interchange for a highway will not directly affect a land area, but indirectly it will draw such establishments as service stations and quick food stores, thus changing land use patterns.

The checklist for a highway project could be divided into three phases: planning, construction and operation. During planning, consideration is given to environmental effects of the highway route and the acquisition and condemnation of property. The construction phase checklist will include displacement of people, noise, soil erosion, water pollution and energy use. Finally, the operation phase will list direct impacts due to noise, air pollution, water pollution due to runoff, energy use, etc., and indirect impacts due to regional development, housing, lifestyle, and economic development.

The checklist technique thus simply lists all of the pertinent factors; then the magnitude and importance of the impacts are estimated. The estimation of impact is quantified by establishing an arbitrary scale , such as

0 = no impact
1 = minimal impact
2 = small impact
3 = moderate impact
4 = significant impact
5 = severe impact

This scale can be used to estimate both the magnitude and the importance of a given item on the checklist. The numbers can then be combined, and a quantitative measurement of the severity of environmental impact for any given alternative estimated.

Example:

A landfill is to be placed in a floodplain of a river. Estimate the impact using the checklist technique.

First the items impacted are listed, then a judgment concerning both importance and magnitude of the impact is made. In this example, the items are only a sample of the impacts one would normally consider. The numbers in this example are then multiplied and the sum obtained. Thus:

| Potential Impact | Importance×Magnitude |
|---|---|
| Groundwater contamination | 5×5 = 25 |
| Surface water contamination | 4×3 = 12 |
| Odor | 1×1 = 1 |
| Noise | 1×2 = 2 |
| Total | 40 |

This total of 40 can then be compared to totals calculated for alternative courses of action.

In the checklist technique most variables must be subjectively judged. Further, it is difficult to predict future conditions such as land-use pattern changes or changes in lifestyle. Even with these drawbacks, however, this method is often used by engineers in governmental agencies and consulting firms, mainly due to its simplicity.

The interaction matrix technique is a two-dimensional listing of existing characteristics and conditions of the environment and detailed proposed actions which may impact the environment. For example, the characteristics of water might be defined as:

* surface
* ocean
* underground
* quantity
* temperature
* groundwater rechange
* snow, ice and permafrost

Similar characteristics must also be defined for air, land, and other important considerations.

Opposite these listings in the matrix are lists of possible actions. In our example, one such action is labeled resource extraction, which could include the following actions:

* blasting and drilling
* surface extraction
* subsurface extraction
* well drilling
* dredging
* timbering
* commercial fishing and hunting

The interactions, as in the checklist technique, are measured in terms of magnitude and the importance. The magnitudes are represented by the extent of the interaction between the environmental characteristics and the proposed actions, and typically can be measured. The importance of the interaction, on the other hand, is often a judgment call on the part of the engineer.

If an interaction is present, for example, between underground water and well drilling, a diagonal line is placed in the block. Numbers can then be assigned to the interaction, with 1 being a small and 5 being a large magnitude or importance, and these placed in the blocks with magnitude above and importance below. Appropriate blocks are filled in, using a great deal of judgment and personal bias, and then are summed over a line, thus giving a numerical grade for either the proposed action or environmental characteristics.

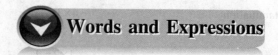

## Words and Expressions

1. adverse ( or beneficial ) effects of a proposed project on the natural and urban

environment 拟建项目对自然环境和城市环境的不利(或有利)的影响
2. interdisciplinary(各)学科(之)间的，跨学科的
3. sensitive environmental setting 敏感的环境定位[设置]
4. conflicting 矛盾的，冲突的，不一致的
5. stormwater runoff 雨水径流(量)
6. soil erosion 土壤的腐蚀[侵蚀]
7. of particular significance 特别重要的
8. prevailing weather conditions 盛行气象条件，主要的天气条件
9. impact of the project on ambient air quality 工程[项目]对环境空气质量的影响
10. woodland 森林地，林区；森林的，林地的
11. runoff pattern 径流模式
12. aquifer recharge rate 含水层回充率，蓄水层回注率
13. checklist method 清单法
14. various project alternatives 各种工程方案
15. negative impacts 负面的影响
16. beneficial or positive impacts 有益的或正面的影响
17. particular project alternative 特别的工程方案
18. visual overview of the assessment 评价的目视观察，评价的直观观察
19. numerical weighting factors 数值权重因子
20. in perspective 展望中(的)，正确地，合乎透视法
21. construction project 建筑项目[工程]
22. juggle time scale 窜改时间标尺
23. perceive 发觉，感觉出，领会，理解
24. conceive 设想，想到，表达
25. circumvent 胜过，击败，绕过，防止……发生，回避，包围
26. inventory 清单，目录，报表；把……编入目录
27. projected effects 预期的影响
28. interaction matrix 交互作用矩阵
29. weighted ranking 加权分类，权重分级
30. establishment 企业，机关，机构
31. quick food store 快餐店
32. highway project 公路项目，公路工程
33. acquisition 获得，取得，采集
34. condemnation 征用；没收，谴责
35. displacement of people 移民，人的移置
36. pertinent 有关的，相干的，恰当的
37. permafrost 永久冻土，永冻地
38. arbitrary 任意的，专制的
39. severity 严厉，严重，刚度
40. blasting 鼓风，环吹，喷砂法，喷砂清理
41. dredge 疏浚，挖掘；挖泥机，挖泥船
42. timber 用木材建造，用木材支撑；木材，原木
43. commercial fishing and hunting 商业垂钓和狩猎
44. personal bias 个人偏见
45. sum 总和，概要；合计，概括，总结

## Notes

(1) They share the basic goal of providing a comprehensive and systematic environmental evaluation of the project, with the greatest degree of objectivity. These procedures range in complexity from simple checklists to more complex "matrix" methods.

他们享有的基本目标是以最大程度的客观性提供对项目的综合和系统的环境评估。这些方法有简单的清单法到较复杂的"矩阵"方法。

(2) In the so-called matrix methods, an attempt is made to quantify or "grade" the relative

impacts of the project alternatives and to provide a numerical basis for comparison and evaluation.

在所谓的矩阵法中，致力于对工程方案的相对影响定量化或"分等级"及为比较和评估提供数值基础。

(3) The purpose of the EIS is not to provide justification for a construction project, but rather to introduce environmental concerns and have them discussed in public before the decision on a project is made.

环境影响报告(EIS)的目的不是要证明建设项目的合理性，而是要介绍环境热点问题并在项目决策前使它们得到公开讨论。

(4) Historically, interest groups in and out of government articulated plans to their liking, the sum of which provided the engineer with a set of alternatives to be evaluated.

从历史上看，政府内外的利益团体都将计划与它们的喜好联系起来，所有的计划都给工程师提供了一系列待评估的方案。

(5) The process of calculating projected effects that a proposed action or construction project will have on environmental quality is called environmental assessment.

预测一项计划的行动或一个拟建项目对环境质量的影响的过程被称为环境评价。

(6) A number of semiquantitative approaches have been used, among them the checklist, the interaction matrix, and the checklist with weighted rankings.

已使用一些半定量的方法，其中有清单法、交互作用矩阵法和带权重分级的清单法。

# Unit 7 Pollution Prevention

## 23. The Pollution Prevention Concept

### Introduction

The amount of waste generated in the United States has reached staggering proportions; according to the United States Environmental Protection Agency (EPA), 250 million tons of solid waste alone are generated annually. Although both the Resource Conservation and Recovery Act (RCRA) and the Hazardous and Solid Waste Act (HSWA) encourage businesses to minimize the wastes they generate, the majority of current environmental protection efforts are centered around treatment and pollution cleanup.

The passage of the Pollution Prevention Act of 1990 has redirected industry's approach to environmental management; pollution prevention has now become the environmental option of this decade and the 21st century(1). Whereas typical waste management strategies concentrate on "end-of-pipe" pollution control, pollution prevention attempts to handle waste at the source (i. e., source reduction)(2). As waste handling and disposal costs increase, the application of pollution prevention measures is becoming more attractive than ever before. Industry is currently exploring the advantages of multimedia waste reduction and developing agendas to *strengthen* environmental design while *lessening* production costs(3).

There are profound opportunities for both the individual and industry to prevent the generation of waste; indeed, pollution prevention is today primarily stimulated by economics, legislation, liability concerns, and the enhanced environmental benefit of managing waste at the source. The EPA's Pollution Prevention Act of 1990 has established pollution prevention as a national policy declaring "waste should be prevented or reduced at the source wherever feasible, while pollution that cannot be prevented should be recycled in an environmentally safe manner" (EPA, 1991). The EPA's policy establishes the following hierarchy of waste management:

1. Source reduction
2. Recycling/reuse
3. Treatment
4. Ultimate disposal

The hierarchy's categories are prioritized so as to promote the examination of each individual alternative prior to the investigation of subsequent options (i. e., the most preferable alternative should be thoroughly evaluated before consideration is given to a less accepted option.) Practices that decrease, avoid, or eliminate the generation of waste are considered source reduction and can include the implementation of procedures as simple and economical as good housekeeping. Recycling is the use, reuse, or reclamation of wastes and/or materials and may involve the incorporation of waste recovery techniques (e. g., distillation, filtration). Recycling can be performed at the facility (i. e., on-site), or at an off-site reclamation facility. Treatment involves the destruction or detoxification of wastes into nontoxic or less toxic materials by chemical, biological or physical methods, or any combination of these methods. Disposal has been included in the hierarchy because it is recognized that residual wastes will exist; the EPA's so-called "ultimate disposal" options include landfilling, land farming, ocean dumping and deep-well injection. However, the term "ultimate disposal" is a misnomer, but is included here because of its adaptation by the EPA. Table 1 provides a rough timetable demonstrating the national approach to waste management. Note how waste management has begun to shift from pollution *control* to pollution *prevention*.

**Table 1 Waste Management Timetable**

| Timeframe | Control |
|---|---|
| Prior to 1945 | No Control |
| 1945-1960 | Little Control |
| 1960-1970 | Some Control |
| 1970-1975 | Greater Control (EPA Founded) |
| 1975-1980 | More Sophisticated Control |
| 1980-1985 | Beginning of Waste Reduction Management |
| 1985-1990 | Waste Reduction Management |
| 1990-1995 | (Pollution Prevention Act) |
| 1995- | ??? |

## Pollution Prevention Hierarchy

As discussed in the Introduction, the hierarchy set forth by the EPA in the Pollution Prevention Act establishes an order in which waste management activities should be employed to reduce the quantity of waste generated. The preferred method is source reduction, as indicated in Figure 1. This approach actually precedes traditional waste management by addressing the source of the problem prior to its occurrence.

Although the EPA's policy does not consider recycling or treatment as actual pollution prevention methods per se, these methods present an opportunity to reduce the amount of waste that might otherwise be discharged into the environment. Clearly, the definition of pollution prevention and its synonyms (e. g., waste minimization) must be understood to fully appreciate and apply these techniques.

Waste minimization generally considers all of the methods in the EPA hierarchy (except for disposal) appropriate to reduce the volume or quantity of waste requiring disposal (e. g., source reduction). The definition of source reduction as applied in the Pollution Prevention Act, however, is "any practice which reduces the amount of any hazardous substance, pollutant or contaminant entering any waste stream or otherwise released into the environment… prior to recycling, treatment or disposal" (EPA, 1991).

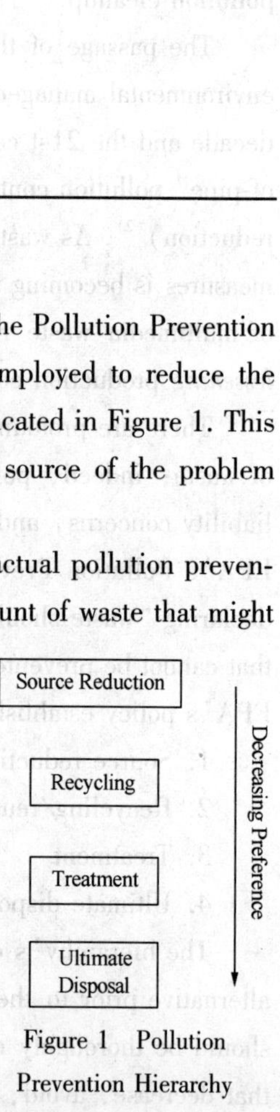


Figure 1 Pollution Prevention Hierarchy

Source reduction reduces the amount of waste generated; it is therefore considered true pollution prevention and has the highest priority in the EPA hierarchy.

Recycling (reuse, reclamation) refers to the use or reuse of materials that would otherwise be disposed of or treated as a waste product. Wastes that cannot be directly reused may often be recovered on-site through methods such as distillation. When on-site recovery or reuse is not feasible due to quality specifications or the inability to perform recovery on-site, off-site recovery at a permitted commercial recovery facility is often a possibility. Such management techniques are considered secondary to source reduction and should only be used when pollution can not be prevented.

The treatment of waste is the third element of the hierarchy and should be utilized only in the absence of feasible source reduction or recycling opportunities. Waste treatment involves the use of chemical, biological, or physical processes to reduce or eliminate waste material. The incineration of wastes is included in this category and is considered "preferable to other treatment methods (i. e., chemical, biological, and physical) because incineration can permanently destroy the hazardous components in waste materials".

Of course, many of these pollution prevention elements are used by industry in combination to achieve the greatest waste reduction. Residual wastes that cannot be prevented or otherwise managed are then disposed of only as a last resort.

## MULTIMEDIA ANALYSIS AND LIFECYCLE COST ANALYSIS

### Multimedia Analysis

In order to properly design and then implement a pollution prevention program, sources of all wastes must be fully understood and evaluated. A multimedia analysis involves a multifaceted approach. It must not only consider one waste stream but all potentially contaminant media (e. g., air, water, land). Past waste management practices have been concerned primarily with treatment. All to often, such methods solve one waste problem by transferring a contaminant from one medium to another (e. g., air-stripping); such waste shifting is *not* pollution prevention or waste reduction.

Pollution prevention techniques must be evaluated through a thorough consideration of all media, hence the term multimedia. This approach is a clear departure from previous pollution treatment or control techniques where it was acceptable to transfer a pollutant from one source to another in order to solve a waste problem. Such strategies merely provide short-term solutions to an ever-increasing problem. As an example, air pollution control equipment prevents or reduces the discharge of waste into the air but at the same time can produce a solid hazardous waste problem.

Lifecycle Analysis

The aforementioned multimedia approach to evaluating a product's waste stream(s) aims to ensure that the treatment of one waste stream does not result in the generation or increase in an additional waste output. Clearly, impacts resulting during the production of a product must be evaluated over its entire history or lifecycle.

A lifecycle analysis, or "Total Systems Approach", is crucial to identifying opportunities for improvement. This type of evaluation identifies "energy use, material inputs, and wastes generated during a product's life: from extraction and processing of raw materials, to manufacture and transport of a product to the marketplace, and, finally, to use and dispose of the product".

During a forum convened by the World Wildlife Fund and the Conservation Foundation in May 1990, various steering committees recommended that a three-part lifecycle model be adopted. This model consists of the following:

1. An inventory of materials and energy used, and environmental releases from all stages in the life of a product or process.

2. An analysis of potential environmental effects related to energy use and material resources and environmental releases.

3. An analysis of the changes needed to bring about environmental improvements for the product or process under evaluation.

Traditional cost analysis often fails to include factors relevant to future damage claims resulting from litigation, the depletion of natural resources, the effects of energy use, and so on(4). Therefore, waste management options such as treatment and disposal may appear preferential if an overall lifecycle cost analysis is not performed. It is evident that environmental costs from "cradle-to-grave" have to be evaluated together with more conventional production costs to accurately ascertain genuine production costs(5). In the future, a total systems approach will most likely involve a more careful evaluation of pollution, energy, and safety issues. For example, if one was to compare the benefits of coal versus oil as a fuel source for an electric power plant, the use of coal might be considered economically favorable. In addition to the cost issues, however, one must be concerned with the environmental effects of coal mining, transportation, and storage prior to use as a fuel. Many have a tendency to overlook the fact that there are serious health and safety matters (e. g., miner exposure) that must be considered, along with the effects of fugitive emissions. When these effects are weighed alongside of standard economic factors, the cost benefits of coal usage may no longer appear valid. Thus, many of the economic benefits associated with pollution prevention are often unrecognized due to inappropriate cost accounting methods. For this reason, economic considerations are detailed in the next chapter.

## POLLUTION PREVENTION ASSESSMENT PROCEDURES

The first step in establishing a pollution prevention program is the obtainment of management commitment. Management commitment is necessary given the inherent need for project structure and control. Management will determine the amount of funding allotted for the program as well as specific program goals. The data collected during the actual evaluation is then used to develop options for reducing the types and amounts of waste generated. Figure 1 depicts a systematic approach that can be used during the procedure. After a particular waste stream or area of concern is identified, feasibility studies are performed involving both economic and technical considerations. Finally, preferred alternatives are implemented. The four phases of the assessment (i. e., planning and organization, assessment, feasibility, and implementation) are introduced in the following subsections. Sources of additional information, as well as information on industrial programs is also provided in this section.

## Planning and Organization

The purpose of this phase is to obtain management commitment, define and develop program goals, and to assemble a project team. Proper planning and organization are crucial to the successful performance of the pollution prevention assessment. Both managers and facility staff play important roles in the assessment procedure by providing the necessary commitment and familiarity with the facility, its processes, and current waste management operations[6]. The benefits of the program, including economic advantages, liability reduction, regulatory compliance and improved public image, often lead to management support.

Once management has made a commitment to the program and goals have been set, a program task force is established. The selection of a team leader will be dependent upon many factors including his or her ability to effectively interface with both the assessment team and management staff.

The task force must be capable of identifying pollution reduction alternatives, as well as be cognizant of inherent obstacles to the process. Barriers frequently arise from the anxiety associated with the belief that the program will negatively affect product quality or result in production losses.

According to an EPA survey, 30 percent of industry comments responded that they were concerned that product quality would decline if waste minimization techniques were implemented (EPA, 1990). Thus, the assessment team, and the team leader in particular, must be ready to react to these and other concerns.

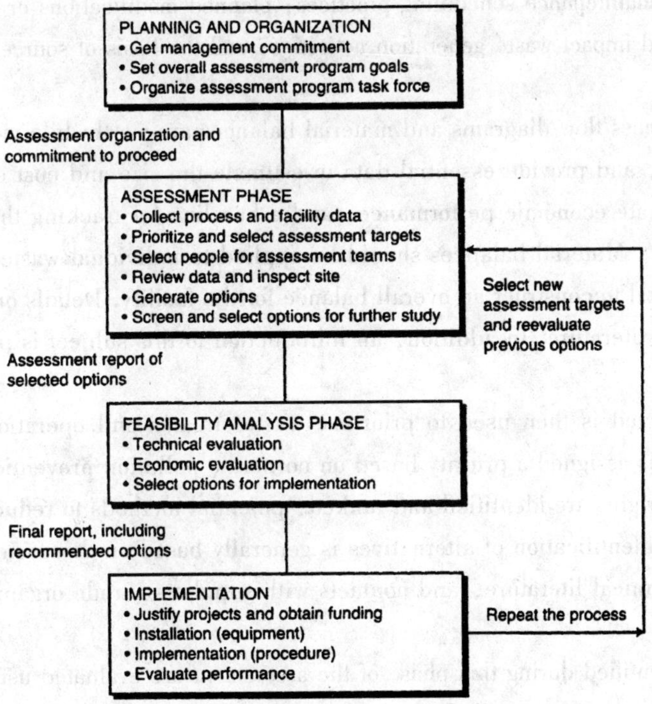


Figure 2 Pollution Prevention Assessment Procedures

## Assessment Phase

The assessment phase aims to collect data needed to identify and analyze pollution prevention opportunities. Assessment of the facility's waste reduction needs includes the examination of hazardous waste streams, process operations, and the identification of techniques that often promise the reduction of waste generation. Information is often derived from observations made during a facility walk-through, interviews with employees (e. g., operators, line workers), and review of site or regulatory records. The American Society of Testing and Materials (ASTM) suggests the following information sources be reviewed, as available.

1. Product design criteria.

2. Process flow diagrams for all solid waste, wastewater, and air emissions sources.

3. Site maps showing the location of all pertinent units (e. g., pollution control devices, points of discharge).

4. Environmental documentation, including: Material Safety Data Sheets (MSDS), military specification data, permits (e. g., NPDES, POTW, RCRA), SARA Title III reports, waste manifests, and any pending permits or application information.

5. Economic data, including: cost of raw material management; cost of air, wastewater, and hazardous waste treatment; waste management operating and maintenance costs; and, waste disposal costs.

6. Managerial information: environmental policies and procedures; prioritization of waste management concerns; automated or computerized waste management systems; inventory and distribution procedures; maintenance scheduling practices; planned modifications or revisions to existing operations that would impact waste generation activities; and the basis of source reduction decisions and policies.

The use of process flow diagrams and material balances are worthwhile methods to "quantify losses or emissions, and provide essential data to estimate the size and cost of additional equipment, data to evaluate economic performance, and a baseline for tracking the progress of minimization efforts" [7]. Material balances should be applied to individual waste streams or processes, and then utilized to construct an overall balance for the facility. Details on these calculations are available in the literature. In addition, an introduction to this subject is provided in the next section.

The data collected is then used to prioritize waste streams and operations for assessment. Each waste stream is assigned a priority based on corporate pollution prevention goals and objectives. Once waste origins are identified and ranked, potential methods to reduce the waste stream are evaluated. The identification of alternatives is generally based on discussions with the facility staff, review of technical literature, and contacts with suppliers, trade organizations, and regulatory agencies.

Alternatives identified during this phase of the assessment are evaluated using screening procedures so as to reduce the number of alternatives requiring further exploration during the feasibility analysis phase. The criteria used during this screening procedure include: cost-effectiveness; implementation time; economic, compliance, safety, and liability concerns; waste reduction potential; and, whether the technology is proven. Options which meet established criteria are then examined further during the feasibility analysis.

### Feasibility Analysis

Preferred alternative selection is performed by an evaluation of technical and economic considerations. The technical evaluation determines whether a given option will work as planned. Some typical considerations follow:

1. Safety concerns
2. Product quality impacts or production delays during implementation
3. Labor and/or training requirements
4. Creation of new environmental concerns
5. Waste reduction potential
6. Utility and budget requirements
7. Space and compatibility concerns

If an option proves to be technically ineffective or inappropriate, it is deleted from the list of potential alternatives. Either following or concurrent with the technical evaluation, an economic study is performed weighing standard measures of profitability such as payback period, investment returns, and net present value[8]. Many of these costs (or more appropriately, cost savings) may be substantial yet are difficult to quantify.

## Implementation

The findings of the overall assessment are used to demonstrate the technical and economic worthiness of program implementation. Once appropriate funding is obtained, the program is implemented not unlike any other project requiring new procedures or equipment. When preferred waste pollution prevention techniques are identified, they are implemented, and should become part of the facility's day-to-day management and operation. Subsequent to the program's execution, its performance should be evaluated in order to demonstrate effectiveness, generate data to further refine and augment waste reduction procedures, and maintain management support[9].

It should be noted that waste reduction, energy conservation, and safety issues are interrelated and often complementary to each other. For example, the reduction in the amount of energy a facility consumes results in reduced emissions associated with the generation of power. Energy expenditures associated with the treatment and transport of waste are similarly reduced when the amount of waste generated is lessened; at the same time worker safety is elevated due to reduced exposure to hazardous materials.

## Sources of information

The successful development and implementation of any pollution prevention program is not only dependent on a thorough understanding of the facility's operations but also requires an intimate knowledge of current opportunities and advances in the field. In fact, 32 percent of industry respondents to an EPA survey identified the lack of technical information as a major factor delaying or preventing the implementation of a waste minimization program (EPA, 1991). Fortunately, the EPA has developed a national Pollution Prevention Information Clearinghouse (PPIC) and the Pollution Prevention Information Exchange System (PIES) to facilitate the exchange of information needed to promote pollution prevention through efficient information transfer (EPA, 1991).

PPIC is operated by the EPA's Office of Research and Development and the Office of Pollution Prevention. The clearinghouse is comprised of four elements:

1. Repository: including a hard copy reference library and collection center and an on-line information retrieval and ordering system.

2. PIES: a computerized conduit to databases and document ordering, accessible via modem and personal computer- (703) 506-1025.

3. Hotline: PPIC uses the RCRA/Superfund and Small Business Ombudsman Hotlines as well as a PPIC technical assistance line to answer pollution prevention questions, access information in the PPIC, and assist in document ordering and searches. To access PPIC by telephone, call:

RCRA/Superfund Hotline (800) 242-9346

Small Business Ombudsman Hotline (800) 368-5888

PPIC Technical Assistance (703) 821-800

4. Networking and Outreach: PPIC compiles and disseminates information packets and bulletins, and initiates networking efforts with other national and international organizations.

Additionally, the EPA publishes a newsletter entitled Pollution Prevention News, which contains information including EPA news, technologies, program updates, and case studies. The EPA's Risk Reduction Engineering Laboratory and the Center for Environmental Research Information has published several guidance documents, developed in cooperation with the California Department of Health Services. The manuals supplement generic waste reduction information presented in the EPA's Waste Minimization Opportunity Assessment Manual (EPA, 1988). Additional information is available through PPIC.

Pollution prevention or waste minimization programs have been established at the State level and as such are good sources of information. Both Federal and State agencies are working with universities and research centers and may also provide assistance. For example, the American Institute of Chemical Engineers has established the Center for Waste Reduction Technologies (CWRT), a program based on targeted research, technology transfer, and enhanced education.

## Industry Programs

A significant pollution prevention resource may very well be found with the "competition". Several large companies have established well-known programs that have successfully incorporated pollution prevention practices into their manufacturing processes. These include, but are not limited to: 3M-Pollution Prevention Pays (3P); Dow Chemical-Waste Reduction Always Pays (WRAP); Chevron-Save Money And Reduce Toxics (SMART); and General Dynamics-Zero Discharge Program.

Smaller companies can benefit by the assistance offered by these larger corporations. It is clear that access to information is of major importance when implementing efficient pollution prevention programs. By adopting such programs, industry is affirming pollution prevention as a good business practice and not simply a noble effort.

## Future Trends

The development of waste management practices in the United States has recently moved toward securing a new pollution prevention ethic. The performance of pollution prevention assessments and their subsequent implementation will encourage increased research into methods that will further aid in the reduction of wastes and pollution.

It is evident that the majority of the present day obstacles to pollution prevention are based on either a lack of information or an anxiety associated with economic concerns. By strengthening the exchange of information among businesses, a better understanding of the unique benefits of pollution prevention will be realized in the future.

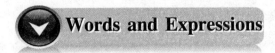

## Words and Expressions

1. staggering 惊人的，交错的，摇摆的
2. solid waste 固体废物
3. end-of-pipe pollution control 末端污染控制
4. waste handling and disposal costs 废物处理[输送]和处置费用
5. multimedia waste reduction 多介质的废物减量(化)
6. lessen production cost 降低生产成本
7. environmentally safe manner 环境安全的方式
8. hierarchy 体系，等级制，分层结构，级别，阶层
9. prioritize 按重点排列，按优先序排列
10. reclamation 再生，回收，恢复，开拓，开垦
11. detoxification 解毒(作用)，去毒
12. ultimate disposal 最终处置
13. preferred method 优先选用的方法，最佳的方法
14. precede traditional waste management 时间上先于传统的废物管理，优于传统的废物管理
15. synonym 同义词
16. priority(在)先，(在)前，优先权，顺序，次序，轻重缓急
17. on-site recovery 就地回收
18. off-site recovery 非现场回收，非就地回收
19. resort 求助，依靠，手段，娱乐场所；使再分开，把……再分类
20. multimedia analysis 多介质分析
21. multifaceted 多方面的
22. life cycle analysis 生命周期分析
23. aforementioned 上述的，前述的，前面提到的
24. ensure 保证，保护，保证得到
25. forum 论坛，讨论会，讲座
26. convene 召集，集会，集合
27. steering 驾驶，掌舵，操纵，指导，领导，引导
28. preferential 优先的，优惠的；优先权
29. cradle-to-grave 从摇篮到坟墓
30. fugitive emissions 易散性排放物，短期[时]排放物
31. alongside 靠在……旁边，横靠；在旁，并排地
32. cost accounting method 成本核算方式
33. management commitment 管理委托(事项)
34. allot for 分配，拨给
35. be cognizant of 认识到，知道，晓得
36. anxiety 担心，忧虑，渴望
37. baseline 基线，底线，原始资料；原始的，开始的，基本的
38. track 轨迹，轨道；跟踪
39. screening procedure 筛选程序，筛选方法
40. preferred alternative selection 优先的方案选择，最优的方案选择
41. utility and budget requirements 公用事业和预算要求
42. payback period 偿还期，报答期

43. safety issue 安全问题
44. complementary to each other 互相补充
45. energy expenditure 能量消耗
46. intimate 密切的，接近的，内部的，熟悉的；宣布，通知，暗示
47. repository 贮藏室，仓库，陈列室，博物馆
48. disseminate 散布，传播，宣传
49. generic 一般的，普通的
50. targeted research 有目标的研究，定向研究
51. technology transfer 技术转移
52. affirm 断言，确认，证实，批准
53. secure 安全的，可靠的；使安全保障，获得

## Notes

(1) The passage of the Pollution Prevention Act of 1990 has redirected industry's approach to environmental management; pollution prevention has now become the environmental option of this decade and the 21st century.

污染预防法案(1990)的通过为工业界环境管理的途径重新指明了方向；污染预防已成为这10年和21世纪环境的选择。

(2) Whereas typical waste management strategies concentrate on "end-of-pipe" pollution control, pollution prevention attempts to handle waste at the source (i. e., source reduction).

尽管一般的废物管理策略致力于"末端"污染控制，但污染预防致力于在源头处理废物(即废物源减量化)。

(3) Industry is currently exploring the advantages of multimedia waste reduction and *developing* agendas to *strengthen* environmental design while *lessening* production costs.

工业界现正在探索多介质废物减量化的优点和形成有关强化环境设计而降低生产成本的议程。

(4) Traditional cost analysis often fails to include factors relevant to future damage claims resulting from litigation, the depletion of natural resources, the effects of energy use, and so on.

传统的成本分析常常不能包括如下相关因素：由诉讼引起的对将来损害的赔偿要求、自然资源的耗竭、能量使用的影响等。

(5) It is evident that environmental costs from "cradle-to-grave" have to be evaluated together with more conventional production costs to accurately ascertain genuine production costs.

显然从"摇篮到坟墓"的环境成本必须与较传统的生产成本一起进行评估以准确确定真正的生产成本。

(6) Both managers and facility staff play important roles in the assessment procedure by providing the necessary commitment and familiarity with the facility, its processes, and current waste management operations.

经理与员工两者在评价过程中都起重要作用，他们会作出必要的承诺，并熟悉设施、工艺过程和现行的废物管理作业。

(7) The use of process flow diagrams and material balances are worthwhile methods to "quantify losses or emissions, and provide essential data to estimate the size and cost of additional equip-

ment, data to evaluate economic performance, and a baseline for tracking the progress of minimization efforts".

运用工艺流程图和物料平衡是合算的方法，该方法能定量损失量或排放量，并提供估算附加设备大小和成本的基本资料，评估经济性能的资料和跟踪废物减量化进展的资料。

(8) Either following or concurrent with the technical evaluation, an economic study is performed weighing standard measures of profitability such as payback period, investment returns, and net present value.

或在技术评估之后，或在技术评估的同时，进行有关经济研究，估量获利性(如偿还期，投资返回和现有净值)的标准大小。

(9) Subsequent to the program's execution, its performance should be evaluated in order to demonstrate effectiveness, generate data to further refine and augment waste reduction procedures, and maintain management support.

在方案执行之后，应对其绩效进行评估以说明效率，形成能进一步改进和增强废物减量化方法的资料及保持管理方的支持。

# 24. The Pollution Prevention Industrial Applications

## Introduction

One of the key elements of the assessment phase of a pollution prevention program involves mass balance equations. These calculations are often referred to as material balances; the calculations are performed via the conservation law for mass. The details of this often-used law are described below.

The conservation law for mass can be applied to any process or system. The general form of the law follows:

$$\textit{mass in} - \textit{mass out} + \textit{mass generated} = \textit{mass accumulated}$$

This equation can be applied to the total mass involved in a process or to a particular species, on either a mole or mass basis. The conservation law for mass can be applied to steady-state or unsteady-state processes and to batch or continuous systems. A steady-state system is one in which there is no change in conditions (e.g., temperature, pressure) or rates of flow with time at any given point in the system; the accumulation term then becomes zero. If there is no chemical reaction, the generation term is zero. All other processes are classified as unsteady-state.

To isolate a system for study, the system is separated from the surroundings by a boundary or envelope that may either be real (e.g., a reactor vessel) or imaginary. Mass crossing the boundary and entering the system is part of the mass-in term. The equation may be used for any compound whose quantity does not change by chemical reaction, or for any chemical element, regardless of whether it has participated in a chemical reaction(1). Furthermore, it may be written for one piece of equipment, several pieces of equipment, or around an entire process (i.e., a total material balance).

The conservation of mass law finds a major application during the performance of pollution prevention assessments[2]. As described in the previous chapter, a pollution prevention assessment is a systematic, planned procedure with the objective of identifying methods to reduce or eliminate waste. The assessment process should characterize the selected waste streams and processes (ICF Technology, 1989)—a necessary ingredient if a material balance is to be performed. Some of the data required for the material balance calculation may be collected during the first review of site-specific data; however, in some instances, the information may not be collected until an actual site walk-through is performed[3].

Simplified mass balances should be developed for each of the important waste-generating operations to identify sources and gain a better understanding of the origins of each waste stream. Since a mass balance is essentially a check to make sure that what goes into a process (i. e., the total mass of all raw materials), what leaves the process (i. e., the total mass of the product(s) and by-products), the material balance should be made individually for all components that enter and leave the process. When chemical reactions take place in a system, there is an advantage to doing "elemental balances" for specific chemical elements in a system. Material balances can assist in determining concentrations of waste constituents where analytical test data are limited. They are particularly useful when there are points in the production process where it is difficult or uneconomical to collect analytical data.

Mass balance calculations are particularly useful for quantifying fugitive emissions, such as evaporative losses. Waste stream data and mass balances will enable one to track flow and characteristics of the waste streams over time[4]. Since in most cases the accumulation equals zero (steady-state operation), it can then be assumed that any buildup is actually leaving the process through fugitive emissions or other means. This will be useful in identifying trends in waste/pollutant generation and will also be critical in the task of measuring the performance of implemented pollution prevention options.

The result of these activities is a catalog of waste streams that provides a description of each waste, including quantities, frequency of discharge, composition, and other important information useful for material balance. Of course, some assumptions or educated estimates will be needed when it is impossible to obtain specific information.

By performing a material balance in conjunction with a pollution prevention assessment, the amount of waste generated becomes known. The success of the pollution prevention program can therefore be measured by using this information on baseline generation rates (i. e., that rate at which waste is generated without pollution prevention considerations).

## Barriers to Pollution Prevention

As discussed previously, industry is beginning to realize that there are profound benefits associated with pollution prevention including cost effectiveness, reduced liability, enhanced public image, and regulatory compliance (these are discussed in the next section). Nevertheless, there are barriers or disincentives identified with pollution prevention. This section will briefly outline barriers that may need to be confronted or considered during the evaluation of a pollution prevention program.

There are numerous reasons why more businesses are not reducing the wastes they generate. The following "dirty dozen" are common disincentives:

1. *Technical limitations.* Given the complexity of present manufacturing processes, waste streams exist that can not be reduced with current technology[5]. The need for continued research and development is evident.

2. *Lack of information.* In some instances, the information needed to make a pollution prevention decision may be confidential or is difficult to obtain. In addition, many decision makers are simply unaware of the potential opportunities available regarding information to aid in the implementation of a pollution prevention program.

3. *Consumer preference obstacles.* Consumer preference strongly affects the manner in which a product is produced, packaged, and marketed. If the implementation of a pollution prevention program results in the increase in the cost of a product, or decreased convenience or availability, consumers might be reluctant to use it.

4. *Concern over product quality decline.* The use of a less hazardous material in a product's manufacturing process may result in decreased life, durability, or competitiveness.

5. *Economic concerns.* Many companies are unaware of the economic advantages associated with pollution prevention. Legitimate concerns may include decreased profit margins or the lack of funds required for the initial capital investment.

6. *Resistance to change.* The unwillingness of many businesses to change is rooted in their reluctance to try technologies that may be unproven, or based on a combination of the barriers discussed in this section.

7. *Regulatory barriers.* Existing regulations that have created incentives for the control and containment of wastes, are at the same time discouraging the exploration of pollution prevention alternatives. Moreover, since regulatory enforcement is often intermittent, current legislation can weaken waste reduction incentives.

8. *Lack of markets.* The implementation of pollution prevention processes and the production of environmentally friendly products will be of no avail if markets do not exist for such goods. As an example, the recycling of newspaper in the United States has resulted in an overabundance of waste paper without markets prepared to take advantage of this "raw" material.

9. *Management apathy.* Many managers capable of making decisions to begin pollution prevention activities, do not realize the potential benefits of pollution prevention and may therefore take on a attitude of passiveness.

10. *Institutional barriers.* In an organization without a strong infrastructure to support pollution prevention plans, waste reduction programs will be difficult to implement. Similarly, if there is no mechanism in place to hold individuals accountable for their actions, the successful implementation of a pollution prevention program will be limited[6].

11. *Lack of awareness of pollution prevention advantages.* As mentioned in *economic concerns*, decision makers may be uninformed of the benefits associated with pollution reduction.

12. *Concern over the dissemination of confidential product information.* If a pollution prevention assessment reveals confidential data pertinent to a company's product, fear may exist that the organization will lose a competitive edge with other businesses in the industry[7].

## Pollution Prevention Advantages

Various means exist to encourage pollution prevention through regulatory measures, economic incentives, and technical assistance programs. Since the benefits of pollution prevention undoubtedly surpass prevention barriers, a baker's dozen incentives is presented below:

1. *Economic benefits.* The most obvious economic benefits associated with pollution prevention are the savings that result from the elimination of waste storage, treatment, handling, transport, and disposal. Additionally, less tangible economic benefits are realized in terms of decreased liability, regulatory compliance costs (e.g., permits), legal and insurance costs, and improved process efficiency. Pollution prevention almost always pays for itself, particularly when the time investment required to comply with regulatory standards is considered. Several of these economic benefits are discussed separately below.

2. *Regulatory compliance.* Quite simply, when wastes are not generated, compliance issues are not a concern. Waste management costs associated with recordkeeping, reporting, and laboratory analysis are reduced or eliminated. Pollution prevention's proactive approach to waste management will better prepare industry for the future regulation of many hazardous substances and wastes that are currently unregulated. Regulations have, and will continue to be, a moving target.

3. *Liability reduction.* Facilities are responsible for their wastes from "cradle-to-grave". By eliminating or reducing waste generation, future liabilities can also be decreased. Additionally, the need for expensive pollution liability insurance requirements may be abated.

4. *Enhanced public image.* Consumers are interested in purchasing goods that are safer for the environment and this demand, depending on how they respond, can mean success or failure for many companies. Business should therefore be sensitive to consumer demands and use pollution prevention efforts to their utmost advantage by producing goods that are environmentally friendly.

5. *Federal and state grants.* Federal and State grant programs have been developed to strengthen pollution prevention programs initiated by states and private entities. The EPA's Pollution Prevention By and For Small Business Grant Program awards grants to small businesses to assist their development and demonstration of new pollution prevention technologies.

6. *Market incentives.* Public demand for environmentally preferred products has generated a market for recycled goods and related products; products can be designed with these environmental characteristics in mind, offering a competitive advantage. In addition, many private and public agencies are beginning to stimulate the market for recycled goods by writing contracts and specifications that call for the use of recycled materials.

7. *Reduced waste treatment costs.* As discussed in *economic benefits*, the increasing costs of traditional end-of-pipe waste management practices are avoided or reduced through the implementation of pollution prevention programs.

8. *Potential tax incentives.* As an effort to promote pollution prevention, taxes may eventually need to be levied to encourage waste generators to consider reduction programs. Conversely, tax breaks to corporations that utilize pollution prevention methods could similarly be developed to foster pollution prevention.

9. *Decreased worker exposure.* By reducing or eliminating chemical exposures, businesses benefit by lessening the potential for chronic workplace exposure, and serious accidents and emergencies. The burden of medical monitoring programs, personal exposure monitoring, and potential damage claims are also reduced.

10. *Decreased energy consumption.* As mentioned previously, energy conservation are often interrelated and complementary to each other. Energy expenditures associated with the treatment and transport of waste are reduced when the amount of waste generated is lessened, while at the same time the pollution associated with energy consumed by these activities is abated.

11. *Increased operating efficiencies.* A potential beneficial side effect of pollution prevention activities is a concurrent increase in operating efficiency. Through a pollution prevention assessment, the assessment team can identify sources of waste that results in hazardous waste generation and loss in process performance. The implementation of a waste reduction program will often rectify such problems through modernization, innovation, and the implementation of good operating practices.

12. *Competitive advantages.* By taking advantage of the many benefits associated with pollution prevention, businesses can gain a competitive edge.

13. *Reduced negative environmental impacts.* Through an evaluation of pollution prevention alternatives, which consider a total systems approach, consideration is given to the negative impact of environmental damage to natural resources and species that occur during raw material procurement and waste disposal. The performance of pollution prevention endeavors will therefore result in enhanced environmental protection.

The development of new markets by means of regulatory and economic incentives will further assist the effective implementation of waste reduction. Various combinations of the pollution prevention barriers provided above have appeared on numerous occasions in the literature, and in many different forms. However, there is one other concern that both industry and the taxpayer should be aware of. Carol Browner, EPA Administrator, has repeatedly claimed that pollution prevention is the organization' s top priority. *Nothing could be further from the truth.*

Despite near unlimited resources, the EPA has contributed little to furthering the pollution prevention effort. The EPA offices in Washington, Research Park Triangle, and Region II have exhibited a level of bureaucratic indifference that has surpassed even the traditional attitudes of many EPA employees. It is virtually impossible to contact any responsible pollution prevention individual at the EPA. Calls are rarely returned. Letters are rarely returned. On the rare occasion when contact is made, the regulatory individual typically passes the caller onto someone else who "really is in a better position to help you", and the cycle starts all over again.

This standard bureaucratic phenomena has been experienced by others in industry and the EPA. Two letters of complaint to the EPA Region II Administrator resulted in a response that was somewhat cynical and suggestive of a reprimand. Ms. Browner chose not to reply to the complaints. Notwithstanding some of the above comments, pollution prevention efforts have been successful in industry because these programs have often either produced profits or reduced costs, or both. The driving force for these successes has primarily been economics and *not* the EPA. Economic considerations are considered in the next section.

## Economic Considerations Associated with Pollution Prevention Programs

The purpose of this section is to outline the basic elements of a pollution prevention cost accounting system that incorporates both traditional and less tangible economic variables[8]. The intent is not to present a detailed discussion of economic analysis but to help identify the more important elements that must be considered to properly quantify pollution prevention options.

The greatest driving force behind any pollution prevention plan is the promise of economic opportunities and cost savings over the long term. Pollution prevention is now recognized as one of the lowest-cost options for waste/pollutant management. Hence, an understanding of the economics involved in pollution prevention programs/options is quite important in making decisions at both the engineering and management levels. Every organization should be able to execute an economic evaluation of a proposed project. If the project cannot be justified economically after *all* factors and considerations have been taken into account, it should obviously not be pursued. The earlier such a project is identified, the fewer resources will be wasted[9].

Before the true cost or profit of a pollution prevention program can be evaluated, the factors contributing to the economics must be recognized. There are two traditional contributing factors-capital costs and operating costs-but there are other important costs and benefits associated with pollution prevention that need to be quantified if a meaningful economic analysis is going to be performed.

The economic evaluation referred to above is usually carried out using standard measures of profitability. Each company and organization has its own economic criteria for selecting projects for implementation. In performing an economic evaluation, various costs and savings must be considered. The economic analysis presented in this section represents a preliminary, rather than a detailed, analysis. For smaller facilities with only a few (and perhaps simple) processes, the entire pollution prevention assessment procedure will tend to be much less formal. In this situation, several obvious pollution prevention options, such as the installation of flow controls and good operating practices, may be implemented with little or no economic evaluation. In these instances, no complicated analyses are necessary to demonstrate the advantages of adopting the selected pollution prevention option. A proper perspective must also be maintained between the magnitude of savings that a potential option may offer and the amount of manpower required to do the technical and economic feasibility analyses.

## Future Trends

The main problem with the traditional type of economic analysis is that it is difficult-nay, in some cases, impossible—to quantify some of the not-so-obvious economic merits of a pollution prevention program. Several considerations, in addition to those provided in the previous section, have just recently surfaced as factors that need to be taken into account in any meaningful economic analysis of a pollution prevention effort. These factors are certain to become an integral part of any pollution prevention analysis in the future. What follows is a listing of these considerations:

Decreased long-term liabilities

Regulatory compliance

Regulatory record keeping

Dealings with the EPA

Dealings with state and local regulatory bodies

Elimination or reduction of fines and penalties

Potential tax benefits

Customer relations

Stockholder support (corporate image)

Improved public image

Reduced technical support

Potential insurance costs and claims

Effect on borrowing power

Improved mental and physical well-being of employees

Reduced health maintenance costs

Employee morale

Other process benefits

Improved worker safety

Avoidance of rising costs of waste treatment and/or disposal

Reduced training costs

Reduced emergency response planning

Many proposed pollution prevention programs have been quenched in their early stages because a comprehensive analysis was not performed. Until the effects described above are included, the true merits of a pollution prevention program may be clouded by incorrect and/or incomplete economic data. Can something be done by industry to remedy this problem? One approach is to use a modified version of the standard Delphi panel that the authors have modestly defined as the WTA (an acronym for the Wainwright Theodore Approach). In order to estimate these "other" factors and/or economic benefits of pollution prevention, several knowledgeable individuals within and perhaps outside the organization are asked to independently provide estimates, with explanatory details, on these benefits. Each individual in the panel is then allowed to independently review all responses. The cycle is then repeated until the group's responses approach convergence.

Finally, pollution prevention measures can provide a company with the opportunity of looking their neighbors in the eye and truthfully saying that all that can reasonably be done to prevent pollution is being done... in effect, the company is doing right by the environment. Is there an advantage to this? It is not only a difficult question to answer quantitatively but also a difficult one to answer. The reader is left with pondering the answer to this question in terms of future activities.

## Words and Expressions

1. mass balance equation 质量平衡方程
2. material balance 物料平衡，物料衡算
3. conservation law for mass 质量守恒定律
4. steady-state process 稳态过程
5. unsteady-state process 非稳态过程
6. batch system 分批系统，分批制度
7. imaginary 想象的，虚构的，假想的
8. find a major application 得到主要的应用
9. ingredient 组成部分，成分，配料，要素
10. actual site walk-through 实际的现场考察[排演]
11. in conjunction with 与……一起，连带
12. disincentive 起抑制或阻碍作用的行为或措施；阻止的，抑制的
13. confront 使面临，使遭遇，迎接(困难)，比较，对照
14. confidential 保密的，机密的
15. reluctant 不愿的，勉强的，难得到的，抵抗的
16. durability 耐久性，持久性，寿命，强度
17. legitimate concern 合法的利害关系
18. initial capital investment 初期的基本投资，初期的基建投资
19. discourage the exploration of pollution prevention alternatives 阻止对污染预防方案的研究，不鼓励对污染预防方案的探索
20. regulatory enforcement 规章的实施，规章的执行
21. environmentally friendly products 环境友好产品
22. apathy 冷淡，无兴趣
23. passiveness 被动，消极，不顺从
24. infrastructure 基本设施，基础设施，基础(结构)，地基
25. competitive edge 竞争优势
26. tangible 有形的，(有)实质的，明确的，确实的
27. comply with regulatory standards 遵守规章标准
28. pollution liability insurance requirements 污染责任保险要求
29. private entities 私人组织，私人机构
30. environmentally preferred products 环境优先的产品，环境最佳的产品
31. levy 征收，征用，征税，征收额
32. foster 鼓励，促进，养育，心怀，抱着
33. potential damage claim 可能的赔偿损失的要求
34. side effect 副作用
35. rectify 纠正，整顿，精馏，蒸馏
36. procurement 获得，收购，采购
37. bureaucratic indifference 官僚(主义)的冷漠
38. cynical 愤世嫉俗的，玩世不恭的
39. reprimand 惩戒，谴责
40. promise 诺言，(有)指望，(有)前途；答应，有……的可能
41. capital cost 基本投资，投资费
42. operating cost 运行成本，操作费用
43. economic analysis 经济分析
44. technical and economic feasibility analyses 技术和经济可行性分析
45. quench 熄灭，抑制，冷却
46. convergence 会聚，集中

## Notes

(1) The equation may be used for any compound whose quantity does not change by chemical reaction, or for any chemical element, regardless of whether it has participated in a chemical reaction.

该方程可用于任何化合物，其数量在化学反应中不变，也可用于任何化学元素，不管其是否参与化学反应。

(2) The conservation of mass law finds a major application during the performance of pollution prevention assessments.

质量守恒定律在污染预防评价中得到重要应用。

(3) Some of the data required for the material balance calculation may be collected during the first review of site-specific data; however, in some instances, the information may not be collected until an actual site walk-through is performed.

物料衡算所需的一些数据可以在第一次审阅具体场地数据时收集；然而，在有些情况下，直到实地考察时才能收集到信息。

(4) Mass balance calculations are particularly useful for quantifying fugitive emissions, such as evaporative losses. Waste stream data and mass balances will enable one to track flow and characteristics of the waste streams over time.

质量衡算对于量化短时排放物(易散性排放物)，如蒸发损失，是特别有用的。废物流数据和质量衡算使人们能跟踪废物的流量和特性随时间的变化。

(5) Given the complexity of present manufacturing processes, waste streams exist that can not be reduced with current technology.

已知目前生产过程的复杂性，那么废物还会存在，它不可能用现有的技术减量化。

(6) Similarly, if there is no mechanism in place to hold individuals accountable for their actions, the successful implementation of a pollution prevention program will be limited.

同样，如果没有要个人对其行为负责的机制，成功实施污染预防计划将受到限制。

(7) If a pollution prevention assessment reveals confidential data pertinent to a company's product, fear may exist that the organization will lose a competitive edge with other businesses in the industry.

如果污染预防评价泄露了与公司产品有关的秘密资料，(评价)机构会担心在同行业中失去竞争优势。

(8) The purpose of this section is to outline the basic elements of a pollution prevention cost accounting system that incorporates both traditional and less tangible economic variables.

本节的目的是概述污染预防费用核算系统的基本原理，该系统包含了传统的和不太有形的经济变量。

(9) If the project cannot be justified economically after all factors and considerations have been taken into account, it should obviously not be pursued. The earlier such a project is identified, the fewer resources will be wasted.

如果一个工程项目在考虑了所有因素后仍不能在经济上证明是合理的，该项目显然不应再继续进行。这样的项目证实得越早，浪费的资源就越少。

# Unit 8 Miscellaneous

## 25. Acid Rain

### Introduction

Acid deposition, popularly known as acid rain, has long been suspected of damaging lakes, streams, forests, and soils, decreasing visibility, corroding monuments and tombstones, and potentially threatening human health in North America and Europe. The National Academy of Sciences and other leading scientific bodies first gave credence to these concerns in the early 1980s when they suggested that emissions of sulfur dioxide from electric power plants were being carried hundreds of miles by prevailing winds, being transformed in the atmosphere into sulfuric acid, falling into pristine lakes, and killing off aquatic life. The process of acid deposition also begins with emissions of nitrogen oxides (primarily from motor vehicles and coal-burning power plants). These pollutants interact with sunlight and water vapor in the upper atmosphere to form acidic compounds. During a storm, these compounds fall to earth as acid rain or snow; the compounds also may join dust or other dry airborne particles and fall as "dry deposition" (EPA, 1987). Regulations have been passed concerning the amount of $SO_2$ and $NO_x$ (oxides of nitrogen) emitted in the air. These regulations have caused the power industries to find ways to cut their emissions. The three ways of lowering emissions-before combustion, during combustion, and after combustion—will be discussed.

Sulfur dioxide, the most important of the two gaseous acid pollutants, is created when the sulfur in coal is released during combustion and reacts with oxygen in the air. The amount of sulfur dioxide created depends on the amount of sulfur in the coal. All coal contains some sulfur, but the amount varies significantly depending on where the coal is mined. Over 80 percent of sulfur dioxide emissions in the United States originate in the 31 states east of or bordering the Mississippi River (EPA, 1987). Most emissions come from the states in or adjacent to the Ohio River Valley.

The extent of damage caused by acid rain depends on the total acidity deposited in a particular area and the sensitivity of the area receiving it. Areas with acid-neutralizing compounds in the soil, for example, can experience years of acid deposition without problems. Such soils are common in much of the United States. But the thin soils of the mountainous and glaciated northeast have very little acid-buffering capacity, making them vulnerable to damage from acid rain(1). Surface waters, soils, and bedrock that have a relatively low buffering capacity are unable to neutralize the acid effectively. Under such conditions, the deposition may increase the acidity of water, reducing much or all of its ability to sustain aquatic life. Forests and agriculture may be vulnerable because acid deposition can leach nutrients from the ground, kill nitrogen-fixing microorganisms that nourish plants, and release toxic metals.

## Regulations

Title IV of the 1990 Clean Air Act Amendments, Acid Deposition Control, which contains comprehensive provisions to control the emissions that cause acid rain, represents a legislative breakthrough in environmental protection. To begin with, it is the first law in the nation' s history to directly address the problem of acid rain. The legislation calls for historic reductions in sulfur dioxide emissions from the burning of fossil fuels, the principal cause of acid rain. It also mandates significant reductions in nitrogen oxide emissions. In addition, the approach embodied in the new provisions represents a radical departure from the traditional "command-and-control" approach to environmental regulation that prevailed in this country during the 1970s and 1980s.

During those years, environmental regulations typically required industry to achieve a particular limit on each pollutant released to the environment by installing specific pollution-control equipment. The acid rain program that the Environmental Protection Agency (EPA) is developing under the Clean Air Act Amendments takes a more flexible approach. It simply sets a national ceiling in sulfur dioxide emissions from electric power plants and allows affected utilities to determine the most cost-effective way to achieve compliance. It is estimated that this approach will result in at least a 20 percent cost savings over a traditional command-and-control program.

The new legislation requires that, by the turn of century, sulfur dioxide emissions must be reduced 10 million tons annually from the levels emitted in 1980; this will amount to roughly a 40 percent reduction from 1980 levels. Because such reductions cannot be achieved overnight, EPA is implementing a two-phase approach that gradually tightens the restrictions placed on power plants that emit sulfur dioxide.

The first phase begins in 1995 and affects 261 units in 110 coal-burning electric utility plants located in 21 eastern and midwestern states. These plants are large and emit high levels of sulfur dioxide. Phase II, which begins in the year 2000, tightens the emissions limits imposed on these large plants and also sets restrictions on smaller, cleaner plants fired by coal, oil, and gas. Approximately 2, 500 units within approximately 1, 000 utility plants will be affected by Phase II. In both phases, affected utilities will be required to install systems that consciously monitor emissions in order to track progress and assure compliance.

The legislation also calls for a 2-million-ton reduction in nitrogen oxide emissions by the year 2000. A significant portion of this reduction will be achieved by utility boilers, which will be required to meet tough new emissions requirements under the acid rain provisions of the act. These requirements will also be implemented in two phases. EPA established emission limitations in mid-1992 for two types of utility boilers (tangentially fired and dry bottom, wall-fired boilers); regulations for all other types of boilers will be issued by 1997. As with the sulfur dioxide emissions, these utilities will be required to install equipment that will continuously monitor emissions.

The acid rain provisions also look to the future by placing a permanent cap on sulfur dioxide emissions and by encouraging energy conservation, the use of renewable and clean alternative technologies, and pollution-prevention practices. These ground-breaking provisions will help ensure that lasting environmental gains are made.

To help bring about the mandated sulfur dioxide emissions reductions in a cost-effective manner, EPA is implementing a market-based allowance-trading system that will provide power plants with maximum flexibility in reducing emissions. Under this system, EPA will allocate allowances to affected utilities each calendar year based upon formulas provided in the legislation. Each allowance permits a utility to emit one ton of sulfur dioxide. To be in compliance with the law, utilities may not emit more sulfur dioxide than they hold allowances for. This means that utilities will have to either reduce emissions to the level of allowances they hold or obtain additional allowances to cover their emissions.

Utilities that reduce their emissions below the number of allowances they hold may elect to trade allowances within their systems, bank allowances for future use, or sell them to other utilities for profit. Allowance trading will be conducted nationwide, so that a utility in North Carolina, for example, will be able to trade with a utility in California. Anyone may hold allowances, including affected utilities, brokers, environmental groups, and private citizens.

The legislation also establishes a permanent cap on the number of allowances EPA issues to utilities. Beginning in the year 2000, EPA will issue 8.95 million allowances to utilities annually. Although the lowest-emitting plants will be able to increase their emissions between 1990 and 2000 by roughly 20 percent, these utilities may not thereafter exceed their year-2000 emission levels.

Utilities that begin operating in 1996 and beyond will not be allocated allowances. Instead, they will have to buy into the system by purchasing allowances. This will effectively limit emissions even as more plants are built and the combustion of fossil fuels increases. These measures will help ensure that the benefits gained from the emissions reductions will not be eroded over time.

The allowance allocation for each unit affected by Phase I of the new law is listed in the legislation. An individual unit's allocation is based on standard formula: the product of a 2.5-pound sulfur dioxide per million Btu emission rate multiplied by the unit's average fuel consumption for 1985—1987.

The allowance system provides incentives for power plants to reduce their emissions substantially more than is required since allowances freed by installing pollution-control equipment can be sold for profit.

EPA's role in allowance trading will be to receive and record allowance transfers and also to ensure at the end of the year that a utility's emissions did not exceed the number of allowances held. When two parties' agree to an allowance transfer, their formally designated representatives will notify EPA in writing to make it official. EPA will record the transaction by entering it into an automated allowance tracking system, but will not otherwise participate in the trading process. The tracking system that will be developed by the Agency over the next two years will monitor compliance by keeping records of allowance holdings and the status of allowances traded. EPA will be writing regulations for such issues as calculating and allocating allowances, for the mechanics of allowance transfers, for allowance tracking, and for the operation of reserves, sales, and auctions.

EPA will maintain a reserve of 300, 000 special allowances that will be allocated to utilities that develop qualifying renewable energy products or use conservation measures. The allowances will be granted to utilities on a first-come, first-served basis starting in 1995 for conservation activities initiated after 1992.

The legislation provides a strong incentive for utilities to comply with the law and not exceed their allowances. Utilities that do exceed their allowances must pay a $ 2000-per-ton excess emissions fee in the following year. Since the excess emissions fee will substantially exceed the expected cost of compliance through the purchases of allowances, EPA expects that the market will do much of the work of ensuring compliance with the mandated reduction requirements.

The idea behind the allowance system was to achieve significant reductions of acid-rain-causing emissions at the lowest possible cost. It happens to be a good system.

## Emissions Reduction -Before Combustion

In an attempt to mitigate the effects of acid rain several proposals have appeared before the U. S. Congress. Reductions in sulfur dioxide emissions, particularly from utility and industrial coal-fired boilers, is the primary target for combating acid rain.

Several means exist to limit the amount of sulfur in the fuel prior to combustion and include the use of lower-sulfur coals and coal cleaning. Reduction of nitrogen oxides emissions cannot be accomplished at this point because it is formed after ( and following) incomplete combustion. The two major reduction procedures described below include coal switching and coal cleaning. Although this development is directed toward coal ( the fuel of primary concern in this nation), it may also be applied, in some instances to other fossil fuels.

## Coal Switching

For many power plants, the most economical strategy for reducing sulfur dioxide emissions tends to be switching from higher-sulfur to lower-sulfur coals. Because the sulfur content in coal varies across regions, this move to consume lower-sulfur coals would result in a major shift in regional coal production from higher-sulfur supply regions, to lower-sulfur supply regions. It could also generate regional hostility by causing shifts in existing coal markets.

Existing coal-fired power plants now burning higher-sulfur coals without scrubbers would be faced with the most stringent requirements for reducing emissions. Most of these plants were initially designed to burn bituminous coals.

Under most circumstances, higher-sulfur plants would face relatively little technical difficulty in shifting to lower-sulfur coals, but there may be some additional costs, especially for upgrading electrostatic precipitators. A few plants, such as cyclone-fired boilers, are not technically well-suited for burning lower-sulfur coals because of the difference in ash-fusion temperatures.

Also, there is some question as to the type of lower-sulfur coals likely to be in demand by power plants shifting to these coals. Since most existing boilers were initially designed to burn biturninous coals, it is not clear whether these units can economically shift to lower-sulfur subbituminous coals. Some of the questions are technical, such as the potential for slagging or fouling when an off-design coal is burned in these boilers.

Consumption of subbituminous coals will entail higher heat rates, higher handling costs, and capacity derates. Together, these economic and technical considerations tend to make the use of subbituminous coals in bituminous boilers a very site-specific issue that does not provide clear economic advantages.

## Coal Cleaning

Another way to control sulfur emission is to clean the coal before burning. This process can reduce sulfur dioxide emission by 20 to 90 percent. Coal cleaning can be done in three ways: physical (gravity separation), chemical (reaction or bioremediation), or electrical. Details on the principal cleaning method, by gravity, are provided below. Details on other separation techniques are available in the literature.

Gravity separation is used by industry. This method depends on the size, shape, density and surface properties of the coal. The process first crushes the coal into small particles and then allows gravity to separate the pyritic sulfur from the coal. The gravity separation process is usually accomplished in a water medium. The coal containing impurities sink to the bottom and the usable coal stays on the top. The benefits of this process is that the coal is easy to handle and the coal can burn better. Although there are some benefits to this process, there are limitations as well.

One of the limitations of gravity separation is that it is ineffective in reducing sulfur content when the coal particles are very fine. When the particles are very fine the metals will not be attached to the sulfur enriched coal and separation will not occur. Another problem is with the medium. The coal must be dried after using water to separate the burning coal from its impurities; drying the coal is very expensive. There is also a major loss of energy release with the purified coal; this means more coal must be burned to get the same amount of energy produced.

## Emissions Reduction -During Combustion

The second method that can be used to reduce the emissions of the precursors of acid rain is during combustion. Both $NO_x$ and $SO_2$ emissions can be reduced during this stage of the combustion process.

The reduction of $NO_x$ emissions is accomplished by primarily three methods: low-$NO_x$ burners, overfire air, and fuel staging or reburning. These methods have the ability to reduce emissions by 80 percent and are cost-effective. The three processes are based on using combustion with stoichiometric air and controlling the temperature. The fuel staging process seems to work the best in reducing the emissions.

In a coal-fired boiler, reburning is accomplished by substituting 15 to 20 percent of the coal with natural gas or low-sulfur oil and burning it at a location downstream of the primary combustion zone of the boiler. Oxides of nitrogen formed in the primary zone are reduced to nitrogen and water vapor as they pass through the reburn zone. Additional air is injected downstream of the reburn zone to complete the combustion process at a lower temperature. In general, NO~ reductions of 50 percent or more are achievable by reburning. When combined with other low-$NO_x$ technologies (such as low-$NO_x$ burners), $NO_x$ reductions of up to 90 percent may be achievable.

Reduction of sulfur dioxide emissions cannot be accomplished easily because many expensive problems occur and these methods are expensive. Two methods that are frequently used in industry are limestone injection multistage burner (LIMB) and fluidized bed combustion (FBC).

LIMB is an emerging control process that can be retrofitted on a large portion of existing coal-fired boilers. In a LIMB system, an $SO_2$ sorbent (limestone) is injected into a boiler equipped with low-$NO_x$ burners. The sorbent absorbs the $SO_2$ and the low-$NO_x$ burners limit the amount of $NO_x$ formed. LIMB is capable of reducing both $SO_2$ and $NO_x$ by about 50 to 60 percent.

The benefit of using the LIMB process is that it is one of the least expensive processes for reducing $SO_2$ emissions during combustion, but the sorbent injected into the boiler tends to increase slagging and fouling, which in turn increase operation and maintenance costs. Because boilers retrofitted with LIMB tend to produce more particulates of smaller sizes, particulate control becomes more difficult. Technical questions remain as to what sorbents are most effective in a LIMB system, and how and where to inject the sorbents.

In an FBC boiler, pulverized coal is burned while suspended over a turbulent cushion of injected air. This technique allows improved combustion efficiencies and reduced boiler fouling and corrosion. Such boilers also are capable of burning different kinds of low-grade fuels like refuse, wood bark, and sewage sludge. In addition, if the coal is mixed with limestone or some other sorbent material during combustion, the $SO_2$ is captured and retained in the ash.

FBC boilers have the potential to control $NO_x$ as well as $SO_2$. FBC boilers must operate within a narrow temperature range (150-1600°F) and lower combustion temperatures inherently limit the formation of $NO_x$. FBC boilers may be able to control $NO_x$ by 50 to 75 percent at the same time as they control $SO_2$ by up to 90 percent. An FBC system does have one major flaw: it requires the construction of a new boiler. The FBC system is more of a replacement technology than a retrofit.

## Emissions Reduction -After Combustion

The final method of reducing emissions that cause acid rain is after combustion has occurred. This method is most frequently used by industry, although it is not preferred by environmentalists because it creates other wastes while reducing emissions.

The most popular process for reducing $NO_x$ after combustion is selective catalyst reduction (SCR)[2]. It is mainly used in Japan. In the SCR system, a mixture of ammonia gas and air is injected upstream of a catalytic reactor chamber. The flue-gas mixture then travels in a vertical, downward-flow direction through a catalytic reactor chamber, where the ammonia gas disassociates $NO_x$ to nitrogen gas and water vapor (EPA, 1987). This process benefits from its high removal rate (80–90 percent) and no retrofitting on the unit is necessary. Catalyst selection is very important for this process. Catalyst selection is based on the following criteria: resistance to toxic materials, abrasion resistance, mechanical strength, resistance to thermal cycling, resistance to the oxidation of $SO_2$, and resistance to plugging.

Industry has mainly chosen the flue-gas desulfurization process (FGD) to combat $SO_2$ emissions. FGD uses sorbents such as limestone to soak up (or scrub) $SO_2$ from exhaust gases. This technology, which is capable of reducing $SO_2$ emissions by up to 95 percent, can be added to existing coal-fired boilers.

FGD has several drawbacks. The control equipment is very expensive and bulky. Smaller facilities do not always have the capital or the space needed for FGD equipment. If, however, the sorbent could be injected into existing ductwork, the cost of the reaction vessel could be eliminated, and it would be much easier to retrofit controls on a wider range of sources.

More recently ETS International, Inc. developed a Limestone Emission Control (LEC) system. The Limestone Emission Control system is a proprietary acid gas control system that has demonstrated high levels of acid gas removals (99 percent) at very competitive costs. A ten-year R&D program included scrutiny by the Ohio Coal Development Office, the U.S. EPA, and DOE as well as major U.S. industry. The system holds great promise for acid gas control from both the chemical and utility industries.

## National Acid Precipitation Assessment Program

In addition to enforcement and monitoring under the provisions of the Clean Air Act, the EPA is actively pursuing a major research effort with other federal agencies under the National Acid Precipitation Assessment Program (NAPAP). This ongoing research project is designed to resolve the critical uncertainties surrounding the causes and effects of acid rain. About $ 300 million have been spent for federal research since NAPAP was initiated in 1980. In September 1987, NAPAP published an interim assessment on the causes and effects of acid deposition.

## Aquatic Effects

One of the most important acid rain research projects being conducted by EPA is the National Surface Water Survey. This survey is designed to provide data on the present and future status of lakes and streams within regions of the United States believed to be susceptible to change as a result of acid deposition. Phase I of the Eastern and Western Lakes Surveys showed that there are essentially no lakes or reservoirs in the mountainous West, northeastern Minnesota, and the Southern Blue Ridge of the Southeast that are considered acidic. The four subregions with the highest percentages of acidic lakes are: the Adirondacks of New York, where 10 percent of the lakes were found to be acidic; the Upper Peninsula of Michigan, where 10 percent of the lakes were also found to be acidic; the Okefenokee Swamp in Florida, which is naturally acidic; and, the lakes in the Florida Panhandle where the cause of acidity is unknown.

The 1988 Stream Survey determined that approximately 2.7 percent of the total stream reaches sampled in the mid-Atlantic and Southeast were acidic. About 10 percent of head waters in the forested ridges of Pennsylvania, Virginia, and West Virginia were found to be acidic. Streams in Florida found to have a low pH are naturally acidic. The study indicated that atmospheric deposition is the major cause of sulfates in streams. Atmospheric deposition was also found to be a major cause of sulfates in the lakes surveyed as part of the National Surface Water Survey.

## Forest Effects

The NAPAP interim assessment reviewed research concerning the effects of acid deposition on forests. It focused on the effects of precursor pollutants (sulfur dioxide and nitrogen oxides) and Volatile Organic Compounds (VOCs) and their oxidants (including ozone and hydrogen peroxides) on eastern spruce-fir, southern pine, eastern hardwood, and western conifer. The assessment found that air pollution is a factor in the decline of both managed and natural forests. The San Bernardino National Forest in California and some types of white pine throughout the eastern United States are seriously affected by ozone.

Forests found to have unknown causes of damage included northeastern spruce-fir, northeastern sugar maple, southeastern yellow pine, and species in the New Jersey Pine Barrens. The high elevation forests such as the spruce-fir in the eastern United States were found to be exposed to severe natural stresses as well as being frequently immersed in clouds containing pollutants at higher concentrations than those observed in rain. Research has shown no direct impacts to seedlings by acidic precipitation or gaseous sulfur dioxide and nitrogen oxides at ambient levels in the United States. Ozone is the leading suspected pollutant that may stress regional forests and reduce growth. Research is underway to resolve the relative importance of physical and natural stresses.

## Crop Effects

The NAPAP assessment indicated that there are no measurable consistent effects on crop yield from the direct effects of simulated acidic rain at ambient levels of acidity. This finding was based on yield measurements of grains, forage, vegetable, and fruit crops exposed to a range of simulated rain acidity levels in controlled exposure studies (EPA, 1987). Continuing research efforts will examine whether stress agents such as drought or insect pests cause crops to be more sensitive to rainfall acidity.

Average ambient concentrations of sulfur dioxide and nitrogen oxides over most agricultural areas in the United States are not high enough or elevated frequently enough to affect crop production on a regional scale. However, crops may be affected locally in areas close to emission sources. Controlled studies also indicate that ambient levels of ozone in the United States are sufficient to reduce the yield of many crops.

## Materials Effects

The NAPAP Interim Report indicated that many uncertainties need to be reduced before a reliable economic assessment could be made of the effects of acid deposition on materials, such as building materials, statues, monuments, and car paint(3). Major areas of uncertainty include inventories of materials at risk, variability of urban air quality, effects on structures, and cost estimates for repair and replacement.

## Human Health Effects

The NAPAP interim assessment reported that there are also many uncertainties associated with assessing the influence of ambient levels of atmospheric pollutants on human health. The primary factors involved are a lack of information on the levels of exposure to acidic aerosols for various population groups across North America; chronic health problems caused by short-term changes in respiratory symptoms and decrease in lung function; and, the effects of repetitive or long-term exposures to air pollutants. Studies on toxicity of drinking water have linked rain acidity to unhealthy levels of toxic metals in drinking water and fish.

## Future Trends

The EPA, in coordination with other federal agencies, is conducting wide-ranging research on the causes and effects of acid deposition. Major research efforts include determining effects on aquatic and forest ecosystems, building materials and human health. In the area of human health EPA is conducting exposure studies on acid aerosols. EPA is also conducting ongoing aquatics research projects that will continue into the future. As part of the National Surface Water Survey, seasonal variability of lakes in the Northeast will be studied.

Over the next several years, major research results are anticipated for improving the basis of decision making on acid rain issues. EPA also expects that Congress and other groups will continue to propose options to reduce acid deposition. As proposals are offered, EPA will provide analyses of costs, consequences, and the feasibility of implementation.

EPA's greatest challenge is to continue to reduce emissions of sulfur dioxide and nitrogen oxides. The Agency must also continue research to reduce the level of scientific and economic uncertainties about acid deposition and work to resolve the regional conflicts related to this problem. In addition to the research efforts, major federal research programs are being funded by the Department of Energy, the Tennessee Valley Authority, and the Argonne, Brookhaven, Lawrence Berkley, and Oak Ridge national laboratories.

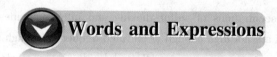

## Words and Expressions

1. acid deposition 酸沉降(物)
2. acid rain 酸雨
3. visibility 能见度
4. monument 纪念碑，纪念馆，界碑
5. tombstone 墓碑，墓石
6. prevailing wind 盛行风
7. pristine 原始(状态)的，早期的，质朴的
8. acid-buffering capacity 酸缓冲能力
9. vulnerable 易损坏的，易受伤的，脆弱的
10. nitrogen-fixing microorganism 固氮微生物
11. nourish 养育，滋养，怀(有)，供给，支持
12. provision 准备，(预防)措施，供给，供应，设备，装置，规定，条款
13. mandate 命令，托管；委托(管理)
14. sulfur dioxide emission 二氧化硫排放(物)
15. nitrogen oxide emission 氮氧化物排放(物)
16. embody 包含，收录，体现，具体化，连接，结合
17. meet tough new emissions requirements

under the act acid rain provisions of the act 满足新的严厉的该法案酸雨条款的排放要求
18. allowance-trading system(排污)允许量交易制度
19. allocate 分配，配给，定位置
20. transaction 办理，处理，(一笔)交易，(具体)事务
21. tracking system 跟踪制度
22. auction 拍卖
23. comply with the law 遵守法律
24. mitigate 使镇静，缓和，减轻，调节
25. industrial coal-fired boiler 工业燃煤锅炉
26. switch from higher-sulfur to lower-sulfur coal 从高硫煤转换到低硫煤
27. hostility 敌意，敌视，敌对(状态)
28. bituminous coal 烟煤
29. cyclone-fired boiler 旋风燃烧锅炉
30. suited for 配做，适合于做
31. subbituminous coal(黑色，褐色)次烟煤
32. pyritic sulfur 黄铁矿硫
33. precursor 预兆，先兆，先驱者，前体，前身
34. retrofit 改型，改进，(式样)翻新，更新
35. pulverized coal 粉煤，煤粉
36. turbulent cushion of injected air 注入空气的湍流缓冲
37. disassociate $NO_x$ to nitrogen gas and water vapor $NO_x$ 解离成氮气和水蒸气
38. abrasion resistance 耐磨性，耐磨度
39. flue-gas desulfurization process 烟气脱硫过程，烟气脱硫工艺
40. exhaust gas 尾气，废气，排气
41. scrutiny 细看，细查，详尽的研究
42. causes and effects of acid rain 酸雨的因果，酸雨的成因和结果
43. interim 中间，暂时，间歇；间歇的，中间的，临时的，阶段的
44. pine 松树，松木
45. hardwood 硬木
46. conifer 针叶树(如松、枞等)
47. sugar maple 糖风树，糖槭树，糖槭木
48. yellow pine 黄松树，黄松木
49. natural stresses 自然压力，自然重压
50. resolve the relative importance 分辨相对重要性
51. measurable consistent effects on crop yield 可测的对作物产量一贯的影响
52. forage 草料，饲料
53. uncertainty 无常，易变，不确定性
54. acidic aerosol 酸性气溶胶
55. chronic 长期的，慢性的
56. respiratory symptom 呼吸症状
57. lung function 肺功能
58. in coordination with 与……协调，与……一致
59. resolve the regional conflict 消除地区冲突，化解地区冲突

## Notes

(1) But the thin soils of the mountainous and glaciated northeast have very little acid-buffering capacity, making them vulnerable to damage from acid rain.

但是，美国东北部山地和冰川地带的瘠薄的土壤对酸的缓冲能力很弱，使其易受酸雨之害。

(2) The most popular process for reducing $NO_x$ after combustion is selective catalyst reduction (SCR).

削减燃烧产生的 $NO_x$ 的最常用方法是选择性催化还原。

(3) The NAPAP Interim Report indicated that many uncertainties need to be reduced before a reliable economic assessment could be made of the effects of acid deposition on materials, such as building materials, statues, monuments, and car paint.

NAPAP(全国酸沉降评价计划)中期报告指出：要作出酸雨对材料(如建筑材料、雕像、纪念碑和汽车油漆)影响的可靠经济评价还需减少许多不确定性。

# 26. Greenhouse Effect and Global Warming

## Introduction

The "greenhouse effect" is a phrase popularly used to describe the increased warming of the earth due to increased levels of carbon dioxide and other atmospheric gases, called greenhouse gases. Just as the glass in a botanical greenhouse traps heat for growing plants, greenhouse gases trap heat and warm the planet. The greenhouse effect, a natural phenomenon, has been an essential part of Earth's history for billions of years. The greenhouse effect is the result of a delicate and non-fixed balance between life and the environment. Yet, the greenhouse effect may be leading the planet to the brink of disaster. Since the Industrial Revolution, the presence of additional quantities of greenhouse gases threatens to affect global climate. However, the predicted effects of this increase are still debated among scientists.

## How the Greenhouse Effect Works

The energy radiated from the sun to the earth is absorbed by the atmosphere, and is balanced by a comparable amount of long-wave energy emitted back to space from the earth's surface. Carbon dioxide molecules (and greenhouse gases) absorb some of the long-wave energy radiating from the planet to the surface of the earth. Because of the greenhouse heat trapping effect, the atmosphere itself radiates a large amount of long wave-length energy downward to the surface of the earth and makes the earth twice as warm as it would have been if warmed by solar radiation alone(1).

The greenhouse gases trap heat because of their chemical makeup and, in particular, their tri-atomic nature. They are relatively transparent to visible sunlight, but they absorb long wave-length, infrared radiation emitted by the earth.

## Cycles of Carbon Dioxide

Carbon dioxide comprises only a very small portion of the atmosphere, a little more than 0.03% by volume. This colorless, odorless gas is of vital importance to life on Earth. Indeed, without it, there probably would not be any plants or animals or human beings on this planet. But too much carbon dioxide could be as harmful as too little.

There are two important natural cycles in the world that play a major part in determining the concentration of carbon dioxide in the atmosphere: terrestrial and oceanic. The terrestrial cycle starts with photosynthetic plants, which use sunlight, water, carbon dioxide, and a pigment called chlorophyll to form glucose and oxygen. It has been estimated that plants consume 500 billion tons of carbon dioxide every year, converting it into organic compounds and oxygen. By respiration, decay, and burning hydrocarbons, carbon dioxide is returned to the atmosphere.

The oceans cover about 70 percent of the Earth' s surface. All the gases that make up the atmosphere are also dissolved in the oceans. Carbon dioxide is exchanged between the atmosphere and the ocean interface until an equilibrium is reached. Carbon dioxide is then transferred to deeper water by convective transport cycles which lower the concentration at the surface, thereby allowing more of the gas from the atmosphere to diffuse in the ocean.

## Greenhouse Gases: Evidence of Climatic Change

The Environmental Protection Agency (EPA) has concluded that energy consumption in general has contributed almost 70 percent to the greenhouse effect in the last 10 years. Four important trace gases are combining to amplify the natural greenhouse effect: carbon dioxide, methane, nitrous oxide, and chlorofluorocarbons (CFCs). Each of these is briefly discussed below.

The primary man-made contributor to the greenhouse effect is carbon dioxide, accounting for approximately 50 percent of the blame for global warming. Carbon dioxide levels have risen at least 25 percent in the last 150 years. For instance, from analysis of air trapped in arctic ice, it is known that pre-Industrial Revolution levels of carbon dioxide were at 280 ppm (ppm = parts per million) and by 1984 the levels had risen to 343 ppm. This rise of carbon dioxide in the atmosphere is largely due to the burning of such fossil fuels as natural gas, coal, and petroleum; another contributor is worldwide deforestation. Deforestation increases carbon dioxide levels in the atmosphere in two basic ways. First, when trees decay or are burned, they release carbon dioxide. Second, without the forest, carbon dioxide that would have been absorbed for photosynthesis remains in the atmosphere. For example, a rainforest can hold 1 or 2 kg of carbon per square meter per year, as compared to a field of crops, which can absorb less than 0. 5 kg of carbon per square meter every year. The current annual rate of atmospheric increase in carbon dioxide is 0. 4%, which is equivalent to $10^{10}$ metric tons per year.

Methane is an odorless, colorless gas found in the atmosphere in traces of less than 2 ppm. Like carbon dioxide, methane is a natural product and is twenty times more effective at trapping heat than carbon dioxide. It is produced by anaerobic bacteria microorganisms that live without oxygen in wetlands, rice fields, cattle, termites, and ocean sediments. Methane' s annual growth rate in the atmosphere is now about 2. 0 percent per year.

Nitrous oxide is produced naturally and artificially. The atmosphere is 79 percent nitrogen and although plants need nitrogen for food, they cannot use it directly from the air. It must first be converted by soil bacteria into ammonium and then into nitrates before plants can absorb it. In the process, bacteria release nitrous oxide gas. In addition, farmers add chemical fertilizers containing nitrogen to the soil. Another way of forming nitrous oxide is by combustion. When anything burns, whether it is a tree in the rainforest, natural gas in a stove, coal in a power plant, or gasoline in a car, nitrogen combines with oxygen to form nitrous oxide. Nitrous oxide also destroys the ozone layer, which filters out dangerous ultraviolet radiation from the sun and protects life on earth. The concentration of nitrous oxide is increasing at the current rate of 0.3% each year.

Chlorofluorocarbons (CFCs) are used as coolants (CFC-12) for refrigerators and air conditioners, as blowing agents (CFC-11) in packing materials and other plastic foams, and as solvents. CFCs trap heat 20,000 times more effectively than carbon dioxide. CFCs are virtually indestructible. They are not destroyed or dissolved by any of the natural processes that normally cleanse the air. In addition, they may stay in the lower atmosphere. These high-power greenhouse gases (CFCs) also attack the ozone layer; each CFC molecule can destroy 10,000 or more molecules of ozone, a gas made up of oxygen. The current rate of atmospheric increase in CFCs is now growing by about 5 to 7 percent annually. Due to this rapid increase in the concentration of CFCs, a worldwide phaseout of chlorofluorocarbons production is being undertaken by the year 2000.

## Greenhouse Debate

Although most scientists are in agreement that higher levels of trace greenhouse gases in the atmosphere are causing global warming and that there is strong evidence to show it has already begun, others feel that the planet is actually entering another ice age. The current trends in weather could be natural fluctuations or could be the result of global climate patterns that run on cycles of thousands of years. Or, on the other hand, they could lead to climatic chaos.

In 1988, James Hansen of the National Aeronautics and Space Administration (NASA) Goddard Institute for Space Studies testified before the Senate Committee on Energy and Natural Resources ("No Way to Cool the Ultimate Greenhouse," 1993). He told the committee he was 99 percent certain that a 1°F rise in world temperatures since the 1850s has been caused by increasing greenhouse effect. "It is time to stop waffling so much and say that the evidence is pretty strong that the greenhouse effect is here." he said. But others argue that the earth has a natural control mechanism that keeps the earth's climate in balance. For instance, higher temperatures due to an increasing greenhouse effect may naturally trigger events that will cool the earth (e.g., increased clouds) and hold the climate in balance.

Scientists in favor claim that the climate models have been reliable enough to conclude that the greenhouse effect is causing global warming. Dr. Hansen, the leading spokesperson on the greenhouse effect, says that "it is just inconceivable that the increase of greenhouse gases in the atmosphere is not affecting our climate."

Other researchers believe that the warming of the earth over the last 100 to 150 years is part of a long-term, natural cycle thathas little to do with the production of greenhouse gases(2). They remain unconvinced that the accumulation in the atmosphere of greenhouse gases is concrete evidence of any rise in the average earth temperature. On this lack of evidence, three scientists from the G. C. Marshall Institute in Washington DC reported that any warming of the earth in the last 100 years is better explained by the variation in natural climate and solar activity. According to this theory, the most probable source of global warming appears to be variations in solar activity. The amount of solar radiation reaching the earth is controlled by three elements that vary cyclically over time(3).

The first element is the tilt of the earth's axis, which varies 22 to 24.5 degrees and back again every 41,000 years. The second element is the month of the year in which the earth is closest to the sun, which varies over cycles of 19,000 and 24,000 years. Finally, the third element is the shape of the orbit of the earth, which, over a period of 100,000 years, changes from being more elliptical to being almost fully circular. They stated that changes in the earth's temperature have followed changes in solar activity over the last 100 years. When solar activity increased from 1880s to 1940s, global temperatures increased. The observed global temperature rise of 1℉ was during this period, before 67 percent of global greenhouse gases emissions had even occurred. When it declined from 1940s to the 1960s, temperatures also declined. When solar activity and sunspot numbers started to move up again in the 1970s and 1980s, temperatures did the same. This is hard to explain as a greenhouse effect phenomenon because the increased emissions of greenhouse gases should have created a period of accelerated temperature rise.

## Effects of Global Warming

A rise in average global temperature is expected to have profound effects. But while scientists are uncertain on the causes of global warming and the degree of impact that it will have, the need to study future situations has led the Environmental Protection Agency (EPA) and other environmental groups to investigate what would happen to the planet after a 3-8℉ warming.

Climate changes will have a significant effect on weather patterns. There will be changes in precipitation, storms, wind direction, and so on. Rising temperatures are expected to increase tropical storm activity. The hurricane season in the Atlantic and Caribbean is expected to start earlier and last longer. Storms will be more severe. Some researchers believe the planet will be a wetter place. Global circulation models (GCM) predicted that a doubling of carbon dioxide could increase humidity 30 to 40 percent. However, such increases will not occur uniformly around the world. Perhaps humid tropical areas will become wetter while semi-arid regions will become drier.

As the earth gets warmer, there will be a rise in the average water level of the oceans. As water is heated, it expands, or increases in volume. According to theory, global warming could cause sea levels to rise. Rising sea levels will gradually flood low lying coastal areas. Beach erosion will be an increasing problem. The EPA has estimated that if the sea level rises 3 feet (0.9 meters), the nation will lose an area the size of Massachusetts, even if it spends more than $100 billion to protect critical shorelines. In addition to lost beaches, houses and other buildings that sit close to the water's edge will be undermined and destroyed.

If polar and temperate zones become warmer, there will be a poleward shift of ecological zones. Animal and plant species that now live in a particular area will no longer be able to survive there. The EPA predicts an increase in extinction rate as well as changes in migration patterns. As the ecological zones shift polewards, there may be a decrease in the amount of area suitable for forests, with a corresponding increase in grasslands and deserts. This means a loss in productive land, both for agriculture and for habitats of a broad range of plants and animals. For example, the nation's major grain crops-corn, wheat, and soybeans-are strongly affected by precipitation and temperatures. The warming trend will cause changes in water quantity and quality in some areas. This will affect drinking supplies as well as the water needs of industry and agriculture. Rising sea levels may contaminate water supplies as seawater migrates up rivers. For example, a 2 feet (0.6 meter) rise in sea level would inundate Philadelphia' s water intakes along the Delaware River, making the water too salty to drink.

Heat waves extract a physical toll on people. The heat puts a strain on the heart, as the body tries to cool itself. Studies have shown that the number of cases of heart and lung disease increases when temperature rises. Contagious diseases, such as influenza and pneumonia, and allergenic diseases, such as asthma, will also be affected by the weather and become more prolific. The lifecycles of mosquitos and other disease-carrying insects are also extended in warmer weather. Climate changes, sea level rises, and other direct effects of greenhouse warming can be expected to have major social, economic, and political effects. For example, if agricultural productivity declines, personal income will fall and jobs will decline.

## COMBATTING THE GREENHOUSE EFFECTS

### National Approach to the Greenhouse Effect

With only 5 percent of the world' s population, the United States today produces nearly 25 percent of global carbon emissions. The past years have yielded a flurry of proposals to deal with global warming. In the United States two comprehensive global warming bills were first introduced to Congress in late 1988, followed by other related legislation. These bills would establish a national goal of cutting carbon dioxide emissions by 20 percent by the year 2000 and include programs to implement national least cost planning, improve automobile fuel economy, develop renewable energy sources, plant trees, and assist developing countries in slowing population growth and deforestation. In the ensuing years, however, the bills were split into sections, each of which will be considered by a different congressional committee. However, key elements of this legislation face severe hurdles due to budget constraints and the opposition of powerful industries. The White House has given mixed signals on its commitment to slowing climate change, but now a recent presidential proposal appears in the offing. President Clinton' s plan is to discourage the use of fossil fuels by raising energy taxes. He has already proposed a tax on various forms of energy that would set the country on a path of reducing carbon dioxide emissions. Combining higher taxes with tougher auto fuel economy standards, increased use of nonfossil energy sources (such as solar and wind power), and investment in energy-efficient technologies, are some of the administration' s strategies to meet the goal. However, it would be unfair to say that Congress has taken no steps on a national level to combat global warming. The new Clean Air Act (of 1990), which represents the first revisions of the original Clean Air Act, is the best American policy regarding the greenhouse effect. The bill sets forth technical standards for 191 toxic air pollutants and calls for significant reductions in tailpipe emissions. It mandates a 10-million-ton two-phase reduction in sulfur dioxide and a two-million-ton reduction in oxides of nitrogen ($NO_x$) emissions, and provides a flexible plan for the 111 affected coal-fired utility plants. The cost for all of this emission control is hefty for industry ( $ 25 to 35 billion per year), but the Clean

Air Act is a triumph for environmentalists.

Having determined that the United States is swiftly becoming more open to taking an approach of active prevention against the greenhouse effect, the field of policy and legislation to combat global warming can be categorized into three main divisions: (1) the reduction of the amount of greenhouse gases emitted to the atmosphere, (2) the total elimination of fossil fuels as energy sources, and (3) the reduction of the effects of greenhouse gas emission by reforestation or alternate methods.

## International Approach to the Greenhouse Effect

Individual efforts and domestic legislation are inadequate to deal with problems that transcend borders. The United States can spend trillions of dollars on reducing greenhouse gas emissions within its borders, but it is wasted money unless the rest of the world follows suit. The greenhouse effect is a global problem, and the world must consent to an international agreement. The most serious obstacle in realizing a worldwide agreement on greenhouse gas emissions is the conflict between developing and developed nations. Developing nations see the problem as a result of industrialized nations' activities and do not want their fledgling economies restrained. As a result of this, they are resistant to international legal control on industry, in which developed nations are anxious to involve them.

As nations try to agree on solutions that are equitable to all, global warming is also likely to be a major world political issue. Some action has already been taken: In 1987, 24 nations and the European Community signed the Montreal Protocol on Substances that Deplete the Ozone Layer. It called for a freeze on the production and use of CFCs. Eleven countries ratified the Protocol by January 1989, when it went into force by the signatories. The Montreal Protocol is considered to be the state of the art agreement among many nations. The question is whether an agreement of this kind can be implemented for carbon dioxide control. One problem is that CFCs are only produced by 37 countries worldwide and that controlling their production is relatively simple compared to trying to control the infinite sources of carbon dioxide. In the case of CFCs, a substitute can be found; but the worldwide control of carbon dioxide is going to require major changes and innovative approaches. The existence of international cooperation on the scale illustrated by the Montreal Protocol is indicative of what can be done between nations to solve long-term environmental problems. An international treaty to solve this problem could include different efforts for different countries. The goal of one nation could be to decrease emissions, whereas that of another could be an intense reforestation plan.

## Future Trends

Regarding future trends, there appears to be three approaches to the response to this problem. The first approach is the "wait-and-see" approach, whose countermeasures may be inappropriate. The second course of action is the "adaptation to incurable changes" approach, which is based on the assumption that there will be plenty of time in which to decide and act on climatic change. The third approach is the "act now" approach, which is the only one that demands an immediate legislative (and industrial) response. But the problem lies in the legal and economic systems, which normally respond only to immediate and certain threats. Which course of action the world will adopt regarding the greenhouse problem is difficult to predict at this time.

## Words and Expressions

1. botanical 植物学的；药材
2. trap heat 吸热，蓄热，捕集热量
3. brink of disaster 灾难的边缘
4. greenhouse heat trapping effect 温室吸热效应
5. chemical makeup 化学组成
6. triatomic nature 三原子性质
7. terrestrial cycle 陆地上的循环
8. photosynthetic plants 光合(作用)植物
9. chlorophyll 叶绿素
10. convective transport 对流输送
11. chlorofluorocarbons 含氯氟烃
12. account for approximately 50 percent 约占50%
13. arctic ice 北极冰
14. deforestation 滥伐森林
15. rainforest 雨林
16. wetland 湿地
17. blowing agent 发泡剂，起泡剂
18. packing materials 填料
19. plastic foam 塑料泡沫
20. climatic chaos 气候的紊乱
21. testify 证明，证实，表明
22. waffle 空话；唠叨，胡扯
23. tilt of the earth' s axis 地轴的倾斜，地轴的不正
24. elliptical 椭圆形的，椭圆的
25. tropical storm activity 热带风暴活动
26. undermine 削弱……的基础，暗中破坏，逐渐损害
27. temperate zone 温带
28. poleward shift 向极转移，向极移动
29. grassland 牧场，草地，草原
30. prolific 多产的，繁殖的，丰富的，富于……的
31. a flurry of proposals 一阵建议
32. in the ensuing years 在以后的几年
33. severe hurdle 严重的障碍
34. budget constraints 预算约束，预算限制
35. hefty 异常大的，很重的，有力的
36. suit 请求，控告，起诉；适合，相配
37. fledgling economy 无经验人的理财
38. ratify 批准，承认，认可
39. reforestation(采伐后)重新造林，更新造林
40. countermeasure 对抗，对抗措施，防范措施，对策

## Notes

(1) Because of the greenhouse heat trapping effect, the atmosphere itself radiates a large amount of long wave-length energy downward to the surface of the earth and makes the earth twice as warm as it would have been if warmed by solar radiation alone.

由于温室吸热效应，大气本身会辐射大量长波能量至地面，使地面比只有太阳辐射时温度高一倍。

(2) Other researchers believe that the warming of the earth over the last 100 to 150 years is part of a long-term, natural cycle that has little to do with the production of greenhouse gases.

其他研究人员认为过去100~150年地球变暖是长期自然循环的一部分，与温室气体产生关系不大。

(3) The amount of solar radiation reaching the earth is controlled by three elements that vary

cyclically over time.

到达地面的太阳辐射量受三个要素的控制，这些要素随时间周期性地变化。

# 27. Economics in Environmental Management

An understanding of the economics involved in environmental management is important in making decisions at both the engineering and management levels. Every engineer or scientist should be able to execute an economic evaluation of a proposed environmental project. If the project is not profitable, it should obviously not be pursued; and, the earlier such a project can be identified, the fewer are the resources that will be wasted.

Economics also plays a role in setting many state and federal air pollution control regulations. The extent of this role varies with the type of regulation. For some types of regulations, cost is explicitly used in determining their stringency. This use may involve a balancing of costs and environmental impacts, costs and dollar valuation of benefits, or environmental impacts and economic consequences of control costs. For other types of regulations, cost analysis is used to choose among alternative regulations with the same level of stringency. For these regulations, the environmental goal is determined by some set of criteria that does not include costs. However, cost-effectiveness analysis is employed to determine the minimum economic way achieving the goal. For some regulations, cost influences enforcement procedures or requirements for demonstration of progress towards compliance with an environmental quality standard. For example, the size of any monetary penalty assessed for noncompliance as part of an enforcement action needs to be set with awareness of the magnitude of the control costs being postponed or bypassed by the noncomplying facility(1). For regulations without a fixed compliance schedule, demonstration of reasonable progress towards the goal is sometimes tied to the cost of attaining the goal on different schedules.

Before the cost of an environmental project can be evaluated, the factors contributing to the cost must be recognized. There are two major contributing factors, namely, capital costs and operating costs; these are discussed in the next two sections. Once the total cost of a project has been estimated, the engineer must determine whether the process (change) will be profitable. This often involves converting all cost contributions to an annualized basis. If more than one project proposal is under study, this method provides a basis for comparing alternate proposals and for choosing the best proposal. Project optimization is covered later in the chapter, where a brief description of a perturbation analysis is presented.

Detailed cost estimates are beyond the scope of this chapter. Such procedures are capable of producing accuracies in the neighborhood of +/- 10%; however, such estimates generally require many months of engineering work. This chapter is designed to give the reader a basis for preliminary cost analysis only.

## Capital Costs

Equipment cost is a function of many variables, one of the most significant of which is capacity. Other important variables include equipment type and location, operating temperature, and degree of equipment sophistication. Preliminary estimates are often made from simple cost-capacity relationships that are valid when other variables are confined to narrow ranges of values; these relationships can be represented by the approximate linear (on log-log coordinates) cost equations of the form

$$C = \alpha (Q)^{\beta}$$

where $C$ represents cost; $Q$ represents some measure of equipment capacity; $\alpha$ and $\beta$ represents empirical "constants" that depend mainly on the equipment type. It should be emphasized that this procedure is suitable for rough estimation only; actual estimates (or quotes) from vendors are more preferable. Only major pieces of equipment are usually included in this analysis; smaller peripheral equipment such as pumps and compressors are not discussed.

If more accurate values are needed and if old price data are available, the use of an indexing method is better, although a bit more time-consuming. The method consists of adjusting the earlier cost data to present values using factors that correct for inflation. A number of such indices are available; one of the most commonly used is the chemical engineering fabricated equipment cost index (FECI), past values of which are listed in Table 1. Other indices for construction, labor, buildings, engineering and so on, are also available in the literature. Generally, it is not wise to use past cost data older than 5 to 10 years, even with the use of the cost indices. Within that time span, the technologies used in the processes have changed drastically. The use of the indices could cause the estimates to be much greater than the actual costs. Such an error might lead to the choice of alternative proposals other than the least costly.

**Table 1 Fabricated Equipment Cost Index**

| Year | Index | Year | Index |
|---|---|---|---|
| 1995 | 362.0 (estimated) | 1984 | 334.1 |
| 1994 | 361.3 | 1983 | 327.4 |
| 1993 | 360.8 | 1982 | 326.0 |
| 1992 | 358.2 | 1981 | 321.8 |
| 1991 | 361.3 | 1980 | 291.6 |
| 1990 | 357.6 | 1979 | 261.7 |
| 1989 | 355.4 | 1978 | 238.6 |
| 1988 | 342.5 | 1977 | 216.6 |
| 1987 | 323.8 | 1976 | 200.8 |
| 1986 | 318.4 | 1975 | 192.2 |
| 1985 | 325.3 | | |

The usual technique for determining the capital costs (i.e., total capital costs, which include equipment design, purchase, and installation) for a project and/or process can be based on the factored method of establishing direct and indirect installation cost as a function of the known equipment costs. This is basically a modified Lang method, whereby cost factors are applied to known equipment costs.

The first step is to obtain from vendors (or, if less accuracy is acceptable, from one of the estimation techniques previously discussed) the purchase prices of primary and auxiliary equipment. The total base price, designated by X-which should include instrumentation, control, taxes, freight costs, and so on-serves as the basis for estimating the direct and indirect installation costs. The installation costs are obtained by multiplying X by the cost factors, which are available in literature. For more refined estimates, the cost factors can be adjusted to more closely model the proposed system by using adjustment factors that take into account the complexity and sensitivity of the system.

The second step is to estimate the direct installation costs by summing all the cost factors involved in the direct installation costs, which include piping, insulation, foundation and supports, and so on(2). The sum of these factors is designated as the DCF (direct installation cost factor). The direct installation costs are the product of the DCF and X.

The third step consists of estimating the indirect installation costs. The procedure here is the same as that for the direct installation cost-that is, all the costs factors for the indirect installation costs (engineering and supervision, startup, construction fees, etc.) are added. The sum is designated as the ICF (indirect installation cost factor). The indirect installation costs are then the product of ICF and X.

Once the direct and indirect installation costs have been calculated, the total capital cost (TCC) may be evaluated as:

$$TCC = X + (DCF)(X) + (ICF)(X)$$

This cost is converted to annualized capital costs with the use of the capital recovery factor (CRF), which is described later. The annualized capital cost (ACC) is the product of the CRF and the TCC and represents the total installed equipment cost distributed over the lifetime of the facility.

Some guidelines in purchasing equipment are listed below:

1. Do not buy or sign any documents unless provided with certified independent test data.
2. Previous clients of the vendor company should be contacted and their facilities visited.
3. Prior approval from the local regulatory officials should be obtained.
4. A guarantee from the vendors involved should be required. Startup assistance is usually needed, and an assurance of prompt technical assistance should be obtained in writing. A complete and coordinated operating manual should be provided.
5. Vendors should provide key replacement parts if necessary.
6. Finally, 10 to 15 percent of the cost should be withheld until the installationis completed.

## Operating Costs

Operating costs can vary from site to site because the costs partly reflect local conditions-for example, staffing practices, labor and utility costs. Operating costs like capital costs may be separated into two categories: direct and indirect costs. Direct costs are those that cover material and labor and are directly involved in operating the facility. These include labor, materials, maintenance and maintenance supplies, replacement parts, waste (e.g., residues after incineration) disposal fees, utilities and laboratory costs. Indirect costs are those operating costs associated with, but not directly involved in operating the facility; costs such as overhead (e.g., building-land leasing and office supplies), administrative fees, local property taxes and insurance fees fall into this category.

The major direct operating costs are usually associated with the labor and materials costs for the project, which involve the cost of the chemicals needed for operation of the process. Labor costs differ greatly, but are a strong function of the degree of controls and/or instrumentation. Typically, there are three working shifts per day with one supervisor per shift. On the other hand, the plants may be manned by a single operator for only one-third or one-half of each shift; that is, usually only operator, supervisor, and site manager are necessary to run the facility. Salary costs vary from state to state and depend significantly on the location of the facility. The cost of utilities generally consists of that of electricity, water, fuel, and steam. The annual costs are estimated with the use of material and energy balances. Cost for waste disposal can be estimated on a per-ton-capital basis. Cost of landfilling ash can run significantly upwards of $ 100/ton if the material is hazardous, and can be as high as $ 10/ton if it is nonhazardous. The cost of handling a scrubber effluent can vary depending on the method of disposal. For example, if the conventional sewer disposal is used, the effluent probably has to be cooled and neutralized before disposal; the cost for this depends on the solids concentration. Annual maintenance costs can be estimated as a percentage of the capital equipment cost. The annual cost of replacement parts can be computed by dividing the cost of the individual part by its expected lifetime. The life expectancies can be found in the literature. Laboratory costs depend on the number of samples tested and the extent of these tests; these costs can be estimated as 10 to 20 percent of the operating labor costs.

The indirect operating costs consist of overhead, local property tax, insurance, and administration, less any credits. The overhead comprises payroll, fringe benefits, social security, unemployment insurance and other compensation that is indirectly paid to plant personnel. This cost can be estimated as 50 to 80 percent of the operating labor, supervision, and maintenance costs. Local property taxes and insurance can be estimated as 1 to 2 percent of the total capital cost (TCC), while administration costs can be estimated as 2 percent of the TCC.

The total operating cost is the sum of the direct operating cost and the indirect operating costs, less any credits that may be recovered (e.g., the value of the recovered steam). Unlike capital costs, operating costs are calculated on an annual basis.

## Hidden Economic Factors

The main problem with the traditional type of economic analysis, discussed above, is that it is difficult—nay, in some cases impossible-to quantify some of the not-so-obvious economic merits of a business and/or environmental program.

Several considerations have just recently surfaced as factors that need to be taken into account in any meaningful economic analysis of a project effort[3]. What follows is a summary of these considerations:

Long-term liabilities

Regulatory compliance

Regulatory recordkeeping

Dealings with the EPA

Dealings with the state and local regulatory bodies

Fines and penalties

Potential tax benefits

Customer relations

Stockholder support (corporate image)

Improved public image

Insurance costs and claims

Effect on borrowing power

Improved mental and physical well being of employees

Reduced health maintenance costs

Employee morale

Worker safety

Rising costs of waste treatment and/or disposal

Training costs

Emergency response planning

Many programs have been quenched in their early states because a comprehensive economic analysis was not performed. Until the effects described above are included, the true merits of a project may be clouded by incorrect and/or incomplete economic data. Can something be done by industry to remedy this problem? One approach is to used a modified version of the standard Delphi Panel. In order to estimate these "other" economic benefits, several knowledgeable individuals within and perhaps outside the organization are asked to independently provide estimates, with explanatory details, on these economic benefits. Each individual in the panel is then allowed to independently review all response. The cycle is then repeated until the groups responses approach convergence.

## Project Evaluation and Optimization

In comparing alternate processes or different options of a particular process from an economic point of view, it is recommended that the total capital cost be converted to an annual basis by distributing it over the projected lifetime of the facility. The sum of both the annualized capital costs (ACCs) and the annual operating costs (AOCs) is known as the total annualized cost (TAC) for the facility. The economic merit of the proposed facility, process, or scheme can be examined once the total annual cost is available. Alternate facilities or options (e.g., a baghouse versus an electrostatic precipitator for particulate control, or two different processes for accomplishing the same degree of waste destruction) may also be compared. Note that a small flaw in this procedure is the assumption that the operating costs will remain constant throughout the lifetime of the facility.

Once a particular process scheme has been selected, it is common practice to optimize the process from a capital cost and O&M (operation and maintenance) standpoint. There are many optimization procedures available, most of them are too detailed for meaningful application for this chapter. These sophisticated optimization techniques, some of which are routinely used in the design of conventional chemical and petroleum plants, invariably involve computer calculations. Use of these techniques in environmental management analysis is usually not warranted, however.

One simple optimization procedure that is recommended is a perturbation study. This involves a systematic change (or perturbation) of variables, one by one, in an attempt to locate the optimum design from a cost and operation viewpoint. To be practical, this often means that the engineer must limit the number of variables by assigning constant values to those process variables that are known beforehand to play an insignificant role. Reasonable guesses and simple short-cut mathematical methods can further simplify the procedure. More information can be gathered from this type of study because it usually identifies those variables that significantly impact on the overall performance of the process and also helps identify the major contributors to the total annualized cost(4).

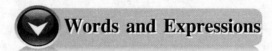

## Words and Expressions

1. make decision 决策
2. economic evaluation 经济评价
3. environmental project 环境项目，环境工程
4. explicitly 明白地，明显地，清楚地
5. stringency 紧急，迫切，缺少，严重，严厉，说服力，强度
6. monetary penalty 金钱上的惩罚，罚款
7. perturbation analysis 摄动分析，扰动分析
8. in the neighborhood of 大约，在……附近，在……上下[左右]
9. vendor 卖方，卖家，自动售货机
10. peripheral equipment 外围设备，辅助设备
11. inflation 膨胀，通货膨胀，夸张
12. replacement parts 备件，替换零件
13. working shifts 工作班(制)
14. cost of handling a scrubber effluent 处理[输送]洗涤器出水的成本[费用]
15. divide the cost of the individual part by its expected lifetime 将单个部件的成本除以其预期寿命
16. life expectancy 概率寿命，估计寿命，预期寿命
17. payroll 工资单，计算报告表
18. fringe benefit 边缘利益
19. explanatory 说明(性)的，解释(性)的
20. convergence 会聚，集中
21. projected lifetime of the facility 设备的预期寿命
22. annualized capital cost(ACCs) 分年的基本(建设)投资
23. annual operating cost(AOCs) 年度运行费用，每年的操作成本
24, perturbation study 摄动研究，扰动研究

## Notes

(1) For example, the size of any monetary penalty assessed for noncompliance as part of an enforcement action needs to be set with awareness of the magnitude of the control costs being postponed or bypassed by the noncomplying facility.

例如，作为执法(强制行为)的一部分，对不达标的罚款需要知道有关延误控制的费用或通过采用不达标设施回避控制的费用。

(2) The second step is to estimate the direct installation costs by summing all the cost factors involved in the direct installation costs, which include piping, insulation, foundation and supports, and so on.

第二步是估算直接安装费用，包括直接安装中的所有费用：管道、隔离、地基和支架等。

(3) Several considerations have just recently surfaced as factors that need to be taken into account in any meaningful economic analysis of a project effort.

作为项目的有意义的经济分析需考虑因素的若干见解最近刚被提出来。

(4) More information can be gathered from this type of study because it usually identifies those variables that significantly impact on the overall performance of the process and also helps identify the major contributors to the total annualized cost.

从这类研究中可以收集更多的信息，因为这类研究通常能识别那些对过程的总效能产生显著影响的变量，也有助于识别总的年度费用的主要构成因素。

# 28. Multimedia Concerns in Pollution Control

The current approach to environmental waste management requires some rethinking. A multimedia approach helps integration of air, water, and land pollution controls and seeks solutions that do not violate the laws of nature. The obvious advantage of a multimedia pollution control approach is its ability to manage the transfer of pollutants so they will not continue to cause pollution problems. Among the possible steps in the multimedia approach are understanding the cross-media nature of pollutants, modifying pollution control methods so as not to shift pollutants from one medium to another, applying available waste reduction technologies, and training environmental professionals in a total environmental concept[1].

A multimedia approach in pollution control is long overdue. As described above, it integrates air, water, and land into a single concern and seeks a solution to pollution that does not endanger society or the environment. The challenges for the future environmental professional include:

1. Conservation of natural resources.

2. Control of air-water-land pollution.

3. Regulation of toxics and disposal of hazardous wastes.

4. Improvement of quality of life.

It is now increasingly clear that some treatment technologies (specific technologies will be discussed in later chapters), while solving one pollution problem, have created others. Most contaminants, particularly toxics, present problems in more than one medium. Since nature does not recognize neat jurisdictional compartments, these same contaminants are often transferred across media. Air pollution control devices and industrial wastewater treatment plants prevent waste from going into the air and water, but the toxic ash and sludge that these systems produce can become hazardous waste problems themselves. For example, removing trace metals from a flue gas usually transfers the products to a liquid or solid phase. Does this exchange an air quality problem for a liquid or solid waste management problem? Waste disposed of on land or in deepwells can contaminate ground water and evaporation from ponds and lagoons can convert solid or liquid waste into air pollution problems. Other examples include acid deposition, residue management, water reuse, and hazardous waste treatment and/or disposal.

Control of cross-media pollutants cycling in the environment is therefore an important step in the management of environmental quality. Pollutants that do not remain where they are released or where they are deposited move from a source to receptors by many routes, including air, water, and land. Unless information is available on how pollutants are transported, transformed, and accumulated after they enter the environment, they cannot effectively be controlled. A better understanding of the cross-media nature of pollutants and their major environmental processes-physical, chemical, and biological-is required.

## Historical Perspective

The EPA' s own single-media offices, often created sequentially as individual environmental problems were identified and responded to in legislation, have played a part in the impeding development of cost-effective multimedia prevention strategies. In the past, innovative cross-media agreements involving or promoting pollution prevention, as well as voluntary arrangements for overall reduction in releases, have not been encouraged. However, new initiatives are characterized by their use of a wide range of tools, including market incentives, public education and information, small business grants, technical assistance, research and technology applications, as well as more traditional regulations and enforcements (see Chapter 5 for additional details).

In the past the responsibility for pollution prevention and/or waste management at the industrial level was delegated to the equivalent of an environmental control department. These individuals were skilled in engineering treatment techniques but, in some instances, had almost no responsibility over what went on in the plant that generated the waste they were supposed to manage. In addition, most engineers are trained to make a product work, not to minimize or prevent pollution[2]. There is still little emphasis (although this is changing) on pollution prevention in the educational arena for engineers. Business school students, the future business managers, also have not had the pollution prevention ethic instilled in them.

The reader should also note that the federal government, through its military arm, is responsible for some major environmental problems. It has further compounded these problems by failing to apply a multimedia or multiagency approach.

This lack of communication and/or willingness to cooperate within the federal government has created a multimedia problem that has just begun to surface. The years of indifference and neglect have allowed pollutants/wastes to contaminate the environment significantly beyond what would have occurred had the responsible parties acted sooner[3].

## Environmental Problems

Environmental problems result from the release of wastes (gaseous, liquid, and solid) that are generated daily by industrial and commercial establishments as well as households[4]. The lack of consciousness regarding conservation of materials, energy, and water has contributed to the wasteful habits of society. The rate of waste generation has been increasing in accordance with the increase in population and the improvement in living standards. With technological advances and changes in lifestyle, the composition of waste has likewise changed. Chemical compounds and products are being manufactured in new forms with different half-lives (time for half of it to react and/or disappear). It has been difficult to manage such compounds and products once they have been discarded. As a result, these wastes have caused many treatment, storage, and disposal problems. Many environmental problems are caused by products that are either misplaced in use or discarded without proper concern of their environmental impacts. Essentially all products are potential wastes, and it is desirable to develop methods to reduce the waste impacts associated with products or to produce environmentally friendly products. Environmental agencies have been lax in promoting and automating tracking mechanisms that identify sources and fate of new products[5].

Solving problems, however, can sometimes create problems. For example, implementation of the Clean Air Act and the Clean Water Act has generated billions of tons of sludge, wastewater and residue that could cause soil contamination and underground water pollution problems. The increased concern over cross-media shifts of pollutants has yet to consistently translate into a systematic understanding of pollution problems and viable changes[6].

As indicated above, environmental protection efforts have emphasized media-specific waste treatment and disposal after the waste has already been created. Many of the pollutants which enter the environment are coming from "area or point sources" such as industrial complexes and land disposal facilities; therefore, they simply cannot be solely controlled by the end-of-pipe solutions.

Furthermore, these end-of-pipe controls that tend to shift pollutants from one medium to another have often caused secondary pollution problems. Therefore, for pollution control purposes, the environment must be perceived as a single integrated system and pollution problems must be viewed holistically. Air quality can hardly be improved if water and land pollution continue to occur. Similarly, water quality cannot be improved if the air and land are polluted.

Many secondary pollution problems today can be traced in part to education, that is, the lack of knowledge and understanding of cross-media principles for the identification and control of pollutants. Neither the Clean Air Act nor the Clean Water Act enacted in the early 1970s adequately addresses the cross-media nature of environmental pollutants. More environmental professionals now realize that pollution legislation is too fragmented and compartmentalized. Only proper education and training will address this situation and hopefully lead to more comprehensive legislation of the total environmental approach.

## Multimedia Approach

The environment is the most important component of life support systems. It is comprised of air, water, soil, and biota through which elements and pollutants cycle. This cycle involves the physical, chemical, or biological processing of pollutants in the environment. It may be short, turning hazardous into nonhazardous substances soon after they are released, or it may continue indefinitely with pollutants posing potential health risks over a long period of time. Physical processes associated with pollutant cycling include leaching from the soil into the ground water, volatilization from water or land to air, and deposition from air to land or water. Chemical processes include decomposition and reaction of pollutants to products with properties that are possibly quite different from those of the original pollutants. Biological processes involve microorganisms that can break down pollutants and convert hazardous pollutants into less toxic forms. However, these microorganisms can also increase the toxicity of a pollutant, for instance by changing mercury into methyl-mercury in soil.

Although pollutants sometimes remain in one medium for a long time, they are most likely mobile. For example, settled pollutants in river sediments can be dislodged by microorganisms, flooding, or dredging. Displacement such as this earlier constituted the PCB problem in New York's Hudson River. Pollutants placed in landfills have been transferred to air and water through volatilization and leaching. About 200 hazardous chemicals were found in the air, water, and soil at the Love Canal land disposal site in New York State. The advantages of applying multimedia approaches lie in their ability:

1. To manage the transfer of pollutants.

2. To avoid duplicating efforts or conflicting activities.

3. To save resources by consolidation of environmental regulations, monitoring, database management, risk assessment, permit issuance and fields inspection.

In recent years, the concept and goals of multimedia pollution prevention have been adopted by many regulatory and other governmental agencies, industries, and the public in the United States and abroad. Multimedia efforts in the United States have been focused on the U.S. Environmental Protection Agency's (EPA) Pollution Prevention Office, which helps coordinate pollution prevention activities across all EPA headquarter offices and regional offices. The current EPA philosophy recognizes that multimedia pollution prevention is best achieved through education and technology transfer rather than through regulatory imposition of mandatory approaches. But the progress of implementing multimedia pollution prevention has been slow.

Recognition of the need for multimedia pollution prevention approaches has extended from the government, industry, and the public to professional societies. The Air Pollution Control Association (APCA) was renamed as the Air and Waste Management Association (AWMA) to incorporate waste management. The American Society of Civil Engineers (ASCE) has established a multimedia management committee under the Environmental Engineering Division. The American Institute of Chemical Engineers (AIChE) has reorganized its Environmental Division to include a section devoted to pollution prevention. The Water Pollution Control Federation (WPCF) has also adopted a set principle addressing pollution prevention.

## Multimedia Application

Perhaps a meaningful understanding of the multimedia approach can be obtained by examining the production and ultimate disposal of a product or service. A flow diagram representing this situation is depicted in Figure 1. Note that each of the ten steps in the overall process has potential inputs of mass and energy, and may produce an environmental pollutant and/or a substance or form of energy that may be used in a subsequent or later step. Traditional approaches to environmental management can provide some environmental relief, but a total systems approach is required if optimum improvements-in terms of pollution/waste reduction-are to be achieved.

One should note that a product and/or service is usually conceived to meet a specific market need with little thought given to the manufacturing parameters. At this stage of consideration, it may be possible to avoid some significant waste generation problems in future operations by answering a few simple questions:

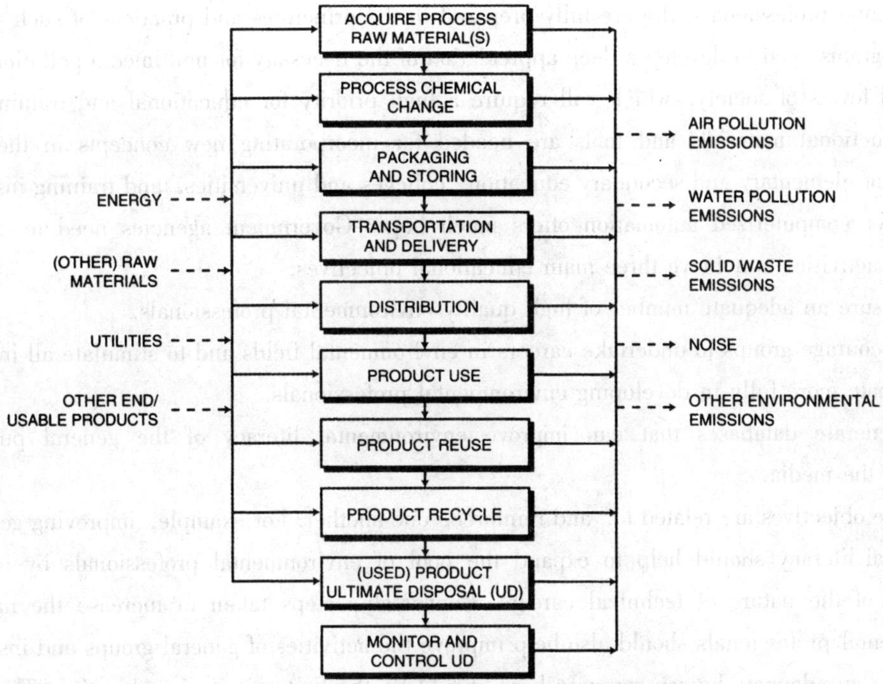


Figure 1 Overall Multimedia Flow Diagram

1. What raw materials are used to manufacture the product?

2. Are any toxic or hazardous chemicals likely to be generated during manufacturing?

3. What performance regulatory specifications must the new product(s) and/or service(s) meet? Is extreme purity required?

4. How reliable will the delivery manufacturing / distribution process be? Are all steps commercially proven? Does the company have experience with the operations required?

5. What types of waste are likely to be generated? What is their physical and chemical form? Are they hazardous? Does the company currently manage these wastes on-site or off-site?

## EDUCATION AND TRAINING

The role of environmental professionals in waste management and pollution control has been changing significantly in recent years. Many talented, dedicated environmental professionals in academia, government, industry, research institutions, and private practice need to cope with this change, and extend their knowledge and experience from media-specific, "end-of-pipe" treatment-and-disposal strategies to multimedia pollution prevention management. The importance of this extension and reorientation in education, however, is such that the effort cannot be further delayed[7]. Many air pollution, water pollution, and solid waste supervisors in government agencies spend their entire careers in just one function because environmental quality supervisors usually work in only one of the media functions. Some may be reluctant to accept such activities. This is understandable given the fact that such a reorientation requires time and energy to learn new concepts and that time is a premium for them[8]. Nevertheless they must support such education and training in order to have well-trained young professionals.

Successful implementation of multimedia pollution prevention programs will require well-trained environmental professionals who are fully prepared in the principles and practices of such programs. These programs need to develop a deep appreciation of the necessity for multimedia pollution prevention in all levels of society, which will require a high priority for educational and training efforts. New instructional materials and tools are needed for incorporating new concepts in the existing curricula of elementary and secondary education, colleges and universities, and training institutions. The use of computerized automation offers much hope. Government agencies need to conduct a variety of activities to achieve three main educational objectives:

1. Ensure an adequate number of high quality environmental professionals.

2. Encourage groups to undertake careers in environmental fields and to stimulate all institutions to participate more fully in developing environmental professionals.

3. Generate databases that can improve environmental literacy of the general public and especially the media.

These objectives are related to, and reinforce, one another. For example, improving general environmental literacy should help to expand the pool of environmental professionals by increasing awareness of the nature of technical careers. Conversely, steps taken to increase the number of environmental professionals should also help improve the activities of general groups and institutions. Developing an adequate human resource base should be the first priority in education. The training environmental professionals receive should be top quality.

There is significant need to provide graduate students with training and experience in more than one discipline. The most important and interesting environmental scientific/technological questions increasingly require interdisciplinary and/or multidisciplinary approaches. Environmental graduate programs must address this aspect. Most practicing environmental professionals face various types of environmental problems that they have not been taught in the universities. Therefore, continuing education opportunities and cross-disciplinary training must be available for them to understand the importance of multimedia pollution prevention principles and strategies, as well as to carry out such principles and strategies.

The education and training plan of multimedia pollution prevention may be divided into technical and nontechnical areas. Technical areas include:

1. Products-Lifecycle analysis methods, trends-in-use patterns, new products, product life-span data, product substitution, and product applicability. (A product's lifecycle includes its design, manufacture, use, maintenance and repair, and final disposal.)

2. Processes-Feedstock substitution, waste minimization, assessment procedures, basic unit process data, unit process waste generation assessment methods, materials handling, cleaning, maintenance and repair.

3. Recycling and Reuse-Market availability, infrastructure capabilities, new processes and product technologies, automated equipment and processes, distribution and marketing, management strategies, automation, waste stream segregation, on-site and off-site, reuse opportunities, close-loop methods, waste recapture, and reuse.

Nontechnical areas include:

1. Educational programs and dissemination of information.

2.Incentive and disincentives.

3.Economic cost and benefits.

4.Sociological human behavioral trends.

5.Management strategies including coordination with various concerned organizations.

## FUTURE TRENDS

Environmental quality and natural resources are under extreme stress in many industrialized nations and in virtually every developing nation as well. Environmental pollution is closely related to population density, energy, transportation demand, and land use patterns, as well as industrial and urban development. The main reason for environmental pollution is the increasing rate of waste generation in terms of quantity and toxicity that has exceeded society' s ability to properly manage it. Another reason is that the management approach has focused on the media-specific and the end-of-pipe strategies. There is increasing reported evidence of socioeconomic and environmental benefits realized from multimedia pollution prevention. The prevention of environmental pollution in the 21st century is going to require not only enforcement of government regulations and controls, but also changes in manufacturing processes and products as well as in lifestyles and behavior throughout society. Education is key in achieving the vital goal of multimedia pollution prevention.

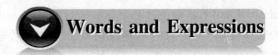

## Words and Expressions

1. environmental waste management 环境废物管理
2. multimedia approach 多介质(控制)方法[途径]
3. overdue 过时的，过期的，迟到的
4. recognize neat jurisdictional compartment 认可整齐的管理分隔，承认整齐的管理区划
5. cross-media pollutant 跨介质污染物
6. receptor 接受器，受纳体，受体
7. impending 逼近的，迫切的，即将来临的
8. military arms 军用武器
9. industrial and commercial establishments 工商企业
10. in accordance with 根据，按照，与……一致
11. single integrated system 单一的综合系统
12. volatilization 挥发(作用)，发散
13. regulatory imposition 按规章的强迫接受，按规章的施加
14. recognition 认出，识别，承认，公认
15. environmental relief 环境减负，环境改善
16. premium 奖励，奖金，佣金，保险费
17. environmental literacy 环境的读写能力
18. multidisciplinary 多学科的
19. cross-disciplinary training 交叉学科培训，跨学科培训
20. multimedia pollution prevention principles 多介质污染预防原理
21. infrastructure capability 基础设施能力[容量]
22. recapture 取回，夺回，恢复，重新俘获
23. incentive 刺激的，鼓励的，诱发的；刺激，鼓励，诱因，动机
24. disincentive 阻止的，抑制的；起阻碍作用的行动或措施

## Notes

(1) Among the possible steps in the multimedia approach are understanding the cross-media nature of pollutants, modifying pollution control methods so as not to shift pollutants from one medium to another, applying available waste reduction technologies, and training environmental professionals in a total environmental concept.

多介质控制方法的可能步骤包括：了解污染物的跨介质特性，改进污染控制方法不使污染物从一个介质转移到另一介质，应用现有的废物减量化技术和以完全的环境概念培训环境专业人员。

(2) In addition, most engineers are trained to make a product work, not to minimize or prevent pollution.

此外，大多数工程师的培训是面向产品生产，而不是最大程度地减少或预防污染。

(3) The years of indifference and neglect have allowed pollutants/wastes to contaminate the environment significantly beyond what would have occurred had the responsible parties acted sooner.

多年的不关心和忽视使污染物/废物污染环境明显超过了责任者早些采取行动会有的程度。

(4) Environmental problems result from the release of wastes (gaseous, liquid, and solid) that are generated daily by industrial and commercial establishments as well as households.

废物(废气、废水和固体废物)的排放引起环境问题，这些废物每天从工业、商业系统和家庭产出。

(5) Environmental agencies have been lax in promoting and automating tracking mechanisms that identify sources and fate of new products.

环保部门在有关识别新产品的来源和归宿的机制的创立和自动跟踪方面是落后了。

(6) The increased concern over cross-media shifts of pollutants has yet to consistently translate into a systematic understanding of pollution problems and viable changes.

对污染物跨介质的转移日益增加的关注要相应地转为对污染问题和相应变化的系统的了解。

(7) The importance of this extension and reorientation in education, however, is such that the effort cannot be further delayed.

然而，教育的这种延伸和重新定向的努力重要到了不能再拖延的程度。

(8) This is understandable given the fact that such a reorientation requires time and energy to learn new concepts and that time is a premium for them.

知道了重新定向需要时间和精力去学习新概念，而时间对他们来说很宝贵，这就可以理解了。

# PART TWO
# Academic Writing

# Unit 1  Abstract Writing
# 英语科技论文摘要的撰写

## 1.1 英语科技论文摘要的含义和类型

### 1.1.1 摘要的含义

摘要(abstract)是论文的梗概，是对论文的简明描述。摘要作为科技论文的重要组成部分，有其特殊的作用和意义。摘要是国际上主要检索机构数据库的重要资源，摘要对于增加期刊和论文的被检索和引用机会、吸引读者和方便交流起着重要的作用。

### 1.1.2 摘要的类型

按内容的不同来分，摘要一般分为两类：报道性摘要和指示性摘要。

1.1.2.1 报道性摘要

报道性摘要(informative abstract)：也称作资料性摘要或信息性摘要，其特点是简要地概括研究目的、研究方法、研究结果和结论。它是论文全文的高度浓缩。报道性摘要不仅为文献工作者和研究人员提供方便，而且为读者提供很多有价值的信息。传统的摘要多为一段式。但20世纪80年代出现了一种“结构式摘要(structrured abstract)”，该摘要是报道性摘要的结构化表达。在该摘要的行文中用醒目的字体直接标出目的、方法、结果和结论等标题。结构式摘要的优点是易于写作(填空式)，便于阅读(摘要要素明晰、具体)。

1.1.2.2 指示性摘要

指示性摘要(indicative abstract)：也称为说明性摘要、描述性摘要(descriptive abstract)或论点摘要(topic abstract)，一般只概括论文的主题、所取得成果的性质及所达到的水平，不涉及研究方法、结果和结论，多用于综述、会议报告等。

摘要的长度应根据文献类型及摘要的用途而变化，报道性摘要通常以150~250words为宜，指示性摘要以100~150words为宜。

当今几乎所有的科技出版物要求的均为报道性摘要，以下除特别说明，讲述的摘要均为报道性摘要。

## 1.2 英语科技论文摘要的要素、基本结构和内容

英语科技论文最常见的结构是所谓的IMRAD结构，即：Introduction(前言)、Methods(方法)、Results and Discussion(结果与讨论)。IMRAD结构的论文简单、清晰、明了并且逻辑性强。摘要从本质上看是一篇高度浓缩的论文，所以其构成与完整论文的IMRAD结构是对应的。因此，摘要一般有四个组成部分：

(1) 研究目的：研究工作的前提、目的和任务、所涉及的主题范围。

(2) 研究方法：研究工作所用的理论、条件、材料、手段、装备和程序等。

(3) 研究结果：研究过程中观察、实验的结果、数据、效果和性能等。

(4) 结论：对结果的分析、比较、评价、应用，提出的问题、今后的课题、假设、启发、建议和预测等。人们通常将这四个组成部分看作为摘要的要素。

目前，有些摘要的主要内容是介绍研究结果，节省了研究目的、研究方法等的说明，此类摘要对一些资深研究人员来说非常实用。

## 1.3 英语科技论文摘要撰写的原则

摘要作为科技论文的重要组成部分，有其自身的规范，为了遵守规范并体现摘要的"独立性"和"自明性"，英语科技论文摘要的撰写必须遵循若干原则。

### 1.3.1 摘要撰写的一般原则

(1) 要简明地表述论文的IMRAD，注重强调研究中的创新之处。表达要准确、清楚，注意表达的逻辑性，尽量使用指示性的词语来表述本论文的属性(研究目的、方法、结果和结论)。

(2) 避免引用文献和使用图表。

(3) 避免使用化学结构式、数学表达式、角标和希腊文等特殊符号。

### 1.3.2 摘要中的时态

摘要写作时所采用的时态大致遵循以下原则：

(1) 介绍背景资料时，如果句子的内容为不受时间影响的普遍事实，应使用现在时；如果句子的内容是对某种研究趋势的概述，则使用现在完成时。

(2) 在叙述研究目的或主要研究活动时，如果采用"论文导向"，多使用现在时(如：This paper presents...)；如果采用"研究导向"，则使用过去时(如：This study investigated...).

(3) 概述实验程序、方法和主要结果时，通常用现在时。

(4) 叙述结论或建议时，可使用现在时、臆测动词或 may、should、could 等助动词。

### 1.3.3 摘要中的人称和语态

科技论文的撰写较多使用第三人称、过去时和被动语态。目前不少期刊都提倡使用主动语态。国际知名期刊 Nature、Cell 等尤其如此，其中第一人称和主动语态的使用十分普遍。可见，为简洁、清楚地表达研究成果，在论文摘要的撰写中不应刻意回避第一人称和主动语态。

在实际工作中，在人称、时态和语态方面应参考拟投期刊的习惯表达或要求，根据实际需要灵活运用。

## 1.4 英语科技论文摘要中的独特句型

英语科技论文摘要有其独特的句型，该句型谓语部分很短，主语部分很长，是一个名词短语，主要信息通过该名词短语表达。该句型有庄重朴实、言简意赅的优点。该句型在摘要中往往用来陈述研究主题，如研究目的、研究内容等。

该名词短语由主语的核心词和其修饰语构成。主语的核心词常常或是一些比较抽象的名词，如 attempt, process, method, effect, result, approach, use, mechanism, principle, technique 等；或是一些表示装置、设备等含义的名词，如 machine, instrument, apparatus, system, device 等。

句子中的谓语动词常常是表示讨论、研究、调查、报告之类的词，如 study, discuss, investigate, describe, develop, present , analyze , report, estimate 等。

句子主语中的修饰语往往为定语，下面就定语的形式和位置作一说明。

### 1.4.1 定语的形式

主语核心词的定语常是介词短语、形容词短语、分词短语、不定式短语和从句等。例句如下：

(1) The general principles of calculating the phase diagram of a system from its thermodynamic properties are discussed.

本文讨论按一个系统的热力特性计算其相图的一般原理。

(2) The results of an investigation on the problems of underwater living are presented.

本文提出对水下生活问题的调查结果。

(3) A hot-acid method for extracting heavy metals from municipal wastewater has been developed.

已经研究出从城市废水中提取重金属的热酸废水处理法。

(4) The techniques suitable for the rapid analysis of some dissolved organic constituents of sewage are presented.

本文提出适宜于对污水中某些溶解的有机成分作快速分析的方法。

(5) The part played by adjustment of chemical composition in obtaining the required properties is also discussed.

本文也讨论调整化学成分对获得所需特性所起的作用。

(6) An attempt to use wind energy more effectively has been made.

对更有效地利用风能进行了尝试。

(7) The parameters which have been reported as affecting the rates of fume formation in welding will be discussed.

本文将讨论据报道说对焊接时烟气生成速度有影响的一些参数。

(8) A new alloy which can resist the action of acids has been developed.

已研制出一种能耐酸的新合金。

### 1.4.2 定语的位置

主语核心词的定语的位置通常都紧跟在主语核心词之后，但定语很长或结构复杂时，也可以采取割裂修饰手段，把句子改为“主语核心词 + 谓语 + 定语 ”。例如：

(1) Results are presented of analyses carried out on the process waters generated during the pyrolysis of oil shale via the Fischer assay retorting process.

本文对用费希尔试料蒸馏法热解油页岩时所产生的工业废水提出分析结果。

(2) An attempt has been made to study the variation of overall oxygen transfer coefficient in relation to the temperature in an aeration system.

对研究氧气传递总系数随曝气系统温度的变化已作过尝试。

(3) Many ways have been found to use small and simple machines to process large and complicated machine parts.

已经找到许多使用小型的简单机器加工复杂的大型机器零件的方法。

(4) In recent years concern has been growing over the pollution of the atmosphere by oxides of sulphur and nitrogen.

近年来(人们)对硫和氮的氧化物造成大气污染越来越担忧。

## 1.5 英语科技论文摘要撰写的要点

一篇好的摘要的写作应重视以下几个方面。

### 1.5.1 摘要特点的充分体现

摘要的写作应反映以下特点：

(1) 完整性——摘要是完全独立的，它应包含论文中的要点，体现各"要素"的组合。

(2) 可读性——用通俗易懂的语言描述复杂的概念和问题。

(3) 科学性——用标准、精练的语言清晰地概述客观事实。

(4) 逻辑性——结构严谨，语意连贯、论点组织合理。

### 1.5.2 首句的撰写

摘要的首句往往表达研究的目的(subject or object).一个简单而有效的方法是开始句用"this paper","this study","this research".首句也可以很简洁地转入研究主旨。由于摘要常紧接标题，若标题已清晰表达出研究的目的，则首句不要重复或解释标题的内容。此时摘要可以直接从方法部分开始写。

### 1.5.3 方法与结果部分的撰写

摘要的"方法"部分的内容依赖于论文的创新性和重要性，重点是介绍新的重要材料与方法，如：新化合物、新设备、新分析方法和新合成方法等。若所用的方法是标准的，众所周知的，或可由方法的名称查到的，在摘要中就可以省略。

研究结果是最被关注的要素，因而常常占摘要的主要部分。摘要应概括论文中的主要和重要研究结果，结果的介绍一般用文字，对于论文中用图表表示的结果，可采用指示性叙述形式。若研究是由一系列实验组成，则尽可能将所得结果合并给出，因为分别给出会显得比较重复。

### 1.5.4 结论部分的撰写

虽然讨论(discussion)部分在论文的正文中很重要，但是在摘要中只需要结论(conclusion)部分，却不需要讨论。讨论一般不写入摘要。

结论通常采用直接陈述的方式。当结论无法很具体时，可采用"The results are discussed in comparison with earlier findings","A mechanism is proposed to explain the results","The implications of the data are discussed in detail"等句子表述研究结果。结论有时可以是对研究发现的解释或者是对有意义结果的建议。

摘要中各要素的顺序一般遵循论文中的次序，这样更有利于读者回想并理解全文的内容。但是，有时由于研究本身的特点或所关注的重点不同，使摘要中各部分的顺序发生变化。

对研究工作的建议和未来的计划一般不写入摘要，因为这两部分的内容不能提供更为有用的科技信息。

## 1.6 英语科技论文摘要实例及简析

### 1.6.1 报道性摘要

实例 1

(1) Formation of oxidation byproducts from ozonation of wastewater (From: Water Research, 2007(41):1481~1490)

Abstract: Disinfection byproduct (DBP) formation in tertiary wastewater was examined after ozonation ($O_3$) and advanced oxidation with $O_3$ and hydrogen peroxide ($O_3/H_2O_2$). $O_3$ and $O_3/H_2O_2$ were applied at multiple dosages to investigate DBP formation during coliform disinfection and trace contaminant oxidation. Results showed $O_3$ provided superior disinfection of fecal and total coliforms compared to $O_3/H_2O_2$. Color, UV absorbance, and SUVA were reduced by $O_3$ and $O_3/H_2O_2$, offering wastewater utilities a few potential surrogates to monitor disinfection or trace contaminant oxidation. At equivalent $O_3$ dosages, $O_3/H_2O_2$ produced greater concentrations of assimilable organic carbon (5%~52%), aldehydes (31%~47%), and carboxylic acids (12%~43%) compared to $O_3$ alone, indicating that organic DBP formation is largely dependent upon hydroxyl radical exposure. Bromate formation occurred when $O_3$ dosages exceeded the $O_3$ demand of the wastewater. Bench-scale tests with free chlorine showed $O_3$ is capable of reducing total organic halide (TOX) formation potential by at least 20%. In summary, $O_3$ provided superior disinfection compared to $O_3/H_2O_2$ while minimizing DBP concentrations. These are important considerations for water reuse, aquifer storage and recovery, and advanced wastewater treatment applications.

简析

摘要首先(用第 1 句话)介绍研究目的(subject or object),句子结构为"... was examined...".主要信息(研究的对象),由名词短语给出。然后,用第 2 句话介绍研究方法,句子结构为"... were applied at...dosages to investigate..."再后,摘要用较大篇幅介绍研究结果,结构信号为"Results showed..."摘要的最后(In summary, ...)提出了"结论"。该摘要全面反映了摘要的要素,重点突出,信息量大。

译文

**废水臭氧消毒中氧化副产物的形成**

摘要:考察了在废水的三级处理中用臭氧($O_3$)消毒和臭氧及过氧化氢($O_3/H_2O_2$)高级氧化后消毒副产物(DBP)的形成。(研究目的)以多剂量投加 $O_3$ 和 $O_3/H_2O_2$ 以研究对大肠杆菌消毒和微量污染物氧化中 DBP 的形成。(研究方法)结果显示,与 $O_3/H_2O_2$ 相比,$O_3$ 对粪便大肠杆菌和总大肠杆菌的消毒效果最佳。通过 $O_3$ 和 $O_3/H_2O_2$ 处理,色度、紫外吸收率和比紫外吸收率都降低,这给废水处理厂提供了一些可能的可监测消毒或微量污染物氧化的代用品。对于等量的 $O_3$ 投加量,$O_3/H_2O_2$ 与单独的 $O_3$ 相比,产生较高浓度的可同化有机碳(5%~52%)、醛(31%~47%)和羧酸(12%~43%),表明有机消毒副产物的形成主要取决于羟基的出现。当 $O_3$ 用量超过废水需要量时,溴酸盐就会形成。采用游离氯的实验室(小型)试验表明 $O_3$ 能减少至少 20%可能产生的总有机卤化物(TOX).(研究结果)总之,与 $O_3/H_2O_2$ 相比,$O_3$ 产生的消毒效果最佳而产生的 DBP 浓度最低。这些是水回用、蓄水层蓄水及恢复和废水高级处理应用中需重要考虑的问题。(结论)

实例 2

(2) Compositional and structural control on anion sorption capability of layered double hydroxides (LDHs) (From: Journal of Colloid and Interface Science, 2006(301): 19~26)

Abstract: Layered double hydroxides (LDHs) have shown great promise as anion getters. In this paper, we demonstrate that the sorption capability of a LDH for a specific oxyanion can be greatly increased by appropriately manipulating material composition and structure. We have synthesized a large set of LDH materials with various combinations of metal cations, interlayer anions, and molar ratios of divalent cation M(II) to trivalent cation M(III). The synthesized materials have then been tested systematically for their sorption capabilities for pertechnetate ( $TcO_4^-$ ). It is discovered that for a given interlayer anion ( either $CO_3^{2-}$ or $NO_3^-$ ) the Ni - Al LDH with a Ni/Al ratio of 3 :1 exhibits the highest sorption capability among all the materials tested. The sorption of $TcO_4^-$ on M(II) - M (III) - $CO_3$ LDHs may be dominated by the edge sites of LDH layers and correlated with the basal spacing d003 of the materials, which increases with the decreasing radii of both divalent and trivalent cations. The sorption reaches its maximum when the layer spacing is just large enough for a pertechnetate anion to fit into a cage space among three adjacent octahedra of metal hydroxides at the edge. Furthermore, the sorption is found to increase with the crystallinity of the materials. For a given combination of metal cations and an interlayer anion, the best crystalline LDH material is obtained generally with a M(II)/M(III) ratio of 3 :1. Synthesis with readily exchangeable nitrate as an interlayer anion greatly increases the sorption capability of a LDH material for pertechnetate. The work reported here will help to establish a general structure - property relationship for the related layered materials.

简析

摘要首先(用第1句话)介绍研究背景。然后，用第2句话清晰地说明了研究目的，句子结构为“In this paper, we demonstrate that...”.之后，用2句话介绍了研究方法(We have synthesized ... with ... have then been tested systematically for... ).再后，用大部分篇幅介绍研究结果，句子结构有“It is discovered that... Furthermore, ... is found to...”. 摘要的最后(用最后1句话)提出了“结论”，句子结构为“The work reported here will help to establish ... relationship for...”.该摘要全面反映了摘要的要素，研究结果表述充分。

译文

**双层氢氧化物(LDHs)对阴离子吸附力的成分和结构控制**

摘要：双层氢氧化物(LDHs)已显示出作为阴离子吸附剂的巨大希望。(研究背景) 本文证明了LDH对特定含氧阴离子的吸附能力可以通过适当控制材料的成分和结构得到显著提高。(研究目的)我们已合成了一批包含不同金属阳离子、层间阴离子，以及不同M(Ⅱ)与M(Ⅲ)摩尔比的LDH材料。对合成材料系统地测试了对高锝酸根($TcO_4^-$)的吸附能力。(研究方法)发现对于一定的层间阴离子($CO_3^{2-}$或$NO_3^-$)，Ni,Al比为3 :1的LDH的吸附能力在所测试的LDH中最强。$TcO_4^-$在M(Ⅱ)-M(Ⅲ)-$CO_3$型LDHs上的吸附可能受LDH层的边缘点位控制并与材料003面的基底间距有关，该吸附随M(Ⅱ)与M(Ⅲ)比的下降而上升。当层间距仅足以使高锝酸根阴离子进入笼空间(该笼处于边缘处三个相邻的八面体金属氢氧化物间)时，吸附达到最大值。此外，还发现吸附随材料的结晶度而上升。对于一定的金属阳离子、层间阴离子的组合，当M(Ⅱ)/M(Ⅲ)为3:1时，LDH的结晶最完善。以易交换的硝酸根为层间阴离子的合成大大增加了LDH材料对高锝酸根离子的吸附能力。(研究结果) 本文工作有助于建立相关的层状材料的结构与性能的一般关系。(结论)

实例3

(3) Persistent organic pollutants (POPs) in the conventional activated sludge treatment process: fate and mass balance(From: Environmental Research, 2005(97): 245~257)

Abstract: The fate and the mass balance of persistent organic pollutants (POPs) during the conventional activated sludge treatment process were investigated in the wastewater treatment plant of the city of Thessaloniki, northern Greece. The POPs of interest were 7 polychlorinated biphenyls and 19 organochlorine pesticides. Target compounds were determined at six different points across the treatment system: the influent, the effluent of the primary sedimentation tank, the effluent of the secondary sedimentation tank, the primary sludge, the activated sludge from the recirculation stream, and the digested/dewatered sludge. The distribution of POPs between the dissolved and the adsorbed phases of wastewater and sludge was investigated. A good linear relationship between the distribution coefficients, $K_d$, and the octanol - water partition coefficients, $K_{ow}$ of the solutes was observed only in raw wastewater, suggesting that other factors affect the phase distribution of organic compounds in treated wastewater. For all POPs, a significant increase in partitioning with a decreasing solids concentration was observed, revealing an effect from non-settling microparticles remaining in the "dissolved" phase during the separation procedure. A good linear relationship was also revealed between $\log K_d$ and the dissolved organic carbon (DOC) content of wastewater, suggesting that DOC favors the advective transport of POPs in the dissolved phase. Almost all POPs showed good mass balance agreements at both the primary and the secondary treatment. The losses observed for some species could be attributed to biodegradation/biotransformation rather than volatilization. The relative distribution between the treated effluent and the waste sludge streams varied largely among different compounds, with p-p-DDE being highly accumulated in the waste sludge (98%) but almost 60% of a-HCH remaining in the treated effluent.

简析

摘要首先(用第1句话)介绍研究目的，句子结构为"... were investigated...".然后，用3句话说明了研究方法，句子结构为"... of interest were... were determined at ... The distribution of ... was investigated...".之后，介绍了研究结果，句子结构有："... was observed, suggesting that...; ... was observed, revealing...; ... was also revealed, suggesting...", 采用-ing短语，使句子结构紧凑。摘要最后部分提出"结论"，句子结构有："... showed..., ... could be attributed to..., ... varied largely among..., with...".摘要的"方法"和"结论"部分叙述充分，反映研究工作特色。

译文

**在传统活性污泥处理过程中的持久性有机污染物(POPs)的归宿和质量平衡**

摘要：研究了希腊北部 Thessaloniki 市污水处理厂的传统活性污泥处理过程中持续性有机污染物(POPs)的归宿及质量平衡。(研究目的)有关的 POPs 是 7 种多氯联苯和 19 种有机氯杀虫剂。在处理系统的如下 6 个不同点测定了目标化合物：进水口、初沉池的出水口、二沉池的出水口、初沉池污泥、回流活性污泥和消化或脱水污泥。研究了在废水及污泥的溶解和吸附相间的 POPs 的分布。(研究方法)可见只有对原废水在分配系数 $K_d$ 与辛醇-水分配系数(溶质的 $K_{ow}$)之间存在良好的线性关系，表明其他因素影响了有机物在处理后的水中的相分布。可以看到，对所有的 POPs, 其分布随固体浓度的减少而显著增加，反映了在分离过程中存在于"溶解"相的非沉降性微颗粒物的影响。还可见在 $\log K_d$ 和废水的溶解有机碳(DOC)含量之间有着良好的线性关系，表明 DOC 有利于 POPs 在溶解相的对流输运。(研究结果)几乎所有的 POPs 在一级和二级处理过程中都显示出良好的质量平衡关系。某些 POPs 的损失可归因于生物降解作用或生物转化作用而不是挥发作用。(POPs)在处理过的出水和废污泥中的相对分布对于不同的化合物变化很大，其中以 p-p-DDE 在废污泥中的积累最高(98%)，而几乎有 60% 的 a-HCH 仍存在于处理过的出水中。(结论)

实例 4

(4) Carbon dioxide and ammonia emissions during composting of mixed paper, yard waste and food waste( From: Waste Management, 2006(26) 62~70)

Abstract: The objective of the work was to provide a method to predict $CO_2$ and $NH_3$ yields during composting of the biodegradable fraction of municipal solid wastes (MSW). The compostable portion of MSW was simulated using three principal biodegradable components, namely mixed paper wastes, yard wastes and food wastes. Twelve laboratory runs were carried out at thermophilic temperatures based on the principles of mixture experimental and full factorial designs. Seeded mixed paper (MXP), seeded yard waste (YW) and seeded food waste (FW), each composed individually, produced 150, 220 and 370g $CO_2$—C, and 2.0, 4.4 and 34g $NH_3$—N per dry kg of initial substrate, respectively. Several experimental runs were also carried out with different mixtures of these three substrates. The effect of seeding was insignificant during composting of food wastes and yard wastes, while seeding was necessary for composting of mixed paper. Polynomial equations were developed to predict $CO_2$ and $NH_3$( in amounts of mass per dry kg of MSW) from mixtures of MSW. No interactions among components were found to be significant when predicting $CO_2$ yields, while the interaction of food wastes and mixed paper was found to be significant when predicting $NH_3$ yields.

简析

摘要首先(用第 1 句话)明确说明研究目的，句子结构为“The objective of the work was to...”.然后，用 2 句话说明了研究方法，句子结构为“... was simulated using..., ... laboratory runs were carried out at... based on...”.之后，用 2 句话介绍了研究结果。摘要最后部分提出“结论”，句子结构有：“The effect of ... was insignificant, while... was necessary; ... equations were developed to predict; No interactions among ... were found to be significant, while... was found to be significant”.摘要的 4 要素反映明晰，可读性强。

译文

**堆肥处理混合废纸、庭院废物和食品废物过程中二氧化碳和氨的排放**

摘要：本工作的目的是提供一种预测城市固体废物(MSW)中可生物降解部分在堆肥处理中二氧化碳和氨产量的方法。(研究目的)MSW 中可堆肥处理的部分用如下三种主要的可生物降解组分模拟，即混合废纸、庭院废物和食品废物。根据混合物实验设计和全因子设计原理，在高温下进行了 12 个实验室试验。(研究方法)接种的混合废纸、接种的庭院废物和接种的食品废物，分别进行单独的堆肥处理，$CO_2$—C 的产量分别为：150g, 220g 和 370g $CO_2$—C/kg 初始基质干重，$NH_3$—N 的产量分别为：2.0g, 4.4g 及 34g $NH_3$—N/kg 初始基质干重。还对这三个基质的不同混合物进行了若干次实验。(研究结果)在食品废物和庭院废物的堆肥处理中，接种的影响是微不足道的，而在混合废纸的堆肥处理中，接种却是必要的。开发出了多项式方程，可预测 MSW 混合物的 $CO_2$和 $NH_3$产量(以每千克 MSW 干重产生的量计)。当预测 $CO_2$产量时，组分间的相互作用都是不重要的，而当预测 $NH_3$产量时，发现食品废物和混合废纸的相互作用是重要的。(结论)

实例 5

(5) Nutrient removal from slaughterhouse wastewater in an intermittently aerated sequencing batch reactor(From: Bioresource Technology, 2008(99): 7644~7650)

Abstract: The performance of a 10 L sequencing batch reactor (SBR) treating slaughterhouse wastewater was examined at ambient temperature. The influent wastewater comprised 4672 ± 952 mg chemical oxygen demand (COD)/L, 356 ± 46 mg total nitrogen (TN)/L and 29 ± 10 mg total phosphorus (TP)/L. The duration of a complete cycle was 8 h and comprised four phases: fill (7 min), react(393 min), settle (30 min) and draw/idle (50 min). During the react phase, the reactor was intermittently aerated with an air supply of 0.8 L/min four times at 50-min intervals, 50 min each time. At an infiuent organic loading rate of 1.2 g COD/(L· d), average efluent concentrations of COD, TN and TP were 150 mg/L, 15 mg/L and 0.8 mg/L, respectively. This represented COD, TN and TP removals of 96%, 96% and 99%, respectively. Phase studies show that biological phosphorus uptake occurred in the first aeration period and nitrogen removal took place in the following reaction time by means of partial nitrification and denitrification. The nitrogen balance analysis indicates that denitrification and biomass synthesis contributed to 66% and 34% of TN removed, respectively.

简析

摘要首先(用第 1 句话)介绍研究目的(subject or object),句子结构为“... was examined...”,主要信息(研究对象)由名词短语给出。然后,用 3 句话说明了研究方法,句子结构简单,信息完整。之后,用 2 句话介绍了研究结果。摘要最后 2 句话提出结论,句子结构为“... studies show that...; ... analysis indicates that...”.摘要的 4 要素反映明晰,句子结构简单,信息完整。

译文

**用间歇式曝气 SBR 去除屠宰场废水中的营养物**

摘要:考察了用 10L 的序批式反应器(SBR)在环境温度下对屠宰场废水的处理。(研究目的)入流废水的 COD 含量为(4672±952)mg/L, 总氮(TN)含量为(356±46)mg/L, 总磷(TP)含量为(29±10)mg/L.一个完整运行周期的时间为 8h, 由如下 4 个阶段组成:进水阶段(7min),反应阶段(393min),沉淀阶段(30min)和排水/闲置阶段(50min).在反应阶段,对反应器进行间歇曝气,曝气量为 0.8L/min,曝气 4 次,间隔 50min, 每次曝气持续时间 50min.(研究方法)当进水的有机负荷为 1.2gCOD/(L·d)时,出水的 COD, TN 和 TP 的平均浓度分别为 150mg/L, 15mg/L 和 0.8mg/L.这意味着 COD, TN 和 TP 的去除率分别为 96%,96%和 99%。(研究结果)阶段研究表明磷的生物吸收发生在第一个曝气阶段,氮的去除以部分硝化与反硝化的方式在随后的反应阶段发生。氮平衡分析表明反硝化作用和生物合成作用产生的 TN 去除率分别为 66%和 34%。(结论)

实例 6

(6) Thermal processing of paper sludge and characterization of its pyrolysis products(From: Waste Management, 2009(29): 1644~1648)

Abstract: Paper sludge is a waste product from the paper and pulp manufacturing industry that is generally disposed of in landfills. Pyrolysis of paper sludge can potentially provide an option for managing this waste by thermal conversion to higher calorific value fuels, bio-gas, bio-oils and charcoal.

This work investigates the properties of paper sludge during pyrolysis and energy required to perform thermal conversion. The products of paper sludge pyrolysis were also investigated to determine their properties and potential energy value. The dominant volatile species of paper sludge pyrolysis at 10℃/min were found to be CO and $CO_2$, contributing to almost 25% of the paper sludge dry weight loss at 500℃. The hydrocarbons ($CH_4$, $C_2H_4$, $C_2H_6$) and hydrogen contributed to only 1% of the total weight loss. The bio-oils collected at 500℃ were primarily comprised of organic acids with the major contribution being linoleic acid, 2, 4-decadienal acid and oleic acid. The high acidic content indicates that in order to convert the paper sludge bio-oil to bio-diesel or petrochemicals, further upgrading would be necessary. The charcoal produced at 500℃ had a calorific value of 13. 3 MJ/kg.

简析

摘要首先用2句话交代研究背景。然后用2句话说明了研究目的，句子结构为“This work investigates...; ... were also investigated...”.之后，用3句话介绍了研究结果，句子结构有“... were found to be..., contributing to...; ... were primarily comprised of ... with...”.摘要最后用2句话概括结论，句子结构为“... indicates that in order to..., further... would be necessary.”.与摘要4要素的一般规范相比，节省了研究方法，增加了研究背景，但总体来看，信息较完整。

译文

**造纸污泥的热处理及其热解产物的表征**

摘要：造纸污泥是造纸和纸浆制造业的废物，通常是经填埋处置的。造纸污泥的热解通过热转化形成较高热值的燃料，包括生物气体、生物油和木炭，为造纸污泥废物的管理可能提供一种选择(方案)。(研究背景)本文研究了造纸污泥在热解中的特性和热转化所需的能量。也研究了造纸污泥热解的产物以确定这些产物的特性和势能。(研究目的)在10℃/min的条件下，造纸污泥热解的主要挥发性产物为CO和$CO_2$，在500℃的条件下，由此产生的造纸污泥干重损失近25%。烃($CH_4$，$C_2H_4$，$C_2H_6$)和氢的产生仅使造纸污泥总重减少1%。在500℃时收集到的生物油主要为有机酸，主要是亚油酸，2,4-十二烯酸和油酸。(研究结果)高的酸含量表明要使来自于造纸污泥的生物油转化成生物柴油或石化产品，需要进一步的改进工作。在500℃产生的木炭的热值为13.3MJ/kg(结论)。

实例7

(7) Facile synthesis of ordered mesoporous silica with high $\gamma$-$Fe_2O_3$ loading via sol-gel process (From: J. Mater. Sci., 2008(43)6359~6365)

Abstract: A series of ordered mesoporous silica loaded with iron oxide was synthesized by facile one-step sol-gel route using Pluronic P123 as the template, tetraethylorthosilicate as the silica source, and hydrated iron nitrite as the precursor under acid conditions. The as-synthesized materials with Fe/Si molar ratio ranging from 0. 1 to 0. 8 were characterized by X-ray diffraction (XRD), transmission electron microscopy (TEM), vibrating sample magnetometry (VSM), and $N_2$ adsorption porosimetry. All samples possess ordered hexagonal mesoporous structure similar to SBA-15, with a high surface area, large pore volume, and uniform pore size. Although higher iron content causes a distortion of hexagonal ordering structure to some extent, the materials still maintain the ordered mesopore structure even with Fe/Si molar ratio as high as 0. 8. Pore structure and TEM data suggest that iron oxide nanoparticles are buried within the silica wall, and increasing the iron oxide loading has little effects on the pore structure of the mesoporous silica. VSM results show as-synthesized samples exhibit superparamagnetic behavior.

简析

摘要首先用 1 句话交代研究目的，兼有研究方法，句子结构为“... was synthesized by...”.然后用 1 句话进一步说明研究方法，句子结构为“... were characterized by...”.之后，用 2 句话介绍了研究结果。摘要最后用 2 句话概括结论，句子结构为“... suggest that...; ... results show...”.摘要完整体现了 4 要素，信息全面。

译文

**用溶胶-凝胶法简易合成负载高含量 $\gamma$-$Fe_2O_3$ 的有序介孔二氧化硅**

摘要：在酸性条件下以嵌段共聚物 P123 为模板剂、正硅酸乙酯为硅源、水合亚硝酸铁为前驱物，通过简易一步溶胶–凝胶法，合成了一系列负载三氧化二铁的有序介孔二氧化硅。(研究目的、研究方法)以上述方法合成的材料(Fe/Si 摩尔比在 0.1~0.8 的范围)通过 X 射线衍射(XRD)、透射电镜(TEM)、振动样品磁力测定(VSM)和氮吸附来表征。(研究方法)所有的样品都拥有与 SBA-15 相似的有序的六方介孔结构，并有高表面积、大的孔体积和均一的孔径。尽管较高的铁含量在一定程度上引起了六方有序结构的扭曲，但材料即使在 Fe/Si摩尔比高达 0.8 时，仍然保持有序的介孔结构。(研究结果)孔结构和 TEM 数据表明：三氧化二铁纳米微颗粒被包埋在二氧化硅骨架中，三氧化二铁负载量的增加对介孔二氧化硅孔结构的影响很小。VSM 结果显示合成的样品表现出超顺磁性能。(结论)

实例 8

(8) Combustion Emissions from Refining Lower Quality Oil: What is the Global Warming Potential? (from: Environ. Sci. Technol., 2010(44): 9584~9589)

The greenhouse gas emission intensity of refining lower quality petroleum was estimated from fuel combustion for energy used by operating plants to process crude oils of varying quality. (研究目的) Refinery crude feed, processing, yield, and fuel data from four regions accounting for 97% of U. S. refining capacity from 1999 to 2008 were compared among regions and years for effects on processing and energy consumption predicted by the processing characteristics of heavier, higher sulfur oils. (研究方法) Crude feed density and sulfur content could predict 94% of processing intensity, 90% of energy intensity, and 85% of carbon dioxide emission intensity differences among regions and years and drove a 39% increase in emissions across regions and years. Fuel combustion energy for processing increased by approximately 61 $MJ/m^3$ crude feed for each 1 $kg/m^3$ sulfur and 44 $MJ/m^3$ for each 1$kg/m^3$ density of crude refined. Differences in products, capacity utilized, and fuels burned were not confounding factors. (研究结果) Fuel combustion increments observed predict that a switch to heavy oil and tar sands could double or triple refinery emissions and add 1.6-3.7 gigatons of carbon dioxide to the atmosphere annually from fuel combustion to process the oil.(结论)

简析

文中已插入中文，说明要素性质。摘要首先(用第 1 句话)介绍研究目的，句子结构为“... was estimated...”.然后，用 1 句话说明了研究方法，句子结构为“... were compared among... for... predicted by...”，较多使用了介词短语，有简洁、明快的表达效果。之后，用 3 个句子介绍了研究结果。摘要最后提出“结论”，句子结构中通过采用宾语从句引出重要结论。本摘要清晰地体现了摘要“4 要素”，英文表达地道。

实例9

(9) Wood Biodegradation in Laboratory-Scale Landfills(from: Environ. Sci. Technol., 2011 (45): 6864~6871)

The objective of this research was to characterize the anaerobic biodegradability of major wood products in municipal waste by measuring methane yields, decay rates, the extent of carbohydrate decomposition, carbon storage, and leachate toxicity. (研究目的兼含研究方法) Tests were conducted in triplicate 8 L reactors operated to obtain maximum yields. (研究方法) Measured methane yields for red oak, eucalyptus, spruce, radiata pine, plywood (PW), oriented strand board (OSB) from hardwood (HW) and softwood (SW), particleboard (PB) and medium-density fiberboard(MDF) were 32.5, 0, 7.5, 0.5, 6.3, 84.5, 0, 5.6, and 4.6 mL $CH_4$ dry $g^{-1}$, respectively. The red oak, a HW, exhibited greater decomposition than either SW (spruce and radiata), a trend that was also measured for the OSB-HW relative to OSB-SW. However, the eucalyptus (HW) exhibited toxicity. (研究结果) Thus, wood species have unique methane yields that should be considered in the development of national inventories of methane production and carbon storage. The current assumption of uniform biodegradability is not appropriate. The ammonia release from urea formaldehyde as present in PB and MDF could contribute to ammonia in landfill leachate. Using the extent of carbon conversion measured in this research, 0-19.9%, predicted methane production from a wood mixture using the Intergovernmental Panel for Climate Change waste model is only 7.9% of that predicted using the 50% carbon conversion default. (结论)

简析

文中已插入中文，说明要素性质。摘要首先用明示的语句介绍研究目的并含有研究方法，句子结构为"The objective of this research was to characterize... by measuring...".然后，再用1句话交代研究方法，句子结构为" Tests were conducted in...".之后，用3个句子介绍了研究结果。摘要最后用较多篇幅 叙述"结论"，反映了研究结果的价值和研究人员的判断力。本摘要要素体现全面，特别是"结论"表达充分，英文表达地道。

实例10

(10) Carbon Dioxide Postcombustion Capture: A Novel Screening Study of the Carbon Dioxide Absorption Performance of 76 Amines(from: Environ. Sci. Technol., 2009(43): 6427~6433)

The significant and rapid reduction of greenhouse gas emissions is recognized as necessary to mitigate the potential climate effects from global warming. The postcombustion capture (PCC) and storage of carbon dioxide ($CO_2$) produced from the use of fossil fuels for electricity generation is a key technology needed to achieve these reductions. The most mature technology for $CO_2$ capture is reversible chemical absorption into an aqueous amine solution. (研究背景) In this study the results from measurements of the $CO_2$ absorption capacity of aqueous amine solutions for 76 different amines are presented. (研究目的) Measurements were made using both a novel isothermal gravimetric analysis (IGA) method and a traditional absorption apparatus. (研究方法) Seven amines, consisting of one primary, three secondary, and three tertiary amines, were identified as exhibiting outstanding absorption capacities. Most have a number of structural features in common including steric hindrance and hydroxyl functionality 2 or 3 carbons from the nitrogen. Initial $CO_2$ absorption rate data from the IGA measurements was also used to indicate relative absorption rates. Most of the outstanding performers in terms of capacity also showed initial absorption rates comparable to the industry standard monoethanolamine (MEA). (研究结果) This indicates, in terms of both absorption capacity and kinetics, that they are promising candidates for further investigation. (结论)

简析

文中已插入中文，说明要素性质。摘要首先用3句话介绍研究背景，句子结构简单，多用名词短语表达信息，背景的描述衬托出本研究的意义。然后，用1句话交代研究目的，句子结构为“In this study ... are presented”.之后，用1个句子介绍了研究方法。再后，用4个句子介绍了研究结果(并非数据)。摘要最后用1个句子概括了“结论”。本摘要要素体现充分，“背景”叙述得体，“结果”表达全面，英文表达地道。

实例11

(11) The use of Mechanical Alloying for the Preparation of Palladized Magnesium Bimetallic Particles for the Remediation of PCBs (from: Journal of Hazardous Materials, 2011(192): 1380~1387)

The kinetic rate of dechlorination of a polychlorinated biphenyl (PCB-151) by mechanically alloyed Mg/Pd was studied for optimization of the bimetallic system. (研究目的) Bimetal production was first carried out in a small scale environment using a SPEX 8000M high-energy ball mill with 4-μm-magnesium and palladium impregnated on graphite, with optimized parameters including milling time and Pd-loading. (研究方法) A 5.57- g sample of bimetal containing 0.1257% Pd and ball milled for 3 min resulted in a degradation rate of 0.00176 $min^{-1}\cdot g^{-1}$catalyst as the most reactive bimetal. (研究结果) The process was then scaled-up, using a Red Devil 5400 Twin-Arm Paint Shaker, fitted with custom plates to hold milling canisters. Optimization parameters tested included milling time, number of ball bearings used, Pd-loading, and total bimetal mass milled. (研究方法) An 85-g sample of bimetal containing 0.1059% Pd and ball-milled for 23 min with 16 ball bearings yielded the most reactive bimetal with a degradation rate of 0.00122 $min^{-1}\cdot g^{-1}$ catalyst. Further testing showed adsorption did not hinder extraction efficiency and that dechlorination products were only seen when using the bimetallic system, as opposed to any of its single components. The bimetallic system was also tested for its ability to degrade a second PCB congener, PCB-45, and a PCB mixture (Arochlor 1254); both contaminants were seen to degrade successfully. (研究结果)

Keywords: Mg/Pd bimetal, Polychlorinated biphenyls, Dechlorination, Mechanical alloying, Ball-milling

简析

文中已插入中文，说明要素性质。摘要首先开门见山，用1句话介绍研究目的，句子结构为“... was studied for...”.然后，用1句话交代研究方法，该句子结构简单、紧凑，句中含有2个“with 结构”。之后，用1个句子介绍了研究结果。再后，用2个句子再次说明研究方法。摘要最后用较多篇幅继续叙述研究结果，摘要中未见“结论”语句。本摘要突出说明“研究方法”和“研究结果”，重要信息表达充分，英文语句组织出色，英文表达地道。

实例12

(12) Application of a Peroxymonosulfate/Cobalt (PMS/Co(Ⅱ)) System to Treat Diseal-con-

taminated Soil (from Chemosphere, 2009(77): 1127~1131)

We investigated the feasibility of using peroxymonosulfate (PMS) with transition metals (PMS/$M^+$ system) for remediation of diesel-contaminated soils. (研究目的) To the best of our knowledge, this is the first attempt to apply a PMS/$M^+$ system for the treatment of diesel-contaminated soils. (研究背景) Two well-known transition metals, Fe(II) and Co(II), used to activate PMS including the effect of co-existence of counter anions ($Cl^-$ and $SO_4^{2-}$) were tested and it revealed that the most effective degradation of diesel was achieved with cobalt chloride. (研究方法与结果) The effect of PMS (i. e. 0~500 mM) indicated that the increasing the molar ratio of PMS/ diesel increased degradation of diesel on soils. The effect of Co(II) (i. e. 0~4 mM) showed that at least 2 mM of Co (II) was needed to degrade above 30% of diesel. Moreover, a maximum diesel degradation of 47% was achieved at a single injection of PMS/Co(II) (i. e. 500 mM/2 mM). Assessments of system pH showed that diesel degradation was higher under acidic conditions (pH 3) possibly due to the dissolution of metal ions from soils that are not possible at other pHs (pH 6 and 9). Sequential injections of both PMS and Co(II) were employed to improve the level of remediation (~90% degradation). The degradation of diesel increased as much as 88% when PMS/Co(II) was sequentially injected. (研究结果) This indicates that PMS/ Co(II) systems are applicable for remediation of soil contaminated with diesel fuel as an aspect of in situ chemical oxidation. (结论)

Keywords: Peroxymonosulfate/cobalt (PMS/Co(II)) system, Diesel fuel, Soils, Sulfate radical

简析

文中已插入中文，说明要素性质。摘要首先开门见山，用1句话介绍研究目的，句子结构采用主动语态 。然后，用1句话交代研究背景。之后，用1个句子介绍了研究方法兼有研究结果。再后用较多篇幅继续叙述研究结果，摘要最后一句为“结论”。本摘要要素体现全面，“研究结果”表达充分，“结论”明确、有力。英文表达规范。

实例13

(13) Supported cobalt oxide on MgO: Highly efficient catalysts for degradation of organic dyes in dilute solutions (from: Applied Catalysis B: Environmental, 2010(95): 93~99)

Cobalt oxide catalysts immobilized on various oxides (MgO, ZnO, $Al_2O_3$, $ZrO_2$, P25, SBA-15) were prepared for degradation of organic dyes in dilute solutions via a sulfate radical approach. (研究目的兼有研究方法) Their efficiency in activation of peroxymonosulfate (PMS) was investigated for the degradation of methylene blue (MB). (研究方法) Among the catalysts employed, the Co/MgO catalyst was found the most active. The complete degradation of MB occurred in <7 min when the Co/MgO catalyst with an optimum $Co_3O_4$ loading of 5 wt% was used. The performance of the Co/MgO catalyst is found better than both the homogeneous cobalt ions and heterogeneous $Co_3O_4$ catalyst. XPS analysis indicates that the surface of the MgO support is extensively covered by the hydroxyl groups. Hence, it is suggested that the alkaline MgO support plays several important roles in (i) dispersing the cobalt oxide nanoparticles well, (ii) minimizing the leaching of cobalt ions into the liquid phase, and (iii) facilitating the formation of surface Co-OH complex which is a critical step for PMS activation. Besides MB, other organic dyes such as orange II and malachite green, can also be degraded within a few minutes using the Co/MgO catalyst. (研究结果) It is believed that the highly efficient and environmentally benign Co/MgO catalyst developed in this work can be widely applied in advanced oxidation technologies towards degradation of organic pollutants.（结论）

简析

文中已插入中文，说明要素性质。摘要首先用1个句子介绍研究目的兼有研究方法。然后，用1句话涉及研究方法。之后，用大部分篇幅介绍“研究结果”，其中包括对重要实验结果的说明。摘要最后一句为“结论”。本摘要要素体现较全面，“研究结果”表达特别充分，“结论”表达明确和简洁。英文表达规范。

实例14：结构式摘要

(14) Evaluation of Spatial Relationships between Health and the Environment: The Rapid Inquiry Facility(from: Environ Health Perspect, 2010(118): 1306~1312)

BACKGROUND: The initiation of environmental public health tracking systems in the United States and the United Kingdom provided an opportunity to advance techniques and tools available for spatial epidemiological analysis integrating both health and environmental data.

OBJECTIVE: The Rapid Inquiry Facility (RIF) allows users to calculate adjusted and unadjusted standardized rates and risks. The RIF is embedded in ArcGIS so that further geographical information system (GIS) spatial functionality can be exploited or results can be exported to statistical packages for further tailored analyses where required. The RIF also links directly to several statistical packages and displays the results in the GIS.

METHODS: The value of the RIF is illustrated here with two case studies: risk of leukemia in areas surrounding oil refineries in the State of Utah (USA) and an analysis of the geographical variation of risk of esophageal cancer in relation to zinc cadmium sulfide exposure in Norwich (United Kingdom).

RESULTS: The risk analysis study in Utah did not suggest any evidence of increased relative risk of leukemia, multiple myeloma, or Hodgkin' s lymphoma in the populations around the five oil-refining facilities but did reveal an excess risk of non-Hodgkin' s lymphoma that might warrant further investigation. The disease-mapping study in Norwich did not reveal any areas with higher relative risks of esophageal cancer common to both males and females, suggesting that a common geographically determined exposure was unlikely to be influencing cancer risk in the area.

CONCLUSION: The RIF offers a tool that allows epidemiologists to quickly carry out ecological environmental epidemiological analysis such as risk assessment or disease mapping.

KEY WORDS: disease mapping, environmental epidemiology, geographical information systems (GIS), risk analysis, spatial epidemiology, tool.

简析

本摘要为一典型的“结构式摘要”，它在传统摘要“4要素”的基础上增加了“BACKGROUND(背景)”这一项。信息量大。该摘要中“目的”，“方法”，“结果”和“结论”等摘要要素表达明确，容易识别。句子结构精练，体现了典型的科技英语句子结构特征。

译文

**健康与环境空间关系的评价：快速调查系统**

背景：在美国和英国发起的环境公众健康跟踪系统提供了改善这样的技术和工具的机会，该技术和工具可用于综合健康和环境数据的空间流行病学分析。

目的：快速调查系统(RIF)使用户能计算调整和未调整的标准化数据和风险。RIF被包含在 ArcGIS 中以使进一步的地理信息系统(GIS)的空间功能性得以开发或结果能被输出至统计包为所需的进一步的特定分析所用。RIF 也直接连接到几个统计包并在 GIS 中显示结果。

方法：RIF 的价值在这里通过两个案例予以说明：在犹他州(美国)炼油厂周围地区的白血病风险，在 Norwich(英国)食管癌相关于锌隔硫化物暴露的风险随地理变化的分析。

结果：在犹他州的风险分析研究表明没有任何证据显示在五个炼油厂周围的人群中白血病、多骨髓瘤，或 Hodgkin 淋巴瘤的相对风险增加，但确实表明有过量的非 Hodgkin 淋巴瘤风险，这使得有理由作进一步的研究。在 Norwich 的疾病-地图标示研究表明没有任何地区有较高的食管癌(对男性和女性人群都是常见的)相对风险，这表明一般的由地理位置决定的暴露不可能影响地区中的患癌风险。

结论：RIF 提供了一个工具使流行病学家能迅速进行生态环境流行病学分析，如风险评价和疾病地图标示。

关键词：疾病地图标示，环境流行病学，地理信息系统(GIS)，风险分析，空间流行病学，工具

实例 15：结构式摘要

(15) Household Exposures to Polybrominated Diphenyl Ethers(PBDEs) in a Wisconsin Cohort (from: Environmental Health Perspectives 2009(117) 1890~1895)

Background: Human exposure to polybrominated diphenyl ethers (PBDEs) is virtually universal in the United States. Although the uses of these chemicals as flame retardants in fabrics, foams, and plastics are well defined, human exposure pathways are not well understood.

Objectives: This study was designed to assess current PBDE body burdens and identify residential sources of exposure among 29 men and 15 women in 38 households.

Methods: Portable X-ray fluorescence (XRF) analyzers were used to measure bromine levels in upholstered furnishings, bedding, vehicle interiors, and electronic devices. Vacuum cleaner contents, indoor air samples, and blood sera were analyzed for PBDE congeners using conventional gas chromatograph methods.

Results: Bromine levels varied widely within similar household items. The greatest range for upholstered items was found among vehicle seat cushions (7–30, 600 ppm). For electronic devices, television sets ranged from 4 ppm to 128, 300 ppm. Based on mixed effects modeling, adjusting for couple households, the bromine content in the participants' sleeping pillows and primary vehicle seat cushions were the strongest predictors of log lipid-adjusted blood serum PBDE concentrations ($p$-values = 0.005 and 0.03, respectively). The total pentaBDE congener levels found in dust samples and in passive air samples were not significant predictors of blood sera levels.

Conclusions: This study demonstrates the usefulness of the portable XRF analyzer in identifying household items that may contribute to human exposure to PBDEs.

Key words: bromine, passive air sample, PBDE, vacuum dust, X-ray fluorescence.

简析

本摘要为又一典型的“结构式摘要”，它在传统摘要“4 要素”的基础上增加了“BACKGROUND(背景)”这一项。“研究结果”介绍充分，“结论”表达简洁。该摘要中各要素表达明确，可读性强。句子结构精练，英语表达地道。

### 1.6.2 指示性摘要

实例 1

A Review of Chemical Warfare Agent Simulants for the Study of Environmental Behavior(from: Envir. Sci. & Tech, 2008(38): 112~136)

Abstract: There is renewed interest in the environmental fate of chemical warfare agents attributable to the intensified threat of chemical weapons use in a terrorist attack. Knowledge of processes that influence the fate of agents in the environment is important for development of disposal strategies and for risk and exposure assessments. However, it is often necessary to conduct studies examining chemical agent behavior using simulants due to the toxicity of the agents and usage restrictions. The objective of this study was to review the physical - chemical properties and mammalian toxicity of compounds that can be used to simulate chemical agents and to identify the most appropriate compounds to simulate specific environmental fate processes, including hydrolysis, sorption, bioavailability, and volatilization.

KEY WORDS: microbial degradation, partition coefficient, simulant, toxicity, warfare agent

简析

摘要开始的 3 句话交代有关背景。之后，用 1 句话简要说明研究目的(综述的内容)，句子结构为“The objective of this study was to review... and to identify…”。摘要体现了指示性摘要的特点。

译文

**综述：用于研究环境行为的化学战剂的模拟剂**

由于在恐怖袭击中使用化学武器的威胁增加，人们重新关注化学战剂在环境中的归宿。了解影响化学战剂归宿的过程对于开发处置策略和进行风险及暴露评价是重要的。然而，由于化学战剂的毒性和使用限制，使用模拟剂研究化学战剂的行为常常是必需的。本研究的目的是综述能用于模拟化学战剂的化合物的物理化学性质和对哺乳动物的毒性，并识别出能模拟具体环境归宿过程(包括水解、吸着、生物利用和挥发)的最适合的化合物。

关键词：微生物降解，分配系数，模拟剂，毒性，战剂

# Unit 2 Research Paper Writing
# 环境类英语科技论文的撰写

## 1 英语科技论文的分类和特点

科技论文是对创造性的科学研究成果进行理论分析和总结，从而揭示某些现象与问题的本质和规律的文章。

### 1.1 英语科技论文的分类

英语科技论文的分类与一般科技论文的分类相似，可有多种分类，在此，按写作目的和按论文内容进行分类。

#### 1.1.1 按写作目的分类

英语科技论文按写作目的可分为期刊论文、学位论文和会议论文等。

#### 1.1.2 按论文内容分类

按论文内容，英语科技论文可分为实验型、理论型和综述型论文三种。

(1) 实验型论文：通过科学实验得到事实，包括新材料、新方法或新工艺条件等，该类论文一般有相当量的数据，由这些数据可得出一些规律和结论。

(2) 理论型论文：可不涉及实验研究，典型的有数学类论文，而更多的是引用文献的实验结果，由此提出新规律、新模型、新计算方法等。该类论文常常可以预测未知系统的现象，有助于较深刻地认识自然规律。

(3) 综述型论文：是对某一主题已做研究的总结和评论。一般是通过文献调查的方法，得到大量有用资料，然后通过归纳、分析等方法，对已有成果作出客观和恰当的评价，包括指出长处与不足和今后的研究方向。该类论文的特点往往具有全面性、再现性和展望性的特点。

### 1.2 英语科技论文的特点

英语科技论文除语言特点之外，与一般科技论文相似。科技论文特别是期刊论文，要有鲜明的论点、充分的论据、严密的论证、确切的论断。因此，英语科技论文一般应体现科学性、创新性、逻辑性、规范性和简洁性。英语科技论文的语言特点是鲜明而丰富的，一篇好的英语科技论文，除了体现一般科技论文的特点外，要充分反映语言特点。

## 2 英语科技论文的主要组成部分

对实验型英语科技论文而言，最常见的是所谓的 IMRAD 结构，即：Introduction(前言)，Methods(方法)，Results and Discussion(结果与讨论)。IMRAD 结构的论文结构紧凑、叙述简明，并且逻辑性强，因此，这一结构现在被广泛采用。该结构论文在正文部分，一般为首先阐述研究的目的，然后，描述研究的方法、试验采用的材料和手段，再后为对结果、讨论的叙述，最后是研究结论的陈述。

# 3 英语科技论文的文体特点

英语科技论文既有与一般科技论文相似的总体特点又有其特有的语言特点。

## 3.1 总体特点

英语科技论文与一般科技论文相似，内容专业，语言文字正规、严谨，论文的结构一般已固定。论文侧重叙事和推理，具有很强的逻辑性，语意连贯，条理清楚，概念明确，判断恰当，推理严谨。从总体上看，英语科技论文有两大显著特点。

### 3.1.1 文体正式

作为严肃的书面文体，论文用词准确，语气正式，语言规范，行文严谨简练。一般不使用带有感情色彩的词句。论文较多使用主要为各种图和表的视觉表现手段(Visual Presentation)。

### 3.1.2 高度的专业性

论文都涉及一个专业范围，论文面对的读者均是本专业的科技人员，作为构成科技论文语言基础的术语，其语义具有严谨性和单一性等特点。

## 3.2 词汇特点

英语科技论文的词汇常常采用纯科技词、通用科技词、派生词、合成词和缩写词等，并多用单个动词，少用动词词组，分述如下。

### 3.2.1 纯科技词

主要指只用于某个学科或专业的词汇或术语。每个学科或专业都有一套含义明确、意义相对狭窄的词语。纯科技词的主要来源为希腊语和拉丁语。掌握特定领域的专业词汇对于阅读和写作英语科技论文很重要。随着新的发明与创造不断出现，新的专业术语也不断产生。

### 3.2.2 通用科技词

指不同专业都经常使用的词汇，但在不同的专业内可有不同的含义，或在同一专业中，又有多种不同的意义。通用科技词语义纷繁，用法灵活，搭配形式多样。

### 3.2.3 派生词

英语的构词法主要有合成、转化和派生三大手段。其中派生词，也就是加前、后缀构成的各种词，出现的频率远远高于其他方法构成的词。

### 3.2.4 合成词

一般指复合名词、复合形容词和复合动词。合成词表意直接、清楚明了，使语言简练，因此，在英语科技论文中经常出现。

### 3.2.5 缩写词

缩写词有短小、简练的优点。一般有如下特征：

(1) 从结构上看，由词组中的首字母构成。还有一种缩写词被称为拼缀词，即一个词失去部分音节，或者各个词都失去部分音节后，连成一个新词，它是复合词的一种缩略形式。

（2）从书写形式上看，有时全用大写，有时全用小写，有时大小写混用。

（3）从词意上看，缩写词往往同形异义，在不同学科专业中具有不同的意义。缩写词在文中第一次出现时要写出全称。

### 3.2.6 多用单个动词，少用动词词组

英语中有大量的词组动词，通常由动词+副词（介词）构成。这类动词使用方便，但意义灵活，因此，在英语科技论文中一般少用，而采用与之相对应的、意义明确的单个动词，这类动词具有多音节、语体庄重和正式的特点，大多源于法语、拉丁语和希腊语。

## 3.3 句子特点

英语科技论文的句子有以下特点。

### 3.3.1 较多使用被动结构

英语科技论文中可见大量使用被动结构。事物发生或形成的主体是人或物。科技论文注重叙事推理，强调客观准确。事物本身是句子叙述的要点，使用被动结构，把描述和研究的对象放在句子的主语位置上，能使其成为句子的焦点，同时省略了不必要的信息。使读者能获得客观重要的信息。下面是一个例子。

Degradation experiments in sediment-slurry phase were performed in 25mL glass centrifuge vials equipped with Teflon cap. Both PMS and PS were evaluated for 2-CB degradation. One gram of sediment spiked with PCB was mixed with 20 mL aqueous phase containing predetermined amount of iron (Fe(Ⅱ)) and/or oxidant (PMS or PS) to obtain 5% slurry. Control experiments using each component (oxidants, Fe(Ⅱ)) alone with spiked sediment were also performed in parallel to catalytic degradation experiments. After addition of oxidant and iron, vials were placed on a tumbler for mixing. At the end of each experiment (24 h runs) aqueous and sediment phases were separated by centrifugation and analyzed for 2-CB. In sediment-slurry experiments, 24 h of reaction time was selected to clearly distinguish between control and degradation experiments. Since most of the sediment degradation processes are desorption controlled, they require longer times (24 h selected here) as compared to aqueous degradation system(4 h).

这一段落为一篇环境类英语科技论文的实验部分的方法描述。该段落有 8 个句子、10 个谓语动词，其中 9 个动词为被动形式，说明了被动结构的高频率使用。

### 3.3.2 较多使用动词非谓语形式

动词非谓语形式是指分词、动词不定式和动名词。英语科技论文常使用非谓语动词形式有两个主要原因：(1)非谓语动词形式能使语言结构紧凑，行文简练，即以最少的篇幅表达最多的信息；(2)非谓语动词形式是对谓语动词的辅助和配合，英语每个简单句中，只能用一个谓语动词，如果同时表达几个动作，就必须选出主要动作动词作谓语，表达主要信息，而将其余动词作为非谓语动词，表达次要信息。下面可以举出一些例子：

（1）Maximum COD reduction observed was 88% at a cell current of 0.2 A (0.90 A/dm$^2$), compared to 82% and 60% at 0.3 A (1.35 A/dm$^2$) and 0.4 A (1.80 A/dm$^2$) respectively.

（句中采用过去分词作定语和过去分词短语作状语。）

（2）Increasing oxidant concentration at fixed Fe(Ⅱ) leads to increase in degradation efficiency until 1∶1 molar ratio of oxidant and Fe(Ⅱ) was achieved; further increase in oxidant concentration does not show any appreciable change in degradation efficiency.

（句中出现了分隔（割裂）现象，划线部分为动名词短语，在句中作主语。）

（3）Some tests require the measurement to be conducted in the stream since the process of obtaining a sample may change the measurement.

（句中采用动词不定式被动式作宾语补足语。）

（4）When indoor air pressure is reduced, pressure-driven radon entry is induced, increasing levels in the home instead of decreasing them.

（该句采用了现在分词短语作状语，表结果，该短语中又含一介词短语，其中动名词短语作介词宾语。）

（5）During the storage step, the role of the precious metals（Pd and Pt）is to oxidize NO to $NO_2$, which then reacts directly with the storage component, forming a mixture of nitrites and nitrates.

（该句用一动词不定式短语作表语，又采用现在分词短语作状语，表结果。）

（6）At a given temperature, increasing the weight hourly space velocity decreases the CO conversion.

（该句采用了动名词短语作主语，过去分词作定语。）

（7）Because of the possibility of some wastes not having the specific gravity of water, the ppm measure has been scrapped in favor of mg/L.

（该句采用了动名词复合结构作介词宾语。）

（8）The amount of perturbation a system is able to absorb without being destroyed is tied to the concept of the ecological niche.

（该句采用动名词被动式作介词宾语。）

（9）One of the best methods to reduce health risks from exposure to organic compounds is for residents or consumers to increase their awareness of the types of toxic chemicals present in household products.

（该句采用一动词不定式短语作定语，又采用动词不定式复合结构作表语。）

（10）The best way to reduce exposure to cigarette smoke in the house is to quit smoking and discourage smoking indoors.

（句中分别采用动词不定式短语作定语和表语。）

（11）The likelihood of an individual developing immediate reactions to indoor air pollutants depends on several factors, including age and pre-existing medical conditions.

（句中采用动名词复合结构作介词宾语，用分词短语作状语。）

（12）The OSHA will also make it mandatory for employers to keep accurate records of employee exposures to harmful agents that are required by safety and health standards.

（句中采用动词不定式复合结构作宾语。）

（13）The legislation provides a strong incentive for utilities to comply with the law and not exceed their allowances.

（动词不定式复合结构作定语。）

（14）The separation process can be facilitated by decreasing the particle size of refuse, thus increasing the number of particles and achieving a greater number of "clean" particles.

（现在分词短语在动名词短语中作结果状语。）

### 3.3.3 陈述句居多，动词时态运用有限，以一般时态为多

科技英语中，作者总是客观地陈述事理和问题，描写过程和状态，说明特性和功能，所述内容多具有一般性、频繁性和特征性。因此，在英语科技论文中以直接叙述的陈述句为多，而感叹和疑问句较少。句中的谓语动词也多采用“零时间(Timeless)”的“一般叙述(General Statement)”。一般现在时用来表述经常发生的并无时限的自然现象、过程和常规等，还用来表述科学定义、定理、方程式或公式的解说以及图表的说明等。一般过去时用来描述在确切的过去时间发生的事情。有专家统计，在科技英语中一般现在时和一般过去时占据时态使用75%以上。下面是几个例子：

(1) For hydrogen peroxide activation in Fenton type processes, iron was found to be the most effective transition metal with highest degradation efficiency at pH around 3.0 (Pignatello et al., 2006). The effectiveness of Fenton process decreases with increasing pH, mainly due to iron speciation and precipitation. In order to stabilize the iron in solution at near neutral pH, many chelating agents were employed (Sun and Pignatello, 1992). Among the chelating agents studied, ethylenediaminetetraacetic acid (EDTA) and nitrilotriacetic acid (NTA) were found effective in activating hydrogen peroxide at near neutral pH. However, the harmful effects of these two chelating agents have recently been highlighted and their applicability for environmental applications has been gradually reduced (Sillanpa a and Pirkanniemi, 2001). Therefore, there is a need for development of environmentally safe and highly effective chelating agents. This investigation aims at evaluating the effectiveness of three different classes (natural, inorganic and synthetic) of environmentally safe chelating agents on common oxidants, namely peroxymonosulfate (PMS), persulfate (PS), and hydrogen peroxide ($H_2O_2$).

(该段落中的动词时态为一般现在时、一般过去时和现在完成时。)

(2) Table 4 summarizes the results from 7-day experiments in which Fe(Ⅱ)/EDDS was used for activation of different oxidants. Fe(Ⅱ)/EDDS/$H_2O_2$ system showed very high reactivity towards 4-CP degradation. In addition to complete 4-CP degradation, overall 25% TOC removal was recorded at the end of 7 day, which is significant since the contribution of EDDS in total TOC was 5 times that of 4-CP. Continuous dissociation of $H_2O_2$ indicates that iron complexes of EDDS are very effective in $H_2O_2$ activation and subsequent 4-CP degradation. The iron chelation capacity of EDDS remained almost constant since almost 95% of iron was available in its soluble form after 7 days. Relatively small drop (w5%) in chelation power of EDDS, even with significant TOC removal (25%), indicates the contribution for TOC removal mainly came from the target contaminant (4-CP) since a significant EDDS degradation would have directly affected the iron chelation power of the system as observed in Fe(Ⅱ)/EDDS/PMS system.

(该段落中出现的动词时态主要为一般现在时和一般过去时。)

### 3.3.4 较多使用名词、名词短语和名词化结构

多用名词、少用动词，是正式英语文体的特点，但在科技英语中更为突出。这是因为科技文体是以事实为根据论述客观事物的，而名词和名词短语正是表达事物的词汇。名词短语是由几个名词或名词与其修饰语构成的名词性词组，包括“名词连用”的情况，即中心名词之前有一个以上其他名词，它们都为中心名词的前置修饰语。与名词短语密切相关的有名词化结构，名词化结构由表示动作或状态的名词与其修饰语构成，还可有起名词作用的非限定

动词(如动名词)与其修饰语构成。名词化结构与名词短语往往能表达一个句子所要表示的内容，因而使表达显现简洁、严密、客观和信息量大的效果。名词化结构的典型结构为：名词后跟 of 短语，引出动作的对象或发出者，该名词化结构中可以有表示地点、时间、原因、目的等通常修饰动词的修饰语，也可以有相当于动词所要求的补语等其他成分。有时，名词化短语中既有动作对象，又有动作的发出者，此时，动作对象用 of 短语表示，动作发出者用 by 短语表示。有时，of 短语表示动作发出者，而动作对象用 on，to 等介词短语来表示。有时，动作的发出者或对象可以用人称代词所有格、"名词+s"或直接用名词表示，放在动作意义的名词之前。

(1) Uncontrolled fast reaction between oxidant and catalyst, elevated temperatures due to exothermic nature of reaction, ineffective utilization of quickly generated hydroxyl radicals and inherent instability of hydrogen peroxide itself are few of the drawbacks of Fenton Reagent, especially when used for ISCO (in-situ chemical oxidation) remediation.

(句中采用 4 个并列的名词短语构成句子的主语，其中包含了丰富的信息。)

(2) The initial slight increase in Co concentration followed by a gradual drop is most probably due to the increase of the pH.

(句中采用名词短语作主语。)

(3) The advantage of Fe(Ⅲ)/PMS system includes: (i) slower radical generation, which facilitate better utilization of generated radials, and (ii) virtually no radical quenching by Fe(Ⅲ), which was predominant in Fe(Ⅱ)/PMS system, especially at higher Fe(Ⅱ) concentrations.

(句中采用 2 个并列的名词短语构成句子的宾语，其中的修饰语提供了丰富的信息。)

(4) Recently, Advanced Oxidation Technologies (AOTs) were successfully used in degradation of a suite of recalcitrant and non-biodegradable contaminants which are resistant to conventional treatment technologies.

(句中采用名词短语作介词宾语。)

(5) The reduced availability of active oxidizing species under high $H_2O_2$ dosages leads to lower COD reduction.

(句中采用名词短语作主语。)

(6) The uniqueness of the toxic substances problem lies in the potential transfer of a chemical to humans with possible attendant public health impacts.

(句中采用名词短语作主语，名词化结构作介词宾语。)

(7) The amount of alum required to achieve a given level of phosphorus removal depends on the amount of phosphorus in the water, as well as other constituents.

(名词短语在句中分别作主语和介词宾语。)

(8) The issue of the release of chemicals into the environment at a level of toxic concentration is an area of intense concern in water quality and ecosystem analyses.

(含名词化结构的名词短语作主语，名词短语作表语。)

(9) The advantages of sludge thickening in reducing the volume of sludge to be handled are substantial.

(名词短语在句中作主语。)

(10) Effluents from tall stacks are often injected at an effective height of several hundred feet

to several thousand feet above the ground because of the added effects of buoyancy and velocity on the plume rise.

（名词化结构在句中作介词宾语。）

（11）Public health is also threatened by infiltration of leachate from MSW disposal into groundwater，and particularly into drinking water supplies.

（名词化结构在句中作介词宾语。）

（12）Nonsmokers' exposure to environmental tobacco smoke is called "passive smoking", "second-hand smoking", and "involuntary smoking".

（名词化结构在句中作主语。）

（13）It is also important to note the location of air-cleaning device inlets in relation to the contaminant sources as an important factor influencing removal efficiencies.

（名词短语在动词不定式短语中作宾语。）

（14）Existing coal-fired power plants now burning higher-sulfur coals without scrubbers would be faced with the most stringent requirements for reducing emissions.

（名词短语在句中作主语。）

（15）The chemical engineers' familiarity with technologies such as spray drying，and activated carbon，membranes（reverse osmosis，ultra-and hyper-filtration），electrolysis，electrodialysis，ion exchange，heat recovery and liquid/solid separation all came in handy as the search continued for technologies to process wastes more economically than before.

（名词短语在句子中作主语。）

### 3.3.5 较多使用长句或复杂句子，但句子结构紧凑

英语科技论文一般描述较复杂的活动与关系。为了清晰、准确地表达科学研究中复杂的现象和现象间的密切联系，论文经常使用结构紧凑的长句和复杂句，如谓语动词简单，其余成分复杂；大量使用 be 和 have 的变化形式做谓语动词，又如使用不同的主从复合句，复合句中从句套从句等。如句子过于简单，会使逻辑关系连接不紧密，论文结构会显得单调和松散。下面是一些例句：

（1）Fe(Ⅱ)/PMS systems were found to be very effective in degrading PCB in a sediment-slurry system with more than 90% PCB removal being observed within 24h.

（句中采用"with 结构"等使句子结构紧凑。）

（2）Fe(Ⅱ) acts as a sulfate radical scavenger at higher concentrations indicating that there is an optimum concentration of Fe(Ⅱ) that leads to most effective degradation of the target contaminant.

（句中采用现在分词短语作状语和从句套从句等手段使句子结构紧凑。）

（3）Even though the rate of reaction for Reaction（4）is relatively lower than those of typical rates of sulfate radical attack on organic compounds，the excess concentration of oxidant will favor Reaction（4）as observed in Fenton system where presence of excess hydrogen peroxide showed detrimental effect on degradation efficiency.

（句中采用多个从句使句子结构虽长但逻辑关系紧密。）

（4）To our knowledge，this is perhaps the first study that documents the heterogeneous activation of peroxymonosulfate with cobalt，the best-known catalyst-activator for this inorganic peroxide.

（句中采用名词短语作同位语，定语从句等使句子结构紧凑 。）

(5) Although further increase in temperature results in a considerable increase in $CO_2$ conversion, reaching 66% at 275℃, the product distribution, as shown in the Fig. 4 inset, suggests that the majority of $CO_2$ was converted to $CH_4$ rather than CO indicating that methanation was more favored than the reverse water gas shift reaction above 200℃.

(句中采用主从复合句结构，内含多个从句和现在分词短语，使句子结构紧凑，信息量大。)

(6) The fact that the $O_2$ and CO signals decrease in an identical manner while the $CO_2$ signal appears as the mirror image of the two suggest that the predominant reaction is the selective oxidation of CO to form $CO_2$, though there appears to be some water formation.

(句中采用多个从句，通过从句表达丰富信息，使句子结构紧凑。)

(7) Though stagnant rivers, smoggy skies, and unsightly dumps were aesthetically displeasing to the citizens of overcrowded cities of earlier centuries, no attempt was made to reverse the negative impact humans had on their environment until it became evident that heavily polluted water, air, and soil could exert an equally negative impact on the health, the aesthetic and cultural pleasures, and the economic opportunities of humans.

(采用复合句结构，内含多个从句和平行结构，使句子虽长但结构紧凑，逻辑联系紧密。)

(8) This method is most frequently used by industry, although it is not preferred by environmentalists because it creates other wastes while reducing emissions.

(句中采用主从复合句、从句套从句等手段使句子结构紧凑。)

# 4 英语科技论文语言表达的基本规范和技巧

为了使英语科技论文体现出科学性、准确性和客观性，其英语表达在措辞和句法结构等方面有一些基本规范和技巧。可从以下几方面来认识和掌握这些基本规范和技巧。

## 4.1 表达准确

准确性是科技论文的基本特征与要求。要做到表达准确应注意以下几点：

### 4.1.1 用词恰当

(1) 正确选词

词汇是语言的基本单位，论文只有选词正确，才能准确地表达意义，传递正确的信息。要选词正确，应区别易混淆的词语，对一组意义相近的词，应选精确度高的词，不使用模棱两可的词汇。

(2) 用含义具体的词

英语科技论文的表达中名词往往有具体与抽象之分。要尽量使用具体词，它能使语言形象、实在。抽象名词的使用往往伴随例证以使其意义具体化。

(3) 少用同义词

科技论文以客观直接的方式揭示事物的本质，较多或不适当地使用同义词会使人产生误解，应当避免。

(4) 不用歧义词语

为了避免对英语表达产生理解的不确定性，应不用歧义词语。在英语科技论文中代词使用频率高，用来代替上文中出现的名词。但必须注意指代清楚，对于易产生理解错误的指代，应重复名词，而不用代词。标点符号的使用不当也易造成歧义，应当正确使用。句子结构安排不当也会造成表达的歧义，所以必须妥善安排。

#### 4.1.2 叙述详略适度

科技论文既要有概述，又要有细节描写。细节描写要做到：(1)体现表达的准确性；(2)提供新信息；(3)突出重点。

## 4.2 表达简洁

科技论文尤其强调表达的简洁，简洁能节省篇幅，使论文易读。要做到简洁，可采取以下方式。

#### 4.2.1 避免赘词和不必要的复杂句子结构

要避免可有可无的词。最常见的赘词是形容词和副词。副词如使用不当，还会削弱其所修饰的词语的意义。避免不必要的复杂句子结构也有利于表达的简洁，一般的准则是：能用简单结构就不用复杂结构。

英语科技论文中常见的冗词列举如下(括号中的词应该省去)：

(already) existing, at (the) present (time), (completely) eliminate, (currently) being, (empty) space, introduced (a new), never (before), now (at this time), start (out), (alternative) choices, (basic) fundamentals, (continue to) remain, (currently) underway, had done (previously), mix (together), none (at all), period (of time), (separate) entities, (still) persists

#### 4.2.2 避免无意义的词语和结构

在英语科技论文中要避免对表达不起作用的词语与结构。因为它们占据了篇幅，没有给出新的信息，还可能产生误导。

可列举出一些常在句中出现的无意义的词语：

as a matter of fact, it should be pointed out that, I might add that, the course of, it is noteworthy that, the fact that, it is significant that, the presence of

#### 4.2.3 用短语替代从句

英语科技论文要做到语言简洁，用短语替代从句是常用的手段。短语类型如下所述：

1) 用名词短语(名词化结构)替代从句

用名词短语可替代时间、条件和原因状语从句，也可替代定语从句和名词性从句等，使复合句变成简单句。名词短语在句子中可作主语、表语、宾语、介词宾语、同位语和定语等。例句如下：

(1) A comparison of the results of the present investigation with those reported in the literature showed that rice husk biomatrix is comparable with other adsorbents.

(If the results of the present investigation were compared with those reported in the literature, it was shown that rice husk biomatrix is comparable with other adsorbents.)

(名词短语替代条件状语从句，在句中作主语。)

(2) We have discussed the passage of an electric current through liquid solutions of acids, bases and salts.

(We have discussed that an electric current passes through liquid solutions of acids, bases and salts.)

(名词短语替代宾语从句，在句中作宾语。)

(3) These sparse data have demonstrated that further studies on the application of ozonation for the degradation of NPs, as a group of quite bio-resistant toxic compounds are required.

(名词短语作介词宾语。)

(4) The release of pollutants into the atmosphere is a traditional technique for disposing of them.

(名词化结构作主语。)

(5) Dilution of air contaminants in the atmosphere is also of prime importance in the prevention of undesirable levels of pollution.

(名词化结构在句中作主语。)

(6) The effects of lead exposure on fetuses and young children include delays in physical and mental development, lower IQ levels, shortened attention spans, and increased behavioral problems.

(名词化结构在句中作主语。)

(7) These short-term effects are usually treatable by some means, oftentimes by eliminating the person's exposure to the source of pollution.

(名词化结构在动名词短语中作宾语。)

(8) The discharge of hazardous substances into the sea or into lakes and rivers often kills fish and other aquatic life.

(名词化结构在句中作主语。)

(9) The dependence of the steady-state conversion of NO to $NO_2$ on the monolith temperature and Pt dispersion are shown in Fig. 2.

(名词短语在句中作主语。)

用名词短语作为前面整个句子或句中一部分的同位语，以避免用一个结构显得松散的并列句或一个非限制性定语从句。其模式为：

(1) ……句子，a (an)+形容词+名词。

(2) ……句子，a (an)+名词+后置定语(过去分词短语，定语从句，形容词短语，同位语从句)。

例句有：

(1) Before such separation can be achieved, however, the material must be in separate and discrete pieces, a condition clearly not met by most components of mixed refuse.

(名词短语作句子同位语。)

(2) The legislation calls for historic reductions in sulfur dioxide emissions from the burning of fossil fuels, the principal cause of acid rain.

(名词短语作为前面句中 sulfur dioxide emissions from the burning of fossil fuels 的同位语。)

(3) The third modification is contact stabilization, or biosorption, a process in which the sorption and bacterial growth phases are separated by a settling bank.

(名词短语作 contact stabilization, or biosorptio 的同位语。)

(4) Carbon dioxide comprises only a very small portion of the atmosphere, a little more than 0.03% by volume.

(名词短语作 a very small portion of the atmosphere 的同位语。)

(5) This usually involves longer detention times in secondary treatment, during which bacteria such as *Nitrobacter* and *Nitrosomonas* convert ammonia nitrogen to $NO_3^-$, a process called *nitrification*.

(名词短语作句子“ bacteria such as *Nitrobacter* and *Nitrosomonas* convert ammonia nitrogen to $NO_3^-$”同位语。)

用一个名词短语放在主语前作主语的同位语，以避免用一个系(词)表(语)结构的句子，使整个句子的结构显得严谨、紧凑和平衡。例句有：

(1) An instrument for measuring dissolved oxygen in a solution, the dissolved oxygen electrode (a probe) is widely used in the water quality analyses.

(2) An air pollution control device, the electrostatic precipitator is able to treat some exhaust gases.

采用表示尺寸、大小等的名词短语作为后置定语，充当一个定语从句的作用。例句有：

Oxygen has a mass about 16 times the mass of a hydrogen atom.

(表示质量大小的名词短语作后置定语。)

2) 用分词短语(包括分词独立结构)替代从句

分词短语可以作状语，放在句首通常可用来表示时间、条件、原因及对主语的附加说明等，放在句末常作附加说明(伴随情况)，或表示结果、方式等，其功用是代替状语从句或并列分句。分词短语还可作后置定语，起定语从句的作用。分词独立结构可放在句首，更常见放在句尾，用作附加说明(即伴随情况)等。例句有：

(1) In urban areas each person produces between five and eight pounds of solid wastes each day, excluding junked automobiles and appliances or industrial solid wastes.

(现在分词短语，状语，作附加说明。)

(2) The residual inorganics then become the building blocks for new life, using the sun as the source of energy.

(现在分词短语，状语，作附加说明。)

(3) This gas is about 60% methane and burns readily, usually being used to heat the digester and answer additional energy needs within a plant.

(现在分词短语(被动式)，状语，作附加说明。)

(4) The cover of the secondary digester often floats up and down, depending on the amount of gas stored.

(现在分词短语，状语，作附加说明。)

(5) Thickening also implies that the process is gravitational, using the difference between particle and fluid densities to achieve greater compacting of solids.

(现在分词短语，状语，作附加说明。)

(6) The wastewater is forced through a semipermeable membrane which acts as a superfilter, rejecting dissolved as well as suspended solids.

(现在分词短语，状语，作附加说明。)

(7) The mice, in turn, spend more time in burrows, thus not eating as much, and allowing the grass to reseed for the following year.

(现在分词短语，作结果状语。)

(8) Having defined the local picture of the Pt/BaO catalyst we can now interpret the storage results in more detail.

(现在分词(完成式)短语，作时间状语。)

(9) This knowledge also went a long way toward initiating better house-keeping procedures within plants, identifying opportunities for raw materials substitution and production process modifications, all resulting in reduced water and waste quantities.

(分词独立结构，作状语，作附加说明。)

(10) The volatile solids parameter is in fact often interpreted as a biological characteristic, the assumption being the VSS is a gross measure of viable biomass.

(分词独立结构，作状语，作附加说明。)

(11) This suggests that $H_2$ is consumed by species other than stored $NO_x$, the obvious species being chemisorbed oxygen, a product of NO decomposition.

(分词独立结构，作状语，作附加说明。)

(12) Consider a situation depicted in Figure 7-2, where two wastewater treatment options are shown—a centralized treatment plant, and several smaller plants— all discharging their effluents into the same river.

(分词独立结构，作状语，作附加说明。)

(13) In countries without such a procedure, guidelines rarely exist, the decision being dependent upon the scale of the proposal, its environmental setting, uniqueness of the project and the likely degree of public opposition.

(分词独立结构，作状语，作附加说明。)

3) 用介词短语替代从句

介词短语在英语科技论文中广泛使用，具有很强的表现力，其主要特点之一就是它能使句子结构简单、精练，具有从句的功能。如，可用介词短语替代时间状语从句和条件状语从句。其中，由 with 构成的“with 短语”和“with 结构”具有特殊的意义。下面是一些例子：

(1) At a temperature greater than 80℃,

(When (or If) the temperature is greater than 80℃,)

(2) In our study of biological wastewater treatment,

(When we studied biological wastewater treatment,)

(3) Because of the inexpensive and high adsorption properties resulted from their ion exchange capabilities,

(Because they are inexpensive and have high adsorption properties resulted from their ion exchange capabilities,)

(4) For a given resistance,

(If the resistance is given,)

(5) Despite their high costs, noble metal catalysts are by far the most studied catalysts in the CO oxidation literature.

(6) In spite of the established importance of the Pt-BaO coupling, the effect of the Pt particle size and dispersion has not been studied in detail for NOx storage and reduction.

(7) After calcination at 500℃ in air, the alumina contained 3.35 wt.% Pt determined by ICP.

(8) For the experiments with an aerobic regeneration feed, the monolith temperature was measured at the front end of the monolith (approximately 0.2 cm from the inlet) where the temperature rise is the largest.

在介词短语中，"with 短语"较为常见。处在句首时，常表示"对于；有了；在……情况下"等的意思，作状语。处在句尾时，常表示结果、附加说明等，也常作状语。类似地，也有"without 短语"。"with (without)短语"也可作后置定语，代替定语从句。例句有：

(1) Table 15-2 summarizes a few specific chemicals that are of special interest in evaluating water quality, with brief descriptions of the problems that they may cause.

("with 短语"，作状语，作附加说明。)

(2) During this time, the focus was entirely on water quality with little concern for the environmental aspects.

("with 短语"，作状语，作附加说明。)

(3) With both excess and stoichiometric $O_2$ concentrations, the CO conversion rate increases with temperature until reaching a maximum at 175℃.

("with 短语"，作条件状语。)

(4) An initial analysis of the process parameters has been made with satisfactory results.

("with 短语"，作结果状语。)

(5) Without free dissolved oxygen, streams and lakes become uninhabitable to most desirable aquatic life.

("without 短语"，作条件状语。)

(6) The latter must be further treated in the plant, but the sand can be dumped as fill without undue odor or other problems.

("without 短语"，作结果状语。)

(7) The objective of a grit chamber is to remove sand and grit without removing the organic material.

("without 短语"，作附加说明。)

(8) It is an inexpensive chemical with excellent bonding characteristics that is produced in high volume throughout the world.

("with 短语"作 chemical 的后置定语。)

在英语科技论文中"with 结构"也广泛使用。其构成为：with (或 without)+名词或代词+分词(短语)、介词短语、形容词(短语)、副词、不定式(短语)、名词(短语)。即，with 后接一个较复杂的结构，也可称为 with 后带"主谓"关系(即由"介词宾语+宾语补语"构成复合宾语)。有人称其为无动词从句，是一个表现力十分丰富的结构，应用广泛。其功用包括作状语，处于句首表示条件、时间和原因等；处于句尾表示结果，附加说明、方式和条件等。也作后置定语，充当定语从句的作用。下面是一些例句：

(1) With the catalyst added into the solution, the oxidizing reaction will be accelerated.

("with+名词+过去分词短语"，作条件状语。)

(2) With friction present, the object has been moved very slowly.

("with+名词+形容词"，作原因状语。)

(3) With the key open, the water in the tank begins to flow.

（“with+名词+形容词”，作时间状语。）

（4）The activated sludge process is a continuous operation，with continous sludge pumping the clean water discharge.

（“with+名词+现在分词短语”，作附加说明。）

（5）The entire food web，or ecosystem，stays in dynamic balance，with adjustments being made as required.

（“with+名词+现在分词(被动式)短语”，作附加说明。）

（6）The storage of $NO_x$ has been studied in great detail with several mechanisms being proposed.

（“with+名词+现在分词(被动式)”，作结果状语。）

（7）Temperature has a pronounced effect on oxygen uptake（usage），with metabolic activity increasing significantly at higher temperature.

（“with+名词+现在分词短语”，作附加说明。）

（8）The logical extension of this idea is to not have any treatment plants at all，but dispose of the wastewater onsite，with each house or building having its own treatment system.

（“with+名词+现在分词短语”，作附加说明。）

（9）The primary clarifier seems to act like a viral and bacteriological concentrator，with a substantial fraction of these microorganisms existing in the sludge instead of the liquid effluent.

（“with+名词+现在分词短语”，作附加说明。）

（10）The pressure of a gas varies inversely as its volume，with temperature being constant.

（“with+名词+现在分词短语”，作条件状语。）

（11）Tubular precipitators consist of cylindrical collection tubes with discharge electrodes located along the axis of the cylinder.

（“with+名词+过去分词短语”，作后置定语。）

（12）The production of $N_2O$ was comparatively low for the three catalysts with the most $N_2O$ produced（ca. 35 ppm）at 125℃ for D50.

（“with+名词+过去分词短语”，作附加说明。）

（13）For streams and rivers with travel times greater than about five days，the ultimate demand for oxygen must include the nitrogenous demand.

（“with+名词+形容词短语”，作后置定语。）

（14）To evaluate its effect，experiments were conducted with pH varying in the range of 8-12.

（“with+名词+现在分词短语”，作条件状语。）

（15）The“Shambles”is both a street and an area in London，and during the eighteenth and nineteenth centuries was a highly commercialized area，with meat packing as a major industry.

（“with+名词+介词短语”，作附加说明。）

（16）Biological sludges are of course continually changing，with the greatest change occurring when the sludge changes from aerobic to anaerobic（or vice versa）.

（“with+名词+现在分词短语”，作附加说明。）

（17）SVI values below 100 are usually considered acceptable，with SVI greater than 200 defined as badly bulking sludges.

（“with+名词+过去分词短语”，作附加说明。）

（18）Numbers can then be assigned to the interaction, with 1 being a small and 5 being a large magnitude or importance, and these placed in the blocks with magnitude above and importance below.

（前一“with+名词(代词)+现在分词短语”，作附加说明，后一“with+名词+副词”，作方式状语。）

（19）Areas with acid-neutralizing compounds in the soil, for example, can experience years of acid deposition without problems.

（“with+名词+介词短语”，作后置定语。）

（20）LDHs are layered materials with positively charged metal hydroxide sheets in brucite layers, balanced by exchangeable charge-compensating anions and water molecules, which are present in the interlayer space.

（“with+名词+介词短语”，作后置定语。）

（21）In this representation formaldehyde（HCOH）and oxygen are produced from carbon dioxide and water, with sunlight the source of energy.

（“with+名词+名词短语”，作条件状语。）

（22）If lead and silver electrodes are put in an electrolyte solution with a microammeter between, the reaction at the lead electrode would be…

（“with+名词+副词”，作附加说明。）

（23）As discussed above, plants（producers）use inorganic chemicals as nutrients and, with sunlight as a source of energy, build high-energy molecules.

（“with+名词+介词短语”，作条件状语。）

（24）Species interaction is common, with the waste product of one species often forming the food supply of another.

（“with+名词+现在分词短语”，作附加说明。）

（25）The mechanism of adsorption on activated carbon is both chemical and physical, with tiny crevices catching and holding colloidal and smaller particles.

（“with+名词+现在分词短语”，作附加说明。）

（26）An activated carbon column is a completely enclosed tube with dirty water pumped up from the bottom and the clear water exiting at the top.

（“with+名词+过去(现在)分词短语”，作后置定语。）

（27）Removal is often continuous, with clean carbon being added at the top of the column.

（“with+名词+现在分词(被动式)短语”，作附加说明。）

（28）Note that this is a staged process, with the solution of organics by extracellular enzymes being followed by the production of organic acids by a large and hearty group of anaerobic microorganisms known, appropriately enough, as the *acid formers*.

（“with+名词+现在分词(被动式)短语”，作附加说明。）

（29）Properly designed units can be constructed and operated with a collection efficiency as high as 90 percent for particulates in the 5 to 10 micron range.

（“with+名词+形容词短语”，作结果状语。）

(30) Particles are collected on a flat parallel collection surface spaced 8 to 12 inches apart, with a series of discharge electrodes located along the centerline of the adjacent plates.

("with+名词+过去分词短语", 作后置定语 。)

(31) The particles settle toward the collection surface at the bottom of the unit with a velocity at or near their settling velocity.

("with+名词+介词短语", 作方式状语。)

(32) Studies have shown concrete to have the highest radon content when compared to all other building materials, with wood having the least.

("with+名词+现在分词短语", 作附加说明。)

(33) Data taken from various states suggest an average indoor radon-222 concentration of 1.5 pCi/L (picoCuries per liter, a concentration of radiation term), and approximately 1 million homes with concentrations exceeding 8 pCi/L.

("with+名词+现在分词短语", 作后置定语。)

(34) Air cleaning systems use high efficiency filters or electronic devices to collect dust and other airborne particles, some with radon products attached to them.

("with+名词+过去分词短语", 作后置定语。)

(35) Respirable suspended particles (generally less than 10 micrometers in diameter), can settle on the tissues of the upper respiratory tract, with the smallest particles (those less than 2.5 micrometers) penetrating the alveoli, the small air sacs in the lungs.

("with+名词+现在分词短语", 作附加说明。)

(36) This apparent discretion has proved to be one of the major areas of difficulty of interpretation, with conflicts of approach between the EU Commission and some Member States.

("with+名词+介词短语", 作附加说明。)

(37) In the United States, scoping is a part of the EIA and is the responsibility of the project proponent, although this is done in close consultation with the federal agency to which the ElS will be submitted, as well as other agencies with a contribution to make to the assessment.

("with+名词+介词短语", 作后置定语。)

(38) As described in the previous chapter, a pollution prevention assessment is a systematic, planned procedure with the objective of identifying methods to reduce or eliminate waste.

("with+名词+动词不定式短语", 作后置定语。)

(39) As the ecological zones shift polewards, there may be a decrease in the amount of area suitable for forests, with a corresponding increase in grasslands and deserts.

("with+名词+介词短语", 作附加说明。)

(40) It may be short, turning hazardous into nonhazardous substances soon after they are released, or it may continue indefinitely with pollutants posing potential health risks over a long period of time.

("with+名词+现在分词短语", 作结果状语。)

(41) One should note that a product and/or service is usually conceived to meet a specific market need with little thought given to the manufacturing parameters.

("with+名词+过去分词短语", 作附加说明。)

4）用形容词短语替代从句

形容词短语由“形容词+介词短语、不定式或状语从句”构成，其功能是作后置定语和状语。作状语，在句首时主要表示原因、条件、让步、对主语的附加说明或对全句的评述，在句尾时，可表示附加说明或对前面句子的评述，更多的是作方式状语。例句有：

（1）Something as small as a worm may be composed of millions of cells.

（形容词短语，作后置定语。）

（2）Small in size and low in price, this device is warmly received by users.

（形容词短语，作原因状语。）

（3）Free from the attack of moisture, a piece of iron will not rust very fast.

（形容词短语，作条件状语。）

（4）A phenomenological picture is proposed that describes the effects of Pt dispersion consistent with the established spatio-temporal behavior of the lean $NO_x$ trap.

（形容词短语，作后置定语。）

（5）From that point the test is colorimetric, using a chemical which when combined with phosphates produces a color directly proportional to the phosphate concentration.

（形容词短语，作后置定语。）

（6）The indicator most often used is a group of microbes called *coliforms* which are organisms normal to the digestive tracts of warm-blooded animals.

（形容词短语，作后置定语。）

（7）The design involves choosing a desired thickened sludge solids concentration, Cu (estimated in laboratory tests) and drawing a line from this value tangent to the underside of the flux curve.

（形容词短语，作后置定语。）

（8）Stacks discharging to the atmosphere have long been one of the methods available to industry for disposing waste gases.

（形容词短语，作后置定语。）

（9）Smaller charged particles pass close to the surface of either the packing material or a scrubbing water droplet.

（形容词短语，作方式状语。）

（10）Population groups at special risk of detrimental effects of CO exposure include fetuses, persons with existing health impairments (especially heart disease), persons under the influence of drugs, and those not adapted to high altitudes who are exposed to both CO and high altitudes.

（形容词短语，作后置定语。）

（11）Several radiation protection groups have approximated the number of annual lung cancer deaths attributable to indoor radon.

（形容词短语，作后置定语。）

（12）This is one of the primary factors responsible for the dramatic increase in landfill tipping fees.

（形容词短语，作后置定语。）

（13）However, a basic disadvantage inherent in any waste processing or recovery system is the additional cost of constructing and operating the facility.

（形容词短语，作后置定语。）

（14）A panel of experts is assembled，responsible for the development of guidelines for the EIA and reviewing the environmental report submitted by the project proponent.

（形容词短语，作状语，作附加说明。）

（15）If a pollution prevention assessment reveals confidential data pertinent to a company's product，fear may exist that the organization will lose a competitive edge with other businesses in the industry.

（形容词短语，作后置定语。）

5）用动词不定式短语替代从句

（1）Both require a large Pt/BaO interface in order for the rate of regeneration to increase.

（动词不定式复合结构作目的状语。）

（2）During the regeneration step，the role of the precious metals（primarily Pt and Rh）is to selectively reduce the stored $NO_x$ to $N_2$.

（动词不定式短语作表语。）

（3）The time for the NO oxidation reaction to reach a steady-state was dependent on the time needed for the barium storage phase to become saturated with $NO_x$.

（两个都为动词不定式复合结构，在句中分别作定语和状语。）

（4）These abilities have made it possible for human population to thrive and flourish beyond natural constraints.

（动词不定式复合结构，在句中作宾语。）

（5）In order to keep each section working at high efficiency，a high degree of sectionalization is recommended.

（动词不定式短语作目的状语。）

### 4.2.4 用词化的手段表意

英语的词化是指通过词的形态的变化或词性的变化来体现词的意义变化或语法功能变化。如英语表达"（向，在）上游（的）"和"（在）下游（的）"概念只需在 stream 词后分别加后缀 up 和 down，就完成了词意和语法功能的变化。要表达"降级、降解"这一概念，只需在 grade 词前加前缀 de，形成 degrade 就完成了词意和语法功能的变化。用词化手段表意，只需一个词就可表达汉语一个较复杂短语的意义，使写出的句子精练。充分运用词化手段能使英语科技论文的语言简洁、明快，更能使文字地道、规范。英语的词化有加缀和转化两种方式。

1）加缀法

英语中有大量的派生词缀，将其加在词根或根词的前部和后部就能表达新的词义，改变其语法功能。

2）转化法

转化法是英语中非常活跃的一种构词方法。许多词可以不经过任何形态变化而用作另一词类，使该词表示新的意义，如由名词向动词的转化等。

现举例说明如下：

（1）Human beings do not have the ability to meet human needs indefinitely without degrading the environment.

句中用 degrading 实现词化手段表意，该词通过 grade 形态变化而形成，整个句子显得简洁。松散的表达形式是：

Human beings do not have the ability to meet human needs indefinitely without making the environment worse.

（2）The deforestation has endangered the survival of many wild animals.

句中用 deforestation 和 endangered 两个词实现词化手段表意，使句子十分简洁。

Deforestation 是通过 forest 加前缀 de-和后缀-ation 形成的，同样，endangered 是由 danger 的形态变化而形成。松散的表达形式是：

The destruction of forest has made many wild animals live in a dangerous situation.

（3）Under different atmospheric conditions，the wind may funnel the diluted effluent down a river valley or between mountain ranges.

句中用 funnel 实现词化手段表意，该词通过转化法，由名词词性转化为动词词性实现词化。

（4）Isolated buildings or the presence of a highrise building in a relatively low area can cause redirection of dispersion patterns and route emissions into an area in which many receptors live.

句中用 route 实现词化手段表意，该词通过转化法，由名词词性转化为动词词性实现词化。

（5）The scrubbing water also can function to absorb gaseous pollutants.

句中用 function 实现词化手段表意。该词通过转化法，由名词词性转化为动词词性实现词化。

（6）This rise of carbon dioxide in the atmosphere is largely due to the burning of such fossil fuels as natural gas，coal，and petroleum；another contributor is worldwide deforestation.

句中用 deforestation 实现词化手段表意，deforestation 通过加缀法形成。

### 4.2.5 用省略手段

为了避免重复使表达简洁，英语句子常采用省略的形式，即省去某个词或某些词。英语中省略的词在译成汉语时有时不能省略。现分述如下：

#### 4.2.5.1 省略前面出现过的词

这种省略较多出现在并列句中，或并列谓语中；有时在带有从句的句子中也有这种省略现象。

1）省略谓语动词，或连同其有关的成分

a. 省略复合谓语动词中的助动词

（1）Natural gas can be sent through pipes to distant towns，and used in houses and factories.

（=… and can be used in houses and factories.）

（2）In a few days the second bottle was retrieved and the DO measured.

（=… and the DO was measured.）

（3）The primary digester is covered，heated and mixed to increase the reaction rate.

（=… is covered，is heated and is mixed…）

（4）The sludge is poured onto the filter，the vacuum exerted，and the volume of the filtrate recorded against time.

（=… the vacuum is exerted，and the volume of the filtrate is recorded against time.）

(5) A great deal of money could be saved, and troubles averted, if sludge could be disposed of as it is drawn off the main process train.

(=… and troubles could be averted, …)

(6) Fireplace flues and chimneys should be inspected and cleaned frequently, and opened completely when in use.

(=… and should be cleaned frequently, and should be opened completely…)

b. 省略复合谓语动词中的主要动词及有关成分，但保留助动词

(1) Oxygen can not burn, but hydrogen can.

(=… but hydrogen can burn.)

(2) A liquid has a definite volume, but a gas does not.

(=… but a gas does not have a definite volume.)

c. 省略整个谓语动词

(1) The cleaned gas then passes up through the center of the unit (inner vortex) and out of the collector.

(=… and passes out of the collector.)

(2) The gas to be cleaned passes horizontally between the plates (horizontal flow type) or vertically up through the plates (vertical flow type).

(=… or passes vertically up through the plates (vertical flow type).)

(3) These are devices in which the contaminant airstream passes around or through a burner and into a refractory-line residence chamber where oxidation occurs.

(=… or passes through a burner and passes into…)

(4) Although gaseous emissions may be controlled by various sorption processes (or by combustion) and particulates (either solid or aerosol) by mechanical collection, filtration, electrostatic precipitators, or wet scrubbers, the effluent from the control device must still be dispersed into the atmosphere.

(=… particulates (either solid or aerosol) may be controlled by mechanical collection, …)

(5) For the aerobic regeneration feed, the most (net) $NH_3$ was generated by the 50% dispersion catalyst at the lowest temperature (125℃), by the 3% dispersion catalyst at the highest temperature (340℃), and by the 8% dispersion catalyst at the intermediate temperatures (170-290℃).

(=… was generated by the 3% dispersion catalyst at the highest temperature (340℃), and was generated by the 8% dispersion catalyst at the intermediate temperatures (170-290℃).

2）省略名词，保留其修饰语

a. 省略形容词后面的名词

Electrical energy is transformed into mechanical by means of electric motors.

(=Electrical energy is transformed into mechanical energy by means of electric motors.)

b. 省略介词短语所修饰的名词

(1) An understanding of the nature of the environment and of human interaction with it is a necessary prerequisite to understanding the work of the environmental engineer.

(= An understanding of the nature of the environment and an understanding of human interaction with it…)

(2) In all instances, the end products of the treatment of polluted water or air or of the disposal of solid wastes must be compatible with the existing environmental resources and must not overtax the assimilative powers of hydrosphere, atmosphere, or lithosphere.

(=… the end products of the treatment of polluted water or air or the end products of the disposal of solid wastes must be compatible with…)

c. 省略 to V 中的 V，保留 to

The development stage lasts as long as it needs to, until the working device has been constructed and tested.

(=The development stage lasts as long as it needs to last,…)

d. 出现多种或其他省略情况

(1) The temperature of melting ice is 0℃, and of boiling water 100℃.

(=… and the temperature of boiling water is 100℃.)

(2) However, it is one thing to declare that water must not be contaminated by pathogens (disease-causing organisms) and another to determine the existence of these organisms.

(=… and it is another thing to determine the existence of these organisms.)

(3) Although information regarding emission rates is limited, in general, the rate of formaldehyde release has been shown to increase with temperature, wood moisture content, humidity, and with decreased formaldehyde concentration in the air.

(=… and increase with decreased formaldehyde concentration in the air.)

(4) Either natural or manmade obstructions alter the atmospheric circulation and with it the dispersion of pollutants.

(=… and with it they alter the dispersion of pollutants.)

(5) The meter is the standard for length, the gram for weights.

(=… the gram is the standard for weights.)

(6) Distances measured from an axis to the right or upward are positive; to the left or downward are negative.

(=…; distances measured from an axis to the left or downward are negative.)

4.2.5.2 省略后面出现的词

为避免重复，有时可省略后面要出现的词。这种省略大多出现在并列成分中，有时在并列句或其他一些场合也可见到。

1) 省略形容词(或其他前位修饰语)后面的名词

(1) The difference between hot and cold water shows up very well if we use a thermometer.

(=The difference between hot water and cold water shows up very well if we use a thermometer.)

(2) Materials may be classified as good or poor conductors.

(=Materials may be classified as good conductors or poor conductors.)

这里要注意识别是否是省略现象，如果涉及指两个事物，而不是并列修饰语，就是省略结构。如，To every action there is an opposite and equal reaction. 这里 opposite 后没有省略名词，an opposite and equal reaction 指一个事物。

2）省略介词后面的名词

如两个介词短语并列，且两个介词后的名词相同时，则前一个介词后面的名词可以省略。省略之后，通常两个介词后面都有逗号。有时，两个介词并列，没有逗号分开。

（1）Electrons may be removed from, or added to, an atom.

（=… removed from an atom or added to an atom.）

（2）Heat may be radiated by, or conducted by, a solid.

（=… radiated by a solid or conducted by a solid.）

（3）Iron filings will be attracted to, and can be picked up by, a magnet.

（=… attracted to a magnet and can be picked up by a magnet.）

（4）Through wires the current is transmitted to and from the lamp.

（=Through wires the current is transmitted to the lamp and from the lamp.）

（5）Further inhalation of or even exposure to smog should be avoided.

（=Further inhalation of smog or even exposure to smog should be avoided.）

（6）Analysis of convective heat transfer requires detailed knowledge of fluid motion in the presence of and adjacent to solid surfaces.

（=… fluid motion in the presence of solid surfaces and adjacent to solid surfaces.）

3）省略修饰名词的介词短语，通常名词后面不用逗号，有时两个名词后面都有逗号

（1）The rate of nuclear reactions is controlled by insertion or removal of the control rods.

（=The rate of nuclear reactions is controlled by insertion of the control rods or removal of the control rods.

（2）Electric generators are really converters, rather than producers, of energy.

（=Electric generators are really converters of energy rather than producers of energy.）

4.2.5.3 不完全从句

在有些连接词引出的从句中，其基本成分不完全，所缺少的部分可从主句中判断。这种从句称作不完全从句。

1）在 when、while、once、if、unless、where、although、as 等引出的从句中，常没有主语及谓语动词中的 be（is, are），而形成"连接词+V-ed$_2$（或 V-ing，或形容词，或介词短语）"结构

（1）When heated, the metal expands, and if allowed to cool it will contract.

（=When it is heated…, and if it is alloweded…）

（2）Before we embark on a discussion on sludge treatment, however, we should be reminded that the processes should be used in wastewater treatment only when necessary.

（=… only when they are necessary.）

（3）When used alone, refrigeration is required to achieve the low temperatures required for condensation.

（=When it is used alone,…）

（4）When compared to the conventional wet scrubber, it uses significantly less liquid.

（=When it is compared to the conventional wet scrubber,…）

（5）When combined with other low-$NO_x$ technologies (such as low-$NO_x$ burners), $NO_x$ reductions of up to 90 percent may be achievable.

( = When the low-$NO_x$ technology is combined with other low-$NO_x$ technologies ( such as low-$NO_x$ burners,…)

(6) Liquids spread out and assume the shapes of their containers while maintaining their volumes constants.

( =… while they maintain their volumes constants.)

(7) Once this is done, the design is a search for the right pipe diameter and grade (slope) which will allow the minimum flow to exceed a velocity necessary for the conveyance of solids, while keeping the velocity at maximum flow less than a limit at which undue erosion and structural damage can occur to the pipes.

( =…, while they keep…)

(8) BOD is not a measure of some specific pollutant. Rather, it is a measure of the amount of oxygen required by bacteria and other microorganisms while stabilizing decomposable organic matter.

( =… while they stabilize decomposable organic matter.)

(9) For complete combustion to occur, each contaminant molecule must come in contact (turbulence) with oxygen at a sufficient temperature, while being maintained at this temperature for an adequate time.

( =…, while they are maintained at this temperature for an adequate time.)

(10) After making preliminary equipment selection, suitable vendors can be contacted for help in arriving at a final answer.

( =After we make preliminary equipment selection,…)

(11) After leaving the spray dryer, the solids-bearings gas passes through a fabric filter (or ESP), where the dry product is collected and where a percentage of unreacted alkali reacts with the $SO_2$for further removal.

( =After it leaves the spray dryer,…)

(12) When considering any of these plume rise equations, it is important to evaluate each in terms of assumptions made and the circumstances existing at the time the particular correlation was formulated.

( =When we consider any of these plume rise equations, )

(13) Fireplace flues and chimneys should be inspected and cleaned frequently, and opened completely when in use.

( =… when they are in use.)

(14) It is possible that, if supported on an oxide, a cobalt catalyst could be stable in a hydrogen atmosphere and retain the benefit of high activity.

( =… if it is supported on an oxide, )

(15) The former method consists of blasting additional air where needed, while step aeration involves the introduction of the waste at several locations, thus evening out the initial oxygen demand.

( =… where it is needed, )

(16) Chemicals from industrial discharges and pathogenic organisms of human origin, if allowed to enter the water distribution system, can cause health problems.

( =… if they are allowed to…)

(17) The BOD, if discharged into a stream with low flow, can still cause damage to aquatic life by depressing the DO.

( =… if it is discharged into a stream with low flow,)

(18) Oxygen, although poorly soluble in water, is fundamental to aquatic life.

( =… although it is poorly soluble in water,)

(19) Although laboratory tests are of some value in estimating centrifuge applicability, tests using continuous models are considerably better and highly recommended whenever possible.

( =… whenever they are possible.)

2) 在 whether、whatever、although 等引出的从句中，谓语动词(有时包括主语)常省略，而形成“连接词+名词”结构

(1) All matter, whether solid, liquid or gas, is made up of atoms.

( =…, whether it is solid, liquid or gas,…)

(2) The specific resistance test, although the closest thing we have to a basic parameter for filtration, should not be used for design.

( =…, although it is the closest thing…)

(3) Although not strictly an air pollution problem, such a plume may be objectionable for aesthetic reasons.

( =…, although it is not strictly an air pollution problem,)

(4) The size reduction step, although not strictly materials separation, is commonly a first step in a solid waste processing facility.

( =…, although it is not strictly materials separation,)

3) 在 as、than 引出的含有比较意义的从句中，常只有相比较的部分

(1) However, for the aerobic regeneration, catalyst D3 produced more net $NH_3$ than the other catalysts under anaerobic conditions.

(2) Sand is about 2.5 times as heavy as most organic solids and thus settles much faster than the light solids.

4) 在“the+比较级……，the+比较级……”句型中常省略谓语动词 be (is, are)

(1) The closer the ratio to 1, the better the degradation efficiency of the system.

(2) The higher the Oxone concentration, the faster the decay rate as well as the higher the performance for the fixed catalyst and oxidant molar ratio.

5) 由 hence 开头的省略句

由 hence 开头的句子常常没有谓语动词，可看成省略了谓语动词 result。

(1) Semiconductors are a class of elements whose electrical properties lie in an area between conductors and insulators, hence their name.

(2) Pollution prevention techniques must be evaluated through a thorough consideration of all media, hence the term multimedia.

(3) In the settling tanks, the microorganisms exist without additional food and become hungry. They are thus activated; hence the term *activated sludge*.

(4) When the blood becomes viscous, it is difficult for the heart to pump it through the capillaries. Hence the increase in blood pressure.

(5) It has been reported that the coupling of transition metal ions like $Fe^{2+}$, $Cu^{2+}$, $Mn^{2+}$ and $Co^{2+}$, to PMS leads to the accelerated generation of SRs and hence higher oxidation efficiencies.

(6) The powder Co/SBA-15 wrapped with polytetrafluoroethylene (PTFE) membranes exhibited excellent operational stability during recycling tests, with high phenol removal and low Co leaching, and hence an easily recyclable catalyst.

## 4.3 表达连贯

在英语语篇中，单词和词组构成句子，句子和段落等构成语篇，所以构成语篇的句子之间必然存在逻辑上的联系。只有语句衔接、语义连贯的语篇才能表达符合逻辑的思想，传达内涵完整的信息。达到语句衔接、语义连贯的主要方式有以下几种：

### 4.3.1 用承接语

承接语反映句子之间逻辑上的联系，起承上启下的作用。承接语不修饰句中的任何成分。承接语大多放在句首，也有放在句中或句末的。但中文表达通常承接语在先。承接语可由副词、数词、介词短语、to V 和句子等担任。连接词 and，but，or 位于句首时也起承接语的作用。承接语也可表示并列句或并列成分之间的逻辑联系。

为了表示不同类型的句际关系，英语科技论文中常用的承接语可分为以下几类：

1）表示顺序或总结

first(ly), second(ly), third(ly)... 第一，第二，第三……

one, two, three... 一，二，三……

on the one hand, on the other hand 一方面，另一方面

to begin with; to start with; in the first place... 首先

next; then... 其次

in turn 随后，又

finally; in conclusion; in a word; to summarize; to sum up; in short; in brief; on the whole... 最后，总之

2）表示补充或进一步说明

and; again; also; besides; further(more); moreover; too; in addition; then; in adding; not only that; what is more... 而且，还有，除此之外

3）表示等同或举例

namely(viz.); that is (i. e. 或 ie); that is to say... 即，也就是说

correspondingly; equally; similarly; in the same way; likewise... 同样，与此类似

or; alternatively; in other words... 或者说，换句话说

better 最好是说

(or) rather 更确切地说

for example (e.g. 或 eg); for instance; thus... 例如

4）表示结果或推理

accordingly; consequently; hence; now; so; then; therefore; thus; as a result; in consequence... 因此，于是，那么

5）表示对比或转折

(or) else; otherwise... 否则

but; however; still; yet; nevertheless… 但是，然而

notwithstanding; in spite of that; for all that… 尽管如此，还是

conversely; instead; rather; on the contrary; in contrast; on the other hand… 相反

in comparison; by comparison 比较起来

6）与某些连接词对应使用的承接语

If, although 等连接词引出的从句在前时，主句开头有时有承接语与连接词呼应。常见的有：

If… then 如果……，那么

although… yet; though… yet 虽然……，但

just as… so 正象……一样，……也

as… so 随着……也就

下面是一些实例：

（1）There is some limited anecdotal evidence of corrosion problems triggered by chloramines and nitrification in at least some circumstances. For example, recent work in Pinellas County, FL, highlighted some concerns related to iron corrosion control and red water. Likewise, elevated copper levels at the tap were suspected to be linked to action of nitrifying bacteria in Willmar, MN, homes. Nitrification also co-occurred with higher lead leaching in Ottawa, Washington DC, and Durham and Greenville, NC, homes. However, any link between nitrification and increased lead contamination was not definitive, nor were mechanisms postulated except for the case of Ottawa for which it was proposed that the higher lead resulted from decreased pH due to nitrification.

本段落中的 For example, Likewise, However 为承接语。

（2）All other factors being equal, a lesser amount of nitrification translated to a higher final pH in the pipe because less acid is produced via nitrification. However, due to the relatively high buffering capacity of the water at 100 mg/L alkalinity, pH values in the pipes with monochloramine/chlorite were only slightly higher ($p \geq 0.04$) relative to the control condition (Figure 1). But, when alkalinity (and buffering) was decreased to 30 mg/L and then 15mg/L, the final pH values with monochloramine/chlorite were much higher ($p<0.001$) than in the control (Figure 1). Thus, the extent of pH reduction by nitrification depends not only on the extent of nitrification but also on the initial alkalinity levels. At 0 alkalinity, where nitrification stopped in the controls, pH values in the pipes with monochloramine/chlorite were actually slightly lower relative to that without inhibitors ($p \geq 0.005$) (Figure 1).

本段落中的 However, But, Thus 为承接语。

（3）The solubility model trends and observed impacts of lower pH on lead release are also consistent with utility experience and other recent data. That is, the effects of pH on lead contamination and lead solubility are relatively weak at the higher alkalinities. For example, the 90%ile lead levels reported by utilities are not a strong function of pH if alkalinity is >30mg/L, but lower pH markedly increases 90%ile lead if alkalinity is <30 mg/L.

本段落中的 That is, For example 作为承接语。

（4）It is worth reiterating that higher alkalinity has an important dual benefit in preventing problems with increased soluble lead due to nitrification. First, as a buffer, the extent of the pH drop due to a given amount of ammonia conversion is reduced. This is obvious based on the average pH (-log of average [ $H^+$ ]) of 7.7, 6.92, and 6.19 at 100, 30, and 15 mg/L alkalinity, respectively (Figure 1). Second, a given pH drop of 0.5 units also has a much lesser impact on soluble lead at higher alkalinity (Figure 4).

本段落中的 First，Second 作为承接语。

(5) This analysis indicates that the initial pH also plays an important dual role. First, systems with pH between 7.5 and 8 are more likely to have active nitrification. Second, a given amount of nitrification activity would induce a much larger pH drop in systems with an initial pH of 8-8.5 since buffer intensity is at minimum at pH = 8.3 in carbonate systems. For example, comparing site 1 at St Paul, MN, and the anonymous utility with similar initial alkalinities (42 and 44 mg/L) and ammonia loss (0.4 and 0.41mg/L) the predicted pH drop due to nitrification is 0.5 unit in St Paul, MN (initial pH of 9), and 0.61 unit in the anonymous utility (initial pH of 8.3) (Table 1). The actual pH drop in St. Paul, MN, was even smaller: 0.12 pH unit (Table 1).

本段落中的 First，Second，For example 作为承接语。

### 4.3.2 用代词

英语科技论文中代词使用频率很高。某事或人第二次或更多次在文中提到，一般就选择代词来表达。使用代词可以避免重复，使行文简洁，同时也使相关句子意义连贯。在英语科技论文中，还可以用指示代词 this，that，these，those 加上一个名词，对上句的内容进行概括，起承上启下作用，使语句的意义指代清楚，内容连贯。举例如下：

(1) In this study, a head to head comparison of pipes with nitrification inhibitors to those without confirmed that nitrification increased lead release, but the extent of the effect is highly dependent on the initial alkalinity level (Figure 2). Specifically, at 100 mg/L alkalinity lead release was not increased by nitrification, as indicated by similar or even higher total and soluble lead release in the pipes with monochloramine/chlorite versus the condition without nitrification inhibitor (Supporting Information, Table S-1, Figure 2). At 30 mg/L alkalinity, nitrification increased total lead release up to 5 times and soluble lead release up to 21 times ($p \leqslant 0.002$, Table S-1) (Figure 2). At 15 mg/L alkalinity, total lead release was increased up to 5.5 times and soluble lead release up to 65 times ($p \leqslant 0.00008$, Table S-1, Supporting Information) (Figure 2). These trends are in agreement with expectations given that the pH was reduced most significantly by nitrification at the lower alkalinity levels (Figure 1).

本段落中用 these 加上名词 trends，对上句的内容进行概括，起承上启下作用。

(2) Although the contribution of calcium from the waste-paper-derived phase is higher for the former sample than for the latter, that difference cannot account for such significant differences in the $H_2S$ breakthrough capacity. The plausible explanation is that those bulky inserts of low-density char mentioned above in the discussion of porosity enhance the capacity. Although they do not provide the high volume of micropores, inside them the space for sulfur deposition exists. They are in close contact with a catalytic phase of sludge origin, and this enables the migration of sulfur. Support for this is the very high capacity of the C-5 sample, where all calcium comes from waste paper. Its high carbon content and limited inorganic phase from sludge lower the "buffer capacity" of the adsorbent. This causes a small fraction of hydrogen sulfide to be oxidized to sulfuric acid as

shown by a decrease in the pH.

本段落中多处使用不同的代词，对前述内容进行概括，起承上启下等作用。

### 4.3.3 重复关键词

这里的关键词是指在英语科技论文中表达重要概念的词，可能会在论文中反复出现。重复关键词既可以使语段衔接紧密、语句意义连贯，又能有效地突出主题，强化语势。在重复关键词时，往往要加上定冠词 the 或像 this，that，these 和 those 这样的指示代词，以便使指代明白无误。现举例如下：

Some information about the speciation of $H_2S$ surface reaction products can be obtained from the analysis of the pH values and their changes after $H_2S$ adsorption. As indicated above, on the basis of the values collected in Table 2, we expect mainly sulfur on the surface with some contribution of salts such as sulfides or sulfates and carbonates. Figure 12 shows DTG curves in nitrogen for the selected initial and exhausted samples. The peaks represent the removal of sulfur-containing species, and their surface areas reflect the weight loss in certain temperature ranges. For all exhausted samples an increase in the intensity of the peaks located at temperatures lower than 200℃, between 200 and 400℃, and between 600 and 800℃ is observed. The first peak may represent the removal of adsorbed water and weakly adsorbed $H_2S$ or $SO_2$ from oxidation of hydrogen sulfide. The second weight loss peak is likely linked to the removal of elemental sulfur. Its intensity increases with an increase in the performance of materials in the process of hydrogen sulfide removal. The lower temperature of that peak than that of the peak found previously for activated carbon is related to deposition of sulfur in big pores, which are present in large volume in these materials as shown from the analysis of pore size distributions. For the samples whose volume of micropores is the highest, a shoulder exists at about 400℃, which represents removal of sulfur from very small pores. The large peak between 600 and 800℃ present for samples with cellulose and calcium is linked to decomposition of calcium sulfate formed as a result of the reaction of sulfuric acid with calcium oxide. That peak is much smaller for the sewage-sludge-derived sample. For the CC sample and its exhausted counterpart there is a sharp weight loss at 400℃. The lack of differences in the intensity of that DTG peak for these two samples suggests its origin in the decomposition of calcium carbonate added to paper as a filler. It is interesting that this peak is not present for other samples containing waste paper. The reason can be the solid state reactions of calcium at high temperature in the presence of various inorganic components of the sewage sludge.

本段落中用 peak 作关键词，有效地突出了主题，强化了语势，使语段衔接紧密、语句意义连贯。

### 4.3.4 适当使用同义词

适当使用同义词与重复关键词一样，也是语篇衔接、语意连贯的手段之一。另外，从修辞的角度来看，英语科技论文中适当使用同义词还可以避免重复，防止语言单调，为论文表达增色。但要防止使用同义词引起误解的情况，兹举例如下：

With respect to the volumetric mass transfer coefficients at different MLSS concentrations, the $k_{La}$ value at a specific SGV decreased steadily as the MLSS concentration raised. A linear relation of $k_{La}$ and SGV was noticed. Such a linear interdependence is not mandatory. Deckwer showed that at a wide range of SGV (between 0 and 15 cm/s) $k_{La}$ flattens at higher SGV. The effect strongly depends on the up flow liquid velocity and is more pronounced in a three phase (water/air/solid) than in a two phase system (water/air). Since the variations in SGV in this study were 0.6 cm/s in reactor A and 0.1cm/s in reactor B a linear dependency of the $k_{La}$ with increasing SGV was noticed.

在以上段落中，relation，interdependence，dependency 3 个名词作为同义词在不同的句子中出现，通过不同的词的表现表达了同一概念，使语句很好衔接，同时增强了修辞色彩。这里，概念名词的选择非常合适。

The intention was to study the influence of the liquid phase and the solid phase on the α-factor separately. The results show that the permeate of the greywater membrane bioreactor had no influence on the oxygen transfer, whereas the greywater supernatant from reactor A operating at an SRT of 80 d had a lower impact on the oxygen transfer than the supernatant of the domestic wastewater plant (SRT=14 d). The washed sludge showed a slightly higher α-factor compared to the unwashed sludge.

在这一段落中，influence，impact 2 个名词作为同义词在不同的句子中出现，增强了修辞色彩，又通过不同的名词的表现表达了同一概念，使语句衔接良好。这里的概念名词是为人熟悉的，因此，它的选择非常合适，不会产生误解的不良结果。应该在阅读中仔细体会"重复关键词"、"尽量少用同义词"和"适当使用同义词"的含义。

### 4.3.5 用主从结构

在英语科技论文中，句子间的意义关系并不都是平行和并重的，往往有主次、轻重之分，有的表示背景与主题的关系，有的表示因果关系。用主从复合句子结构可将这种意义的分层和偏正体现出来。当一个语段内的几个句子都使用同一主语时，就表明这些句子意义相关，如果其意义有主从之分，用主从结构表达会使衔接更紧密和连贯。兹举例如下：

(1) Another commercial carbonaceous adsorbent worth mentioning here is catalytic carbon, which converts hydrogen sulfide to elemental sulfur. An example is US Filter's Midas on which about 60wt% sulfur can be stored before a breakthrough of hydrogen sulfide occurs. This material is prepared by physical mixing of alkaline-earth-metal oxides (calcium and/or magnesium) with coal-based activated carbons. As a result of this process high dispersion of the inorganic phase is achieved. That basic inorganic phase acts as a catalyst for hydrogen sulfide oxidation to sulfur. On oxide particles, in the presence of water, $H_2S$ dissociates and $HS^-$ ions are oxidized to sulfur, which immediately migrates to small pores of a neighboring carbon particle.

(2) The results described above demonstrate the importance of the composition and arrangement of inorganic/organic phases for the removal of hydrogen sulfide. By mixing sewage sludge and waste paper, exceptionally good adsorbents were obtained. Their capacities are comparable to those of the best activated carbons existing on the market. The interesting finding is that although some microporosity is necessary to increase the storage area for oxidation products, the carbonaceous phase does not need to be highly microporous. It is important that it provides space for deposition of sulfur, which is formed on the inorganic-phase catalyst. That space can be in meso-and macropores as shown in the case of char derived from the waste paper.

以上 2 个段落中的划线部分均是主从复合句，它们使语片结构紧凑，语意衔接更紧密和连贯。

### 4.3.6 用平行结构

平行是指将结构相同或类似、意义相关或并重的语言成分并列在一起。平行结构具有结构整齐、表述简练、语义突出的特点。由于结构一致，所以能够表达概念间的联系，达到意义的连贯。平行包括句子间的平行，也包括句子内，包括从句间或短语及单词间的平行。为了清楚地表明平行关系，在平行结构中，有时要重复该结构的标记。下面是一些例句：

(1) Engineers adapt the principles of natural mechanisms to engineered systems for pollution control when they construct tall stacks to disperse and dilute air pollutants, design biological treatment facilities for the removal of organics from wastewater, use chemicals to oxidize and precipitate out the iron and manganese in drinking-water supplies, or bury solid wastes in controlled landfill operations.（单词平行）

(2) The assumption here is that concentration can vary with length, width and depth.（单词平行）

(3) Techniques used to control gaseous emissions are absorption, adsorption, combustion, and condensation.（单词平行）

(4) Countries contemplating regulation of chemical wastes must consider for each chemical not only how toxic it is but also how flammable, corrosive, and explosive it is, and whether it will produce mutations or cause cancer.（单词平行）

(5) Particle removal can be accomplished in a variety of ways, including shaking the bags, blowing a jet of air on the bags, or rapidly expanding the bags by a pulse of compressed air.（单词平行）

(6) In general, the various types of bag cleaning methods can be divided into those involving fabric flexing and those involving a reverse flow of clean air.（单词平行）

(7) In general, contact condensers are more flexible, simpler, and less expensive than surface condensers.（单词平行）

(8) Forests and agriculture may be vulnerable because acid deposition can leach nutrients from the ground, kill nitrogen-fixing microorganisms that nourish plants, and release toxic metals.（单词平行）

(9) After the material has been collected from consumers, it must be cleaned, sold to an industry, transported, remanufactured and (most importantly) sold once again to consumers.（单词平行）

(10) The ideal solution, economically, energetically, and environmentally, would be to recover and reuse many of the solid wastes.（单词平行）

(11) Unless information is available on how pollutants are transported, transformed, and accumulated after they enter the environment, they cannot effectively be controlled.（单词平行）

(12) Environmental engineering has been defined as the branch of engineering that is concerned with protecting the environment from the potentially deleterious effects of human activity, protecting human populations from the effects of adverse environmental factors, and improving environmental quality for human health and well-being.（短语平行）

(13) Natural systems have been ever active, dispersing smoke from forest fires, diluting animal wastes washed into streams and rivers, and converting debris of past generations of plant and animal

life into soil rich enough to support future populations.（短语平行）

（14）For every natural act of pollution, for every undesirable alteration in the physical, chemical, or biological characteristics of the environment, for every incident that eroded the quality of the immediate, or local, environment, there were natural actions that restored that quality.（短语平行）

（15）The delicate balance of our biosphere has been disturbed and the state in which we now find ourselves is a direct consequence of our having ignored the limits of the earth's ability to overcome heavy pollution loads, and of our having been ignorant of the constraints imposed by the limits of the self-cleansing mechanisms of our biosphere.（短语平行）

（16）The first characteristic, solids concentration, is perhaps the most important variable, defining the volume of sludge to be handled, and determining whether the sludge behaves as a liquid or a solid.（短语平行）

（17）Models are classified by the number of dimensions modeled and by the type of model employed.（短语平行）

（18）Chlorofluorocarbons（CFCs）are used as coolants（CFC-12）for refrigerators and air conditioners, as blowing agents（CFC-11）in packing materials and other plastic foams, and as solvents .（短语平行）

（19）Among the possible steps in the multimedia approach are understanding the cross-media nature of pollutants, modifying pollution control methods so as not to shift pollutants from one medium to another, applying available waste reduction technologies, and training environmental professionals in a total environmental concept.（短语平行）

（20）Other processes related to algal growth and nutrient recycling are sorption and desorption of inorganic material, settling and deposition of phytoplankton, uptake of nutrients and growth of phytoplankton, death of phytoplankton, mineralization of organic nutrients, and nutrient generation from the sediment.（短语平行）

（21）Wrong or variable F/M ratios, fluctuations in temperature, high concentrations of heavy metals, and deficiencies in nutrients in the incoming wastewater have all been blamed for bulking.（短语平行）

（22）Typical applications are found in mining and metallurgical operations, the cement and plastics industries, pulp and paper mill operations, chemical and pharmaceutical processes, petroleum production and combustion operations.（短语平行）

（23）For activated carbon, the amount of hydrocarbon vapors that can be adsorbed depends on the physical and chemical characteristics of the vapors, their concentration in the gas stream, system temperature, system pressure, humidity of the gas stream, and the molecular weight of the vapor.（短语平行）

（24）Physical adsorption is a reversible process; the adsorbed vapors can be released（desorbed）by increasing the temperature, decreasing the pressure or using a combination of both.（短语平行）

（25）In the spray dryer, the sorbent solution, or slurry, is atomized into the incoming flue gas stream to increase the liquid-gas interface and to promote the mass transfer of the $SO_2$ from the gas to the slurry droplets where it is absorbed.（短语平行）

（26）The three ways of lowering emissions- before combustion, during combustion, and after combustion—will be discussed.（短语平行）

（27）EPA will be writing regulations for such issues as calculating and allocating allowances, for the mechanics of allowance transfers, for allowance tracking, and for the operation of reserves, sales, and auctions.（短语平行）

（28）The Noise Control Act of 1972 provides the EPA with the authority to require labels on all products that generate noise capable of adversely affecting public health or welfare and on those products sold wholly or in part for their effectiveness in reducing noise.（短语平行）

（29）In general the objectives would be to identify the concerns and issues that warrant attention, to provide for public involvement and to prepare a detailed brief for the investigation of specific issues associated with the development.（短语平行）

（30）In the so-called matrix methods, an attempt is made to quantify or "grade" the relative impacts of the project alternatives and to provide a numerical basis for comparison and evaluation.（短语平行）

（31）Environmental engineering in particular has been defined as that branch of engineering which is concerned with (a) the protection of human populations from the effects of adverse environmental factors; (b) the protection of environments, both local and global, from the potentially deleterious effects of human activities; and (c) the improvement of environmental quality for people's health and well being.（短语平行）

（32）The industries use the results to determine if they are in compliance with laws and regulations, to design treatment units, and to defend themselves in law suits.（短语平行）

（33）The conservation law for mass can be applied to steady-state or unsteady-state processes and to batch or continuous systems.（短语平行）

（34）The equation may be used for any compound whose quantity does not change by chemical reaction, or for any chemical element, regardless of whether it has participated in a chemical reaction.（短语平行）

（35）The National Academy of Sciences and other leading scientific bodies first gave credence to these concerns in the early 1980s when they suggested that emissions of sulfur dioxide from electric power plants were being carried hundreds of miles by prevailing winds, being transformed in the atmosphere into sulfuric acid, falling into pristine lakes, and killing off aquatic life.（短语平行）

（36）Advantages accounting for their use are simple construction, low initial cost and maintenance, low pressure losses, and simple disposal of waste materials.（短语平行）

（37）Impurities enter the water as it moves through the atmosphere, across the earth's surface, and between soil particles in the ground.（短语平行）

（38）Unfortunately, the sludges have three characteristics which make such a simple solution unlikely: they are aesthetically displeasing, they are potentially harmful, and they have too much water.（句间平行）

（39）Most engineers agree that although it would be most practical to treat the sludge at the plant in order to achieve the removal of toxic components, there are no effective methods available for removing heavy metals, pesticides and other potential toxins from the sludge, and that the control must be over the influent. (句间平行)

(40) For example, if water or a waste stream treatment is not available at the site, this may preclude use of a wet scrubber system and instead focus on particulate removal by dry systems, such as cyclones or baghouses and/or ESP. If auxiliary fuel is unavailable on a continuous basis, it may not be possible to combust organic pollutant vapors in an incineration system. If the particle-size distribution in the gas stream is relatively fine, cyclone collectors would probably not be considered. If the pollutant vapors can be reused in the process, control efforts may be directed to adsorption systems. (句间平行)

(41) In providing detailed information concerning specific methods or instruments, emphasis is placed on those that are readily available, easy to use, reasonably priced, and that provide the required levels of sensitivity and accuracy. (句间平行)

(42) If the plume is caught in the turbulent wake of the stack or of buildings in the vicinity of the source or stack, the effluent will be mixed rapidly downward toward the ground. If the plume is emitted free of these turbulent zones, a number of emission factors and meteorological factors will influence the rise of the plume. (句间平行)

(43) Where these particles are deposited and how long they are retained depends on their size, chemical composition, and density. (句间平行)

(44) Plastic recycling is still a relatively new field. Industry is researching new technologies that promise to increase the quantity of plastics recycled and that will allow mixing of different plastics. (句间平行)

(45) The common theme of environmental engineering is a basic understanding of environments, how they function, how they can be damaged, what hazards they present to people, and how people and environments can be protected from such effects. (句间平行)

(46) Air quality can hardly be improved if water and land pollution continue to occur. Similarly, water quality cannot be improved if the air and land are polluted. (句间平行)

(47) This is understandable given the fact that such a reorientation requires time and energy to learn new concepts and that time is a premium for them. (句间平行)

(48) If the project site is the last remaining woodland in an urban community, or if some of the trees are among an endangered species, the impacts would be considered more severe than otherwise. (句间平行)

## 4.4 英语科技论文常用句型结构

在英语科技论文中，句子是表意的基本单位，句子结构对于表达起着重要作用。对于非母语(外语)的写作，人们往往会根据语法规则自拟句子，这是一件费力不小的事情，但这常常是事倍功半，造成论文的表达有失地道、规范。因此，有必要熟悉和掌握相当多的常用句型，在写作中加以模仿和灵活运用，这是一条提高英语科技论文写作质量的有效途径。英语科技论文的常用句型一般可分为两类：第一类按论文语言表达的功能分类；第二类按语句所处的论文结构来划分。

### 4.4.1 表达不同功能的常用句型

1）定义

定义是对某一事物的本质特征或某一概念的内涵与外延的准确而简要的说明。在英语科技论文中，常见的表达方式和句型有：

a. 不定冠词+单数名词(或不可数名词)+is+不定冠词+类别词+后置定语，例句有：

(1) Iron is an element which has an atomic weight of 55. 85.

(Iron is an element with an atomic weight of 55. 85).

(2) Environmental science is an interdisciplinary area of study that includes both applied and theoretical aspects of human impact on the world.

(3) Seeding is a process in which the microorganisms which create the oxygen intake are added to the BOD bottle.

(4) Sound is a disturbance that propagates through a medium having the properties of inertia (mass) and elasticity.

(5) By definition, noise is a sound that is annoying and has a long-term physiological effect on an individual.

b. C is (may be/can be) defined as D.

c. By C is meant D (By C we mean D).

d. In this paper(In this context/For this purpose), C will be taken to mean (will be used in the/will be considered to be/will refer to/will be used in)…

e. The term C means (signifies/is considered to be/is taken to be/refers to)….

2）分类

在英语科技论文中常需要对叙述的对象从不同的角度进行分类。用于表达分类的常用句型有：

(1) C break down(divide/subdivide/classify)+总体+into(as/in)+部分

(2) C be (can be/may be/might be) broken down (divided/subdivided/classified/categorized/grouped) into(as/in)….

(3) C fall into (can be divided into/may be classified into) two major(general/broad/main) groups (classes/categories).

(4) C be of two major(general/broad/main) groups (classes/categories).

(5) There are two major (general/broad/main) groups(classes/categories) of….

3）比例关系

比例关系是英语科技论文中常见的描述内容。用于表达比例关系的常用句型有：

(1) A be big in proportion to the size of B but small in proportion to the size of C.

(2) The ratio between A and B is C:D.

(3) A be directly (inversely) proportional to B.

(4) A and B be mixed in a proportion of C to D.

(5) Three A(复数): B, C and D will, if they be mixed together in the right proportion, give us E.

(6) …, A and B are usually reduced in scale by a factor of two or three.

(7) A be made to scale.

(8) …, A be proportionally increased.

(9) A be formed by mixing B and C in a ratio of D E to F G.

(10) A be relative to B.

(11) A varies directly (inversely) as B.

(12) A varies directly (inversely) with B.

(13) A varies in direct relation to B.

(14) A(复数) depend inversely as B.

4) 用途与功能

在英语科技论文中常见的表达用途和功能的常用句型有：

(1) A (复数) apply only to B.

(2) A be applicable to (adapted to/not adaptable to/particularly useful for/fit for/more suitable for/well suited for) B.

(3) A holds for (is perfectly used for/is fit for) B.

(4) A can fit into B.

(5) A be not fitted to B.

(6) A (复数) lend themselves readily to B.

5) 行为与结果

在英语科技论文中常见的表达行为与结果的常用句型有：

(1) This causes (gives rise to/leads to/results in/brings about)….

(2) The result obtained agrees with (is in agreement with/is in line with/is consistent with/fits into)….

(3) Thus,….

(4) Hence,….

(5) Therefore,….

(6) Consequently,….

(7) As a result,….

(8) It follows that….

(9) … so that….

(10) … such that….

6) 目的

在英语科技论文中常常要阐述研究目的，常用句型有：

(1) The chief(main/primary/major/principal) aim (purpose/object/objective/goal) of the experiment was….

(2) The experiment has been conducted in the hope of (that)….

(3) The experiment has been conducted with a view to do….

(4) The experiment has been conducted in order to do….

(5) In order to do…,….

(6) To do…,….

(7) The experiment has been conducted to the end that….

(8) The experiment has been conducted so that….

(9) Performing the study we hope (expect)....

(10) Performing the study we intend to do....

(11) This article (paper) aims at....

(12) The designer(author) aims to....

(13) It is the aim of....

(14) The experiment was designed to do....

(15) The experiment was intended to show....

(16) The objective of... is to do....

7) 相关性

在英语科技论文中表达问题的相关性的常用句型有：

(1) This is a problem concerned with (concerning/related to/relating to/bearing on/dealing with)....

(2) A be associated with (directly bound up with/mainly concerned with/linked closely to/directly related to/in connection with) B.

(3) A be strongly dependent on (relative to/relevant to) B.

(4) A (复数) have to do with B.

8) 举例

在英语科技论文中表达举例的常用句型有：

(1) This can be illustrated (demonstrated) by (through)....

(2) A specific case can be provided to do....

(3) An example of this involves....

(4) Another typical example is....

(5) One example will suffice....

9) 数量大小、数量减少和数量增加

在英语科技论文中常需要表达数量大小、数量减少和数量增加，而英语的习惯表达方式与中文表达方式差别很大，因此，在使用过程中容易出错。这里进行详细说明，以引起重视。

a. 数量大小

... as+形容词 (large/heavy/many/much/high/low/great)+as+具体数字

表示“大到(至)/重达……/多达……/高达……/低到(至)……/……达……”

b. 数量减少

表示“减少”意义的动词有 reduce，shorten，fall，drop，decrease，step down，cut down 等。

(1) 数字+比较级+than....

表示净增加(减少)的数量。

A is two less than B.

(A 比 B 少两个。)

(2) 倍数 N+比较级+than....

表示前者是后者的 N 分之一。

The substance reacts four times slower than the other one.

(这种物质反应速度是另一种物质的 1/4。)

(3) 倍数 N+as+形容词/副词+as....

表示前者是后者的 N 分之一。

The hydrogen atom is nearly 16 times as light as the oxygen atom.

(氢原子的重量约为氧原子的 1/16。)

(4) 表示减少的动词+(to)+数字/倍数 N....

表示减少到多少或者减少到原来的 1/N。

The pressure will be reduced to one-fourth of its original value.

(压力将减少到原来数值的四分之一。)

(5) 表示减少的动词+by+百分数/数字....

表示净减少了多少。

Last year the sanitary wastewater discharge of this town decreased by 25%.

(去年该城镇的生活污水排放量减少了 25%。)

(6) 表示减少的动词+by+倍数 N....

表示减少到原来的 N 分之一[即减少了(1-1/N)]。

The pollutant concentration is stepped down by ten times.

(污染物浓度降了十分之九。)

(7) 表示减少的动词+by a factor of+数字 N....

表示减少到原来的 N 分之一[即减少了(1-1/N)]。

The metal concentration in the effluent is reduced by a factor of four.

(出水中的金属浓度减少了 3/4。)

(8) N-fold decrease/reduction...,

表示减少到原来的 N 分之一[即减少了(1-1/N)]。

The principal advantage over the old-fashioned wastewater treatment equipment is a four-fold reduction (decrease) in weight.

(与老式的废水处理设备相比，主要优点是重量减轻了四分之三。)

c. 数量增加

常见表示"增加"意义的动词有 increase, rise, exceed, grow, raise, expand, go up 等。

(1) 数字+比较级+than....

表示净增加的数量。

A is three more than B.

(A 比 B 多三个。)

(2) 倍数 N+比较级+than....

表示前者是后者的 N 倍。

The oxygen atom is 16 times heavier than the hydrogen atom.

(氧原子的重量是氢原子的 16 倍。)

(3) 倍数 N+as+形容词/副词+as....

表示前者是后者的 N 倍。

In water sound travels nearly five times as fast as in air.

(声音在水中的传播速度几乎是在空气中传播速度的五倍。)

(4) 表示增加的动词+to+数字/倍数 N....

表示增加到多少或增到 N 倍。

The members of the association have increased to 1500.

(这个协会的会员增加到了1500名。)

(5) 表示增加的动词+倍数N....

表示增加到N倍[即增加了(N-1)倍]。

The production of A has been increased four times as against 1993.

(A的产量比1993年增加了3倍。)

(6) 表示增加的动词+by+百分数/数字....

表示净增加了多少。

C rose by 5 percent, compared with the same period of last year.

(与去年同期相比,C增加了5%。)

(7) 表示增加的动词+by+倍数N....

表示增加到N倍[即增加了(N-1)倍]。

The strength of the attraction or repulsion increased by 4 times if the distance between the original charges is halved.

(如果原电荷的距离缩短一半,则吸引力或排斥力就增加三倍(即增加到原来的四倍)。)

(8) 表示增减的动词+by a factor of+数字N....

表示增加到N倍(即增加了(N-1)倍),或者减少到原来的N分之一[即减少了(1-1/N)]。

The sludge volume index(SVI) is reduced by a factor of four.

(污泥体积指数减少了3/4。)

(9) ... as+much/many/fast/long+again as 或 again as+much/many/fast/long+as....

表示"是……的两倍"或"比……多(快)一倍"。

The amount left was estimated to be again as much as all the zinc that has been mined.

(当时估计,剩余的锌储量是已开采量的两倍。)

Wheel A turns as fast again as wheel B.

(A轮转动比B轮快一倍。)

(10) half+as+much/many/fast/long+again as 或 half again as+much/many/fast/long+as....

表示"是……倍半","比……多一半"。

The resistance of aluminum is approximately half again as great as that of copper for the same dimensions.

(尺寸相同时,铝的电阻约为铜的一倍半。)

(11) 主语+谓语+double/treble/quadruple+...

表示增加一倍(翻一番)/增加两倍(增加到三倍)/增加三倍(翻两番)。

The new facility will double the capacity of the existing one.

(新的设施是现有设施容量的两倍。)

The growth rate of A will be quadrupled by 2000.

(到2000年,A的增长率将翻两番。)

10) 近似值

在英语科技论文中表达近似值的常用句型有:

a. 使用某些形容词、副词或介词,如approximate, approximately, nearly, almost, a little, slightly, under, over等。

（1）The pipeline is approximately 200 meters long.

（2）That wire has a length of slightly（a little）over 3 meters.

b. 使用一些被动语态句

（1）A be thought（believed/estimated/said/known/assumed/shown）to do….

（2）The temperature in the reactor is thought to reach 60℃.

11）状态的转化或变化

在英语科技论文中表达状态的转化或变化的常用句型有：

（1）... change(transfer/convert/transform/translate) B(变化前状态) to（into）C(变化后状态).

（2）B(变化前状态) be changed（transferred/converted/transformed/translated）to（into）C（变化后状态).

12）比较

比较对象要一致：

（1）Unlike that of alkali-metal or alkaline-earth-metal cations, hydrolysis of trivalent lanthanides proceeds significantly at this pH.

（2）In contrast to bromide anion, the free fluoride anion is strongly distorted on the vibrational spectroscopy time scale.

在英语科技论文中表达比较的常用句型有：

a. "compare to"用于表示相似性，" compare with"用于表示差异性。

（1）Compared to compound 3, compound 4 shows an NMR spectrum with corresponding peaks.

（2）Compared with compound 3, compound 4 shows a more complex NMR spectrum.

（3）A comparison of M and N（M with N）shows that....

（4）A comparison of between M and N shows that....

b. "than"和表示比较的形容词或副词不可省略

（1）The alkyne stretching bands for the complexes are all lower than those for the uncoordinated alkyne ligands.

（2）The alkyne stretching bands are all lower for the complexes than for the uncoordinated alkyne ligands.

（3）The decrease in isomer shift for compound 1 is greater in a given pressure increment than that for compound 2.

（4）The decrease in isomer shift in a given pressure increment is greater for compound 1 than for compound 2.

c. 避免将"different from"，"similar to"，"identical to"，"identical with"等分开，但中间插入介词短语时例外。

（1）The complex shows an NMR resonance significantly different from that of compound 1.

（2）Compound 5 does not catalyze hydrogenation under conditions similar to those for compound 6.

（3）The single crystals are all similar in structure to the crystals of compound 7.

（4）Solution A is identical in appearance with solution B.

d.“relative to”，“as compared to”，“as compared with”及“versus”等也可用于引导比较。

（1）The greater acidity of nitric acid relative to nitrous acid is due to the initial-state charge distribution in the molecules.

（2）The lowering of the vibronic coupling constants for Ni as compared with Cu is due to configuration interaction.

（3）This behavior is analogous to the reduced A-like reactivity in thiolate versus phenoxide complexes.

e. By comparison，very little is known about....

f. As a comparison（by contrast/in contrast/by way of contrast），....

比较这个功能范畴所含的内容较丰富，可再细分为以下几类：

g. 区分

（1）… be to separate A from B.

（2）… distinguish A from B.

（3）… know A from B by C.

（4）One should not confuse A with B.

h. 不同于

（1）A be different（distinct/distinguished）from B.

（2）A differs from B in C....

（3）A be distinguished from B by C.

（4）A unlike B，is C.

i. 相当于

（1）A be equivalent to B.

（2）A corresponds to B.

j. 优越于

（1）A has（possesses/offers）advantage over B.

（2）A be superior to B.

（3）A be used in preference to B for C.

（4）A be preferred to B for C.

（5）A has certain advantages over B.

k. 相似

（1）A be analogous to B in C.

（2）A（复数）have many features similar to B.

（3）A be very similar to B.

（4）The two A resemble one another in some ways.

（5）A bears little resemblance to B.

l. 与……一致

（1）A be compatible（in conformity/consistent/in good agreement/in accord/in line/uniform）with B.

（2）A accords with（agrees quite well with）B.

(3) A be in good agreement with B.

(4) A (复数) illustrate the agreement of B with C.

(5) A and B coincide.

(6) A coincides with B.

(7) A shall conform to B.

(8) A corresponds to B.

(9) A matches exactly B.

(10) The result obtained agrees with (is in agreement with/is in line with/is consistent with/fits into) the computer simulation.

13) 使用特殊连系动词

在英语科技论文的句子中常使用特殊连系动词(由少数实义动词变成的连系动词)，包括get，turn，go，stay，appear，look，prove 等，例句如下：

(1) This curve looks (appears) puzzling.

(2) The results prove correct.

(3) When the heat is transferred to the solution, the solution will get hot.

(4) In this case, the output stays high.

14) 使用半助动词

在英语科技论文的句子中时常使用半助动词(remain，seem，appear，happen 等)与动词不定式合成谓语，例句如下：

(1) The problem remains to be solved.

(2) These experiments seem (appear) to indicate that there are only two types of ions in the solution.

15) 采用动词不定式作主语

在英语科技论文中常采用动词不定式作主语，常见的句型有：

(1) It is necessary (for us) to do....

(2) It takes... for... to do....

(3) It remains now (for us) to do....

(4) It is desired to do....

16) 采用动词不定式作定语

在英语科技论文中常采用动词不定式作定语，常见的句型有：

(1) D is the solution concentration to be measured.

(2) There are many problems for us to consider.

(3) "介词+which+动词不定式"的句型，例句有：

We shall use such a principle on which to base our discussion of the catalytic activity.

(4) "the ability (tendency, capacity 等)+of A to do B"的句型

17) 采用动词不定式复合结构

在英语科技论文中常采用动词不定式复合结构，常见的句型有：

(1) 不定式有其自己的逻辑主语，表达"为了使……"

For... to do..., it is necessary to do....

(2) "too... to (do)"结构，例句有：

The velocity of light is too great for us to measure in simple units.

18）采用动词不定式的其他句型

在英语科技论文中还常采用动词不定式的其他句型，常见的句型有：

（1）"主语+系动词+形容词+不定式主动形式(来自及物动词)"，例句为：

This device is easy to operate.

（2）"主语+及物动词+it+形容词+不定式短语"，例句为：

We find it very easy to solve this problem.

（3）"主语+及物动词+名词+形容词+不定式主动形式(来自及物动词)"，例句为：

We find this parameter difficult to measure.

19）采用分词短语作状语

在英语科技论文中十分常见使用分词短语作状语，处于句首时主要表示原因、条件、时间、对主语的附加说明等。处于句尾时，可表示附加说明、结果及方式等。例句如下：

（1）This gas is about 60% methane and burns readily, usually being used to heat the digester and answer additional energy needs within a plant.

（2）The cover of the secondary digester often floats up and down, depending on the amount of gas stored.

（3）Thickening also implies that the process is gravitational, using the difference between particle and fluid densities to achieve greater compacting of solids.

（4）The mice, in turn, spend more time in burrows, thus not eating as much, and allowing the grass to reseed for the following year.

（5）Having defined the local picture of the Pt/BaO catalyst we can now interpret the storage results in more detail.

20）采用分词独立结构作状语

在英语科技论文中常见使用分词独立结构作状语，常常处于句尾，用作附加说明(即伴随状况)等。例句如下：

（1）This knowledge also went a long way toward initiating better house-keeping procedures within plants, identifying opportunities for raw materials substitution and production process modifications, all resulting in reduced water and waste quantities.

（2）The volatile solids parameter is in fact often interpreted as a biological characteristic, the assumption being the VSS is a gross measure of viable biomass.

（3）This suggests that $H_2$ is consumed by species other than stored $NO_x$, the obvious species being chemisorbed oxygen, a product of NO decomposition.

21）采用形容词短语作状语

在英语科技论文中时常使用形容词短语作状语，处于句首时主要表示原因、条件、让步、对主语的附加说明或对全句的评注。处于句尾时，可表示附加说明或对前面句子的评述或作方式状语。例句如下：

（1）Simple in structure and low in price, this device is in great demand.

（2）Free from the attack of moisture, a piece of iron will not rust very fast.

（3）Contrary to our knowledge, the reaction does not take place under this condition.

（4）Analogous to an expert, the robot is capable of undertaking some dangerous assignments.

(5) Smaller charged particles pass close to the surface of either the packing material or a scrubbing water droplet.

(6) A panel of experts is assembled, responsible for the development of guidelines for the EIA and reviewing the environmental report submitted by the project proponent.

22) 采用动名词短语与动名词复合结构

在英语科技论文中常用动名词短语和动名词复合结构，它们在句中常作介词宾语。

a. 动名词复合结构作介词宾语，例句有：

(1) Automation is not a question of machines replacing man.

(2) This is an example of magnetism being converted into electricity.

(3) The regular array of atoms in the lattice results in there being certain sets of paealled and equally spaced planes in the crystal.

b."介词+动名词"，常见句型有：

(1)"by+动名词"，意为"通过……"

(2)"on(upon)+动名词"，意为"一……就"，"在……之后"

(3)"in+动名词"，意为"在……时候(期间)"，"在……过程中"，"在……方面"

23) 采用名词短语作句子的同位语

在英语科技论文中常采用名词短语作为前面整个句子或句中一部分的同位语，以避免用一个结构显得松散的并列句或一个非限制性定语从句。其模式为：

……句子，a (an)+形容词+名词。

……句子，a (an)+名词+后置定语(过去分词短语，定语从句，形容词短语，同位语从句)。例句有：

(1) Before such separation can be achieved, however, the material must be in separate and discrete pieces, a condition clearly not met by most components of mixed refuse.

(2) The legislation calls for historic reductions in sulfur dioxide emissions from the burning of fossil fuels, the principal cause of acid rain.

24) 采用名词短语作主语的同位语

在英语科技论文中常采用名词短语放在主语前作主语的同位语，以避免使用一个系表结构的句子。例句有：

(1) An instrument for measuring dissolved oxygen in a solution, the dissolved oxygen electrode (a probe) is widely used in the water quality analyses.

(2) An air pollution control device, the electrostatic precipitator is able to treat some exhaust gases.

25) 采用"with 结构"

在英语科技论文中广泛使用"with 结构。有人称其为无动词从句，是一个表现力十分丰富的结构。其功用包括作状语，处于句首表示条件、时间和原因等；处于句尾表示结果，附加说明、方式和条件等。也作后置定语，充当定语从句的作用。下面是一些例句：

(1) With the catalyst added into the solution, the oxidizing reaction will be accelerated.

(2) With friction present, the object has been moved very slowly.

(3) With the key open, the water in the tank begins to flow.

(4) The activated sludge process is a continuous operation, with continous sludge pumping the

clean water discharge.

(5) The entire food web, or ecosystem, stays in dynamic balance, with adjustments being made as required.

(6) The storage of $NO_x$ has been studied in great detail with several mechanisms being proposed.

(7) Temperature has a pronounced effect on oxygen uptake (usage), with metabolic activity increasing significantly at higher temperature.

(8) The logical extension of this idea is to not have any treatment plants at all, but dispose of the wastewater onsite, with each house or building having its own treatment system.

(9) For streams and rivers with travel times greater than about five days, the ultimate demand for oxygen must include the nitrogenous demand.

(10) To evaluate its effect, experiments were conducted with pH varying in the range of 8-12.

(11) As described in the previous chapter, a pollution prevention assessment is a systematic, planned procedure with the objective of identifying methods to reduce or eliminate waste.

26) 采用同位语从句

在英语科技论文中常采用同位语从句，常用的句型有：

a. 毫无疑问……

There is no doubt (question) that….

b. 有证据表明……

There is evidence that….

c. 由名词从句转变成的同位语从句，例句有：

(1) The question whether there is water on that planet will be discussed.

(2) The users have no guarantee how long this kind of device will be operating.

27) 采用"What 从句"

在英语科技论文中常采用"What 从句"，常用句型有：

a. 表示"什么、多大、哪个、哪种，……"(即疑问代词的词义)，例句有：

(1) We must understand what is meant by the concept.

(2) Now you should determine what the reaction rate is in the reactor.

b. 表示"……的东西，……的内容"，常用句型有：

(1) What this paper describes…. (What is described in this paper….)

(2) What I have said above….

(3) What we need….

(4) What has been assumed (observed)….

(5) What is holding it….

c. 表示"所谓的、所称的，通常所说的"，常用句型有：

… what is called (termed/named/described as/known as/referred to as/spoken of as)….

28) 采用由 as 引导的定语从句

在英语科技论文中常采用由 as 引导的定语从句。

a. as 引导非限制性定语从句，修饰整个主句，位置灵活。常用句型有：

(1) As we mentioned earlier,….

(2) As we shall see,....
(3) As these examples illustrate,....
(4) As will be explained in Section 2,
(5) ..., as has been pointed out,
(6) ..., as has been described,
(7) ..., as will be shown later.

b. 表示“顾名思义”，常用句型有：

..., as the (its/their) name shows (indicates/implies/suggests)...

c. as 引导的修饰某个名词的限制性定语从句(往往与 such 和 the same 连用)，例句有：

(1) Variable 1 produces the same effect as is produced by the simultaneous action of other variables.

(2) Such instruments as we use to measure DO in water are called dissolved oxygen electrodes.

d. 与 as 有关的一种特殊结构

..., as+过去分词(介词短语/副词)，常用句型有：

(1) As pointed out in the literature,....
(2) As with the circular settling tanks,....
(3) As before,....
(4) ..., as shown in Fig. 1-4.
(5) ... as stated above (as described above)....

29) 采用强调句型“it is(was)... that (which/who)...”

在英语科技论文中常用强调句型“it is(was)... that (which/who)...”，例句有：

(1) It is the biodegradation, or the gradual use of this energy, by a chain of organisms which causes many of the water pollution problems.

(2) It was not until 1914 that the first workable pilot plant was constructed.

(3) It is when an object is heated that the average speed of molecules is increased.

30) 采用全倒装句型

在英语科技论文中常用全倒装句型，句型结构如下：

a. “表语+连系动词+主语”的句型，例句有：

(1) Of wider application is the fact....
(2) Among the most noteworthy achievements at that time was the realization that....
(3) Basic to understanding…is....
(4) Most interesting has been the fact that....
(5) Here is a.... (Below is a....)

b. “状语+不及物动词(或被动态谓语)+主语”的句型，例句有：

(1) First comes the outer covering, called the cell membrane.

(2) On an even larger scale looms the possibility of climatic alterations from increased levels of carbon dioxide produced in fossil fuel combustion.

(3) In the brain occour such higher fuctions as memory and reasoning.

(4) To this, of course, must be added the production of solids in secondary treatment.

c. “分词(过去分词为多)+助动词 be 的时态形式+主语”的句型，例句有：

(1) Shown on this page is a block diagram of a treatment process.

(2) Also omitted, to prevent the article from becoming unduly long, are discussions of design procedures and experimental techniques.

(3) Underlying the various techniques are the fundamental disciplines of physics, chemistry and biology.

d."介词短语(或引导词 there)+被动态谓语+主语"的句型，例句有：

(1) A stereoscope is an instrument through which can be seen two pictures of the same scene.(倒装句型出现在定语从句中。)

(2) There is shown a block diagram of a radio receiver in Fig. 4.

(3) Along the axis of the cylinder in the tubular precipitators are located discharge electrodes.

31) 采用部分倒装句型

在英语科技论文中常用部分倒装句型，句型结构如下：

a. 谓语的第一部分放在主语前的情况

"only+状语"开头的句型，例句有：

(1) Only in such a case shall we take this measure.

(2) Only via transition-metal catalysis is it possible to form sulfate radicals as the major oxidizing species.

(3) Only under a mached condition is there a maximum yield.

否定性含义副词或具有否定含义的介词短语处于句首，例句有：

(1) Seldom was the need to treat industrial wastewaters on a separate basis identified.

(2) A thermometer indicates temperature, but in no way does it show the amount of heat.

(3) Not until 1914 was the first workable pilot plant constructed.

由"so 或 neither, nor"开头的句型，例句有：

(1) Two electrons will be repelled from each other and so will two nuclei.

(2) All experiments do not necessarily function simultaneously, nor do they require information from the environment.

(3) As we will see later, not all oxygen atoms are alike, nor are all carbon atoms.

(4) In the absence of friction, the driving wheel would not run the belt, neither would the belt run the wheel to be driven.

b. 把强调的成分或较短的成分提前

"宾语+主语+谓语"句型，例句有：

(1) This process we shall discuss in detail.

(2) Substances that behave like metal in that kind of experiment we call conductors.

(3) A long list can we make of the products derived from petroleum.

"状语或表语+主语+谓语"句型，例句有：

(1) With pressure the gaps decreases.

(2) Certain it is that all essential processes of plant growth and development occur in water.

(3) Of one thing we can be sure: ….

"主语+谓语+宾补(或状语)+宾语"句型，例句有：

(1) Friction makes necessary a good lubrication system.

(2) We can take as an example the case shown in Fig. 1.

(3) Every measurement we make carries with it a degree of uncertainty, or error.

32) 全否定与部分否定

在英语科技论文中常使用 no, none, neither 等词表示全否定，使用 every, both, all 等与 not 连用来表示部分否定。例句有：

(1) None of these problems can be solved immediately.

(2) No treatment process is perfect.

(3) Neither of the devices is good in quality.

(4) All these analytical data are not correct.

(5) Both techniques are not suitable.

(6) Environmental engineers are not familiar with all these problems.

33) 多用"have"句型以及其他主谓宾结构

在英语科技论文中常采用"have"句型以及其他主谓宾结构来代替中文式英语中的"be"句型。例句如下：

(1) This wire has a length of ten meters.

(2) This device has a more complicated structure than that one.

(3) Tin does not have as high a melting point as lead does.

(4) That type of equipment has the advantages of simple structure and high efficiency.

(5) Silver possesses more free electrons than iron does.

(6) This technique requires a far smaller amount of operation than that one.

(7) We should use as simple a structure as possible.

(8) This device consumes much less energy than that one does.

(9) This device has a considerable advantage in removal efficiency over that one.

34) 采用静态结构

在英语科技论文中常采用静态结构来表示状态、特征、情况以替代采用行为或过程的结构。最常见的有以下两种模式。

a. 使用"连系动词+表语"结构(该表语主要为形容词，有时也可以是名词或介词短语)，例句有：

(1) ... is applicable to....

(2) This phenomenon is ignorable.

(3) ... are descriptive of.... (=describe)

(4) ... are symptomatic of.... (=indicate 或 suggest)

(5) ... stays (remains) constant (unchanged/the same/fixed/unaltered) (=is kept/held/left/maintained)

(6) ... is indicative of.... (=indicates)

(7) ... is an indication of.... (=indicates)

(8) ... is an illustration of.... (=illustrates)

(9) The problem is now under consideration. (=is now being considered)

(10) This specification is in great demand. (=is needed by a great many people)

b. 使用某些静态动词

(1) ... leads to....

（2）… results in....

（3）… fall into....

35）采用代词代后

在英语科技论文中常常采用 it，its，they，their 等代词来代替后面的人或物，即代词代后问题。例句如下：

（1）When they get hot，all metals melt.

（2）Because of its capacity to handle large volumes of data in a very short time，a computer may be the only means of resolving problems when time is limited.

36）故障

在英语科技论文中表达故障的常用句型有：

（1）… be out of order,....

（2）There be something wrong with A,....

（3）A malfunction,....

（4）… A fails,....

（5）… thanks to…failure.

（6）In case of… failure,....

（7）Trouble develop in....

（8）There should be… broken down,....

37）优点(长处)

在英语科技论文中常常要表达优点、长处，典型例句有：

（1）One of the greatest advantages of this method is its great simplicity.

（2）One of the greatest advantages of this method is that it is very simple.

（3）Among the greatest advantages of this method is its great simplicity.

（4）This method has many advantages over those available (over the existing ones).

（5）This method is advantageous in many respects (as) compared with those available (with the existing ones).

（6）This new method has a few advantages over the existing ones (over those available).

38）符合(达到)标准、满足(符合)条件和要求

在英语科技论文中表达符合(达到)标准、满足(符合)条件和要求的常用句型有：

a. 符合(达到)标准，例句有：

（1）A (复数) answer to (meet/reach/attain) the highest technical standards.

（2）A be fit for the national standard.

（3）A be in accordance with B.

（4）A be in full conformity with the standard.

（5）A be quite up to the standard.

（6）A conforms with (complies to) the standard.

（7）A is (comes) up to the standard.

（8）A (复数) make the grade.

（9）A high degree of accuracy can be achieved by....

（10）A can be made to an accuracy of B.

b. 满足(符合)条件和要求，例句有：

(1) A be in accordance with the requirements of....

(2) A complies fully with (conforms to) the requirements of....

(3) A should meet all conditions (demand/need).

(4) ... the conditions(demand/need) must be satisfied.

(5) A(复数)conform to (cover/fulfil/meet/satisfy) requirements.

39) 方式与手段

在英语科技论文中表达方式与手段的常用句型有：

(1) A can be done by means of (by using/with/through) B.

(2) With the help (aid) of A....

(3) A be achieved by....

(4) A be obtained with (be accomplished with) B.

(5) By this means,....

(6) ... by way of (with the assistance of) B.

(7) ... in a more orderly (very similar) fashion.

(8) ... in a detachable (reverse/violent) manner.

(9) ... in quite a complex(different/normal/following) way.

40) 物体相对位置关系

在英语科技论文中表达物体相对位置关系的常用句型有：

(1) ... in reference to....

(2) ... relative to (in relation to)....

(3) ... with (in) respect to....

41) 物体形状

在英语科技论文中表达物体形状的常用句型有：

(1) A is shaped like B. (或者：A is+表形状的形容词 B+in shape.)(A 的形状是 B。)

(2) A is circular in shape.

(3) A appeared in the shape of B.

(4) It is shaped into a basin-like form.

(5) A was made into the shape of B.

(6) A is B-shaped.

42) 作出结论

在英语科技论文中表达作出结论的常用句型有：

(1) A can arrive at (come to) the conclusion.

(2) A leads to a number of conclusions about B.

(3) ... conclusion follows....

(4) A cannot draw (reach) any definite conclusion from B.

(5) It can be concluded that....

(6) These data lead us to a conclusion (to conclude) that...

(7) These data enable us to conclude that....

(8) On the basis of these data (From these data) one (we) can conclude....

(9) On the basis of these data (From these data) it can be concluded that...

43) ……得到应用。

在英语科技论文中表达“……得到应用”的常用句型有:

(1) ... be widely used.

(2) ... be in wide use.

(3) ... be widely in use.

(4) A finds an important application (use) in B.

(5) A (复数) have found their place in B.

(6) A will be put (go) into mass production.

(7) ... put A into operation,....

(8) A may be put into service (use).

(9) A comes into use.

44) ……得多

常用句型有:

... much (far/well/considerably/greatly/significantly/a great deal) +比较级

45) 现有的……

常用句型有:

(1) ... available.

(2) ... we have (possess)....

(3) the existing....

(4) the current....

46) 所产生的……

常用句型有:

(1) The resultant....

(2) The resulting....

47) 有……

常用句型有:

(1) There be....

(2) ... be available.

(3) We have (possess/are in possession of)....

(4) There be available....

48) 在下面……

常用句型有:

(1) In what follows,....

(2) In what is to follow,....

(3) In the following,....

49) 在下一节中……

常用句型有:

(1) In the following (next/subsequent) section,....

(2) In the section which follows (to follow)....

50）A 随 B 的变化曲线图……

常用句型有：

（1）The graph（plot）of A as a function of B….

（2）The graph（plot）of A versus（against）B….

（3）The graph（plot）of the dependence of A on（upon）B….

（4）The graph（plot）of the variation of A with B….

51）本章论述……

常用句型有：

（1）This chapter deals with（treats/covers/describes/discusses/involves/bears on）….

（2）This chapter is confined to（devoted to/concerned with）….

52）……保持不变。

常用句型有：

（1）… remain（stay）constant（unchanged/fixed/the same/unaltered）.

（2）… be kept（held/maintained/left）constant（unchanged/fixed/the same/unaltered）.

53）毫无疑问……

常用句型有：

（1）There is no doubt that….

（2）There is no question that….

（3）Beyond doubt（question），….

（4）Without doubt（question），….

（5）No（Out of）doubt，….

54）……在……范围之内(外)。

常用句型有：

… is（lies/goes/comes/falls）within（beyond）….

55）……的体积各不相同。

常用句型有：

（1）… are different in size.

（2）are of different sizes.

（3）vary（differ）in size.

（4）come in different（various）sizes.

56）有证据表明……

常用句型有：

（1）There is evidence that….

（2）There is evidence to show（indicate/suggest）that….

57）该理论认为……

常用句型有：

The theory holds（maintains/claims/implies）that….

58）该理论来源于……

常用句型有：

（1）The theory comes（stems/emerges/originates）from….

(2) The theory is obtained (provided/furnished) from....

59) X 表示(指示/象征)了……

常用句型有:

(1) X is an indication of....

(2) X gives an indication of....

(3) X indicates....

(4) X is indicative of....

60) 把 C 代入 D 就可得到……

常用句型有:

(1) Substituting C in (into) D, we obtain (have/get)....

(2) Substituting (Substitution of C) in (into) D gives (yields/produces/results in/leads to)....

61) 该实验未能说明……

常用句型有:

(1) This (The) experiment failed to show (demonstrate)....

(2) This (The) experiment has not shown (demonstrated)....

62) 我们做了许多实验来检验这一理论。

常用句型有:

We have done (performed/made/conducted/carried out) a number of experiments to test (verify/prove/check) the theory.

63) 该式(方程式/反应式)适用于……[或该式(方程式/反应式)对……是成立的。]

常用句型有:

This (The) equation holds for (holds true for/is true for/is valid for/apples to)....

64) 我们成功地完成了……

常用句型有:

We have succeeded in doing.... (been successful in doing.... /successfully done....)

65) 这与……没有(几乎没有, 有一点, 有很多)共同之处。

常用句型有:

This has nothing (little/something/much) in common with....

66) 最好是……

常用句型有:

(1) It is desirable (that) to do....

(2) It is most desirable (that) to do....

(3) It would be best if....

67) 前面的例子表明……

常用句型有:

The previous (preceding/forgoing) example shows....

68) (本文)对……进行了研究(分析)。

常用句型有:

(1) A study (An analysis) has been made of....

(2) A study (An analysis) of… has been made.

69) 这引起了……

常用句型有：

(1) This causes (gives rise to/leads to/results in/brings about)….

(2) This is responsible for….

70) 这是由……引起的。

常用句型有：

(1) This is duo to….

(2) This is caused by….

(3) This results from….

(4) This is the result of….

(5) This arises from….

(6) The reason for this be that….

71) ……在一定范围内变动。

常用句型有：

… range (vary) from… to….

72) 这一点应受到(引起)我们的重视(关注)。

常用句型有：

This should arouse (attract/gain/have/receive) our attention.

73) 问题的关键是……

常用句型有：

(1) The key to the problem is….

(2) The crux (heart/essence/main aspects) of the problem is….

74) 这需要进一步的研究。

常用句型有：

This demands (requires/bears/deserves/calls for) further research (effort/study/work/investigation).

75) ……曾提出了这个问题。

常用句型有：

… put forward (raised/posed/advanced/brought up/formulated) this problem.

76) 他们根据……而提出了这一理论。

常用句型有：

They advanced (put forward/developed/proposed/suggested/created/constructed/formulated/elaborated) this theory, which is based on (rests on/proceeds from)….

77) 结果表明……

常用句型有：

The results show (indicate/imply/suggest/demonstrate/reveal/establish/bear out/confirm/support/favor/provide)….

78) ……特点是……(……是以……为特征的。)

常用句型有：

(1) … be characterized by….

(2) The features of… are….

79) 已知……就能做……

常用句型有:

(1) Given (Knowing)…, one (we) can do….

(2) Given (Knowing)…, it is possible to do….

80) 在这种情况下,C 不会发生,D 也不会发生。

常用句型有:

In this case, C does not take place, neither (nor) does D.

81) 这个问题有待解决。

常用句型有:

This problem remains to be solved.

82) 似乎只存在(有)……[好像只存在(有)……]

常用句型有:

There seem to be only….

83) C 与 D 相互作用。

常用句型有:

C interacts with D.

84) 现有的论文没有提到这一点。

常用句型有:

The papers(articles) available have made no mention of this point.

85) 并不像参考文献[15]所叙述的那样……"

常用句型有:

Unlike what was described in literature (reference) [15],….

86) 为了做……,必须做……

常用句型有:

To (In order to) do…, it is necessary to do….

87) 这条曲线显示了 C 随 D 的变化情况。

常用句型有:

The curve shows the variation of the C with D.

88) ……对这一论题很感兴趣。

常用句型有:

… be very interested in this topic.

89) 水是一种物质,而空气也是一种物质。

常用句型有:

Water is a kind of matter and so is air.

90) ……被称为……

常用句型有:

… be called(termed/named/known as/referred to as/spoken of)….

91) 为了做到所有这一切,对……必须要有一个清晰的了解。

常用句型有：

To do all this, a sound knowledge of… is necessary.

92）与目前国际文献中提出的……的理论相比，本文提出的理论可以导致….

常用句型有：

Compared with the theory presented in the international literature published in recent years that…, the theory given in this paper will lead to….

93）所有这一切表明采用……方法制备催化剂B的工艺是完全可行的。

常用句型有：

All this illustrates that the technology (process) of the preparation of the catalyst B by… is quite workable.

94）本文介绍(讨论/论述)了……

常用句型有：

(1) This paper(article) describes (presents/discusses/introduces/deals with)….

(2) … is (are) described (presented/discussed/introduced)….

95）在这种情况下，……是可以忽略不计的。

常用句型有：

In this case… is negligible.

96）这些值似乎不能满足该标准。

常用句型有：

It does not appear that these values meet the standard.

97）在这种情况下，必须尽量利用水体的自净能力。

常用句型有：

In this case it is necessary to use as much self-purification capacity of the water body as possible.

98）M的变化对P的性能的影响比N的变化产生的影响大。

常用句型有：

The variation of M has a greater effect on the performance of P than the variation of N.

99）根据对……分析，就得到了一种新的产品构型。

常用句型有：

On the basis of the analysis of…, a novel product configuration results.

100）本文基于理论和经验检验方法对……的选择提出建议。

常用句型有：

Based on the theoretical and empirical test methods, this paper offers (advances) a suggestion for selecting….

101）本文论述了该环境系统的数学模型，深入探讨了模型的参数估计和验证。

常用句型为：

This paper discusses the mathematical model of the environmental system, with a deep inquiry into the parameter estimation and model validation.

102）……由……组成。

常用句型为：

T contains (comprises/consists of/is composed of/is made up of) M, N, O, P, etc.

103) 越来越……

常用句型有:

(1) The applications of… are wider and wider.

(2) The applications of… are ever wider.

(3) … have become more and more complicated.

(4) … have become increasingly complicated.

104) 越……，越……

常用句型有:

(1) The higher the wave, the greater its power; the greater the number of waves a sound has, the larger is its frequency or pitch.

(2) The more chromate used, the more organics were in the sample, and hence the higher the COD.

105) 如果下面两个条件中任意一个成立，那么我们就得到一个新的表达式。

常用句型有:

If either of the following two conditions holds (applys/holds true/is true/is valid), then we have a new expression.

106) 模拟结果表明，所有这两种方案都是容易实现的。

常用句型有:

The simulation results show that both the schemes are easy to implement.

107) ……保持…… 优点。

常用句型有:

… maintain (keep up/preserve) the advantages of….

108) ……值得……

常用句型有:

(1) … deserves to be further studied.

(2) … be worth doing….

(3) … be worthy to be done.

(4) It is worthwhile….

109) 还有其他一些因素应该仔细考虑，如政策，不可预见的自然灾害以及有关工厂的经济效益。

常用句型为:

There are also several other factors, such as policies, unpredictable natural disasters, and the economic benefits of the factory involved, which should be carefully considered.

110) 进行了……试验，结果令人满意。

常用句型有:

The test of… was conducted (carried out/performed/done) with satisfactory results.

111) ……具有头等重要性。

常用句型有:

… be of prime importance.

112）……具有重要意义。

常用句型有：

... be of great significance.

113）这个结果已被 Plewa 和我们的试验证明。

常用句型有：

The result has been proved by both Plewa's tests and ours.

114）这些参数在数值上没有很大的差别。

常用句型有：

（1）These parameters do not differ greatly in magnitude.

（2）There is not much difference in magnitude between these parameters.

115）几次试验就足以获得令人满意的结果了。

常用句型有：

Several tests are enough for us to obtain a satisfactory result.

116）这一点在文献[10]~[12]中有论述。

常用句型有：

This point is described in the literature[10]~[12].

117）这种装置的体积小。

常用句型有：

The device is small in size.

118）重点放在该文所讲的内容上。

常用句型有：

Emphasis is put on what is described in this paper.

119）这种设备的特点是体积小、重量轻、效率高和成本低。

常用句型有：

（1）The equipment is characterized by small size，light weight，high efficiency and low cost.

（2）The equipment is small in size，light in weight，high in efficiency and low in cost.

（3）The features of this device are small size，light weight，high efficiency and low cost.

120）x 与 y 之间的关系可用下式表示。

常用句型有：

x and y are related by the following equation.

121）A 与 B 之间的关系由 C 来表示。

常用句型有：

（1）A and B are related by C.

（2）A is related to B by C.

122）……的方法是……

常用句型有：

... is found（determined/increased/achieved/solved/calculated/reduced/obtained/satisfied/separated/constructed/analyzed/adjusted/minimized）by....

123）近年来……有了非常迅速的发展。

常用句型有：

Recent years have witnessed a very rapid growth of….

### 4.4.2 论文主要组成部分中的常用句型

英语科技论文有若干主要组成部分。每部分都有其特定的功能。要实现这些特定功能，论文的各组成部分都必须有组成要素，句子是反映论文组成要素的载体，规范、地道的句子对于表达具有重要的作用。掌握并能灵活运用常用句型对于保证写作的质量是一条有效措施。这部分的内容将在下面的有关章节进行分述。

# 5 环境类英语科技论文正文的撰写

正文是科技论文的主体部分，环境类英语科技论文正文与一般英语科技论文正文一样，包括引言(Introduction)，材料与方法(Materials and Methods)(或实验部分 Experimental)，(研究)结果(Results)，讨论(Discussion)和结论(Conclusion)等方面(即 IMRAD 结构)。当今，较多的期刊论文结构将结果与讨论归并在一节中，冠名为“Results and discussion”，也有一些期刊论文结构中不单列“Conclusion”。对于涵盖信息较多、篇幅较长的研究论文，在正文中常采用更次一级的标题，以分出层次。

本章节依次介绍英语科技论文正文中引言、材料与方法(或实验部分)、(研究)结果、讨论及结论各部分的功用、组成要素、时态、语态与常用句型，并提供实例与简析。

## 5.1 引言部分

### 5.1.1 功用

引言(Introduction)又称前言、序言，是论文正文的第一大组成部分。

引言的功用是有效提供研究背景、相关研究工作的评述、研究主题、目的和意义等。

### 5.1.2 组成要素

引言部分通常包括以下几个方面的信息，即组成要素。

(1) 研究领域的一般概况与进展(一般背景)

引言部分的第一句话或开头几句话往往从一个大主题出发，然后逐步缩小到论文和研究工作所涉及的具体问题。这样一方面交代背景知识和论文所涉及研究工作所属的大领域，另一方面也可以说明研究工作的重要性。

(2) 已有研究工作的回顾和评论(简要文献评论)

论述国内外已发表的相关研究，一定要涉及相关性好、时效性强、经典和重要的研究工作，重要的是不仅要“叙述”，而且要“评价”，即经过分析，指出研究的不足、共同点和差异、研究发展趋势等，不能仅“点到为止”。当然，论述一定要准确、完整和简练。

(3) 有待研究的问题或现象

根据已有研究的成果与不足，结合现有的需要和可能，提出有待研究问题或现象。

(4) 本研究的主题、目的和意义等

阐明本研究的主题、目的、意义和本研究的创新性等。

(5) 概念与术语的定义(可选)

需要明确定义的概念和术语。

上述内容为一篇论文引言的基本组成部分，具体要素组合可视情况而变。

在撰写引言时应注意以下问题：

（1）避免内容空泛或过于简要；

（2）不要重述摘要和解释摘要；

（3）不应对实验的理论、方法做详尽叙述，不提前使用结果、结论和建议。

### 5.1.3 时态、语态与常用句型

1）时态

引言中时态运用的通用规则为：介绍已有的认识和普遍的、不受时间影响的事实时，使用一般现在时；叙述本人或他人已有或近期的工作或认识和特定的、过去的行为或事件时，采用一般过去时；描述的是从过去到现在的一种研究趋势，或对某个问题的研究程度，可用现在完成时。对于主从复合句，如主句说明提出方法或技术是他人过去的行为，从句的内容是描述他人曾提出的方法或技术，且该方法或技术已成为一种标准的方法或技术，或者现在仍被使用时，则主句用一般过去时，从句用一般现在时；若从句所引述的他人的研究结果只在特定的条件下有效，并非不受时间影响的普遍事实，那么在从句中应使用一般过去时。

a. 研究背景

（1）介绍一般性资料、现象或普遍事实时，多使用一般现在时；

（2）引述他人过去的研究行为时，通常采用一般过去时，常以“that 从句”叙述被引作者的研究成果，从句中动词的时态因随所表达信息的性质而定。

（3）描述特定研究领域中最近的某种趋势，或者强调表示某些“最近”发生的事件对现在的影响时，常采用现在完成时。

（4）对存在问题，如果叙述的是普遍事实，就用一般现在时。

（5）如果叙述过去已开始并持续到现在的趋势或事件，则使用现在完成时。

b. 研究目的

在叙述研究目的时，可采取两种方式，即论文导向和研究导向。

（1）论文导向

用 paper，report，article 等特征名词，直接提及研究论文本身，表示论文提供信息的行为，重点在于介绍新的技术或方法、分析某个问题或提出某个论证。由于论文提供信息的行为是不受时间影响的事实，所以通常使用一般现在时。

（2）研究导向

用 study，research，investigation 或 experiment 等特征名词来介绍研究活动，重点在于提出某些调查或实验结果，由于句中所涉及的是已经过去的事情，因而多使用一般过去时。

2）语态

在英语科技论文的引言部分语态的使用较灵活，根据需要主动态和被动态兼用。有时要适当地使用“I”、“We”、“Our...”为主语的主动态，以明确地表示是作者本人的工作。

3）常用句型

a. 研究背景

（1）A is a serious problem in many areas of....

（2）A showed（found/reported/noted/suggested/observed/pointed out）that B does....

（3）A found（reported/suggested/observed）that B did C....

（4）如果“that 从句”中的信息不是很确定的研究成果（如建议、假设等），则主句中的动词应使用 suggested 或 hypothesized 之类的臆测动词，从句中则使用“may+现在时”。

A suggested（hypothesized/proposed/argued）that B may help C to do....

(5) In recent years, a variety of standards have been proposed in the literature.

(6) Many (A number of/Several/Few) studies (experiments) have been conducted (done/performed/published) on....

(7) Several researchers (A number of authors/Many investigators/Few writers) have studied (investigated/examined/explored/reported/discussed/considered)....

(8) The problem of... has been the subject of much research (the subject of few studies/the focus of a great deal of research) in recent years (in the last decade/since 1990).

(9) There has been much research (little research) on A.

(10) ... put forward (raised/posed/advanced/brought up/formulated) this problem.

b. 存在的问题

(1) 以 however, few, little 和 no 等表示过去研究的不足。

(Ⅰ) However, few studies have been done (published) on....

(Ⅱ) However, little research has been devoted to (little attention has been paid to/little is known about) A.

(Ⅲ) However, no studies have investigated (no work has been done) on A.

(Ⅳ) However, insufficient data are available on A.

(2) 以 although, while 引导或以 but, yet 转折的复合句来提出问题。

(Ⅰ) Although much research has been devoted to A (While much work has been done on A/Although many studies have been published concerning A/While many researches have investigated A/Although much literature is available on A), little research has been done on B (little attention has been paid to B/little information is available on B/little work has been published on B/few researches have studied B/few studies have investigated B).

(Ⅱ) Although much research has been done on A, little work has been done on B.

(Ⅲ) A (复数) have been studied extensively, but little attention has been devoted to B.

(3) Little literature is available on A.

(4) Few studies have been done on A.

(5) Little attention has been paid (devoted) to A.

(6) Little is known about A.

(7) The key to the problem is....

(8) The crux (heart/main aspect/essence) of the problem is....

c. 研究目的

在叙述研究目的时，可采取两种方式，即论文导向和研究导向。

(1) 论文导向

用 paper, report, article 等特征名词，直接提及研究论文本身，表示论文提供信息的行为。常用的句型有：

(Ⅰ) The purpose (aim/goal/objective/object) of this (present) paper (article/report) is to do....

(Ⅱ) This paper (article/report) describes (introduces/presents/discusses/concerns/reports/explains/examines/analyses/proposes/evaluates/demonstrates)....

(Ⅲ) In this paper, we propose a new algorithm of....

(Ⅳ) In this paper, experimental results are presented to show that….

(Ⅴ) The primary focus of this paper is on….

(2) 研究导向

用 study, research, investigation 或 experiment 等特征名词来介绍研究活动，重点在于提出某些调查或实验结果。常用的句型有：

(Ⅰ) The main (chief/primary/major/principal) purpose (aim/goal/objective/object) of the (this) experiment (study/research/investigation) reported here was to do….

(Ⅱ) The study has been started in the hope of (that)….

(Ⅲ) The study has been started with the view to do (in order to do/to the end that/so that)….

(Ⅳ) The study is intended (designed) to do….

(Ⅴ) Performing the study we hope that (intend to do/expect to do)….

(Ⅵ) In the research described here (In this study/In the present investigation/In this research/In the experiments reported here), we did… [A was (were) done].

(Ⅶ) This study (research/investigation/experiment/project/survey) did (tested/investigated/measured)….

(Ⅷ) The study was designed to evaluate….

应注意的是，在阐述作者本人研究目的的句子中应有类似 This paper, The experiment reported here 等词，以表示所涉及的内容是作者的工作，而不是其他学者过去的研究。

d. 研究的意义、研究结果的实际应用和进一步研究的建议

在叙述本研究的意义时，可在动词 be 前加上情态动词 should。在指出本研究结果的实际应用和进一步研究的建议时，可用 might, may, would, could 等情态动词。但用 should 则表示作者对自己研究的价值相当肯定或所提的建议比较强烈。

其他实际典型例句(单句或连续语句)如下所示：

(1) Recently, sulfate radical (SR) based-advanced oxidation technologies (SR-AOTs) have attracted great scientific and technological interest in the area of water treatment and other environmental applications.

(2) Consequently, many studies have stated that the type of support and cobalt precursor must be very carefully chosen to attain highly active cobalt catalysts.

(3) As a result, it is critical to investigate the effect of synthesis variables of the supported cobalt on the hetero-PMS-act.

(4) To the best of our knowledge, this is the first systematic study dealing with the effect of synthesis variables and UV radiation on the hetero-PMS-act in supported cobalt systems.

(5) The primary objective of this study is to investigate the effectiveness of hetero-PMS-act systems for 2,4-DCP degradation and find ways to minimize cobalt leaching. Meanwhile, we wanted to explore the role of UV radiation in PMS-Co/$TiO_2$ systems for 2,4-DCP degradation.

(6) This investigation aims at evaluating the effectiveness of three different classes (natural, inorganic and synthetic) of environmentally safe chelating agents on common oxidants, namely peroxymonosulfate (PMS), persulfate (PS), and hydrogenperoxide ($H_2O_2$).

(7) So far, the use of EDDS in Fenton type systems has not been reported.

(8) The main objective of the present study was to evaluate the effectiveness of three representative chelating agents (Citrate, EDDS, and Pyrophosphate) on Fe(Ⅱ)-mediated activation of three commonly used peroxides (PMS, PS, $H_2O_2$) at neutral pH conditions.

(9) To the authors' best knowledge, this is the first study to evaluate the effectiveness of all three oxidants, PMS, PS and $H_2O_2$, for chlorophenol degradation in the presence of the commonly used activation agent Fe (Ⅱ), using environmentally safe chelating agents such as citrate, pyrophosphate and EDDS at neutral pH conditions.

(10) However, aesthetic aspects and health concerns associated with the adverse effect of dissolved cobalt species in water still need to be addressed.

(11) Consequently, it is critical to develop a highly efficient nanoscale metal catalyst for the heterogeneous PMS activation, where the adverse impact of metal leaching on aqueous environments is notably minimized and the catalyst is readily recovered after its application.

(12) In this study, we report the development of a novel approach of using Fe-Co mixed oxide nanocatalysts as a mediator for the heterogeneous PMS activation.

(13) We focus on the effect of several factors such as cobalt content, calcination temperature, and $TiO_2$ support for the Fe-Co catalysts on cobalt leaching, Fe-Co interaction, and heterogeneous activation of PMS.

(14) The catalyst performance was evaluated in terms of the degradation of highly toxic and poorly biodegradable 2,4-DCP, which mainly originates from the environmental transformation of herbicides such as 2,4-dichlorophenoxyacetic acid.

(15) Moreover, we discuss the environmentally friendly aspects of the Fe-Co nanocatalysts, including reduced cobalt leaching and ferromagnetic-based separation properties.

(16) Recently, the growing interests in sulfate radical-based advanced oxidation technologies (AOTs) are driven by the increasing demand of cost-effective and environmentally benign routes for wastewater treatment.

(17) There are a few studies on the generation mechanism of sulfate radicals by cobalt-catalyzed decomposition of PMS in the homogeneous system.

(18) Consequently, the purpose of this study is to improve the catalytic activity of cobalt for the heterogeneous PMS activation by controlling the physicochemical properties of cobalt or $Co_3O_4$ material at the nanolevel.

(19) In this study, we focused on immobilizing cobalt catalyst at the surface of $TiO_2$.

(20) This paper deals with the development of a novel Fe(Ⅱ)-PMS advanced oxidation technology for PCBs degradation in aqueous and sediment systems.

(21) The effectiveness of the proposed process was evaluated based on the degradation and total organic carbon (TOC) removal efficiencies of a model PCB (2-chlorobiphenyl, 2-CB).

(22) The catalytic process was optimized for soluble iron species and oxidant concentration.

(23) The role of both iron species, Fe(Ⅱ) and Fe(Ⅲ), for oxidant activation and subsequent 2-CB degradation was evaluated.

(24) Dominating reactive radicals responsible for the degradation process were identified by quenching studies using a variety of alcohols with markedly different reactivity towards hydroxyl and

sulfate radicals.

(25) A low molecular weight biodegradable iron chelating agent, sodium citrate, was used to control the amount of soluble iron at neutral pH.

(26) Finally, the degradation of 2-CB in sediment-slurry phase was studied.

(27) Several conventional processes, including adsorption and biological treatment, are commonly applied for the removal of 2,4-D in contaminated water.

(28) Some alternative processes have also been explored in the destruction of 2,4-D, other pesticides, and a variety of other organic contaminants.

(29) Among emerging treatment approaches, advanced oxidation processes (AOPs) are considered among the most effective and are currently gaining significant importance in water treatment applications.

(30) Nevertheless, only little has been reported on the actual mechanism of how the light enhances the reaction rate in this process when both the catalyst and the oxidant are present.

(31) To our knowledge, no studies on the use of solar radiation as the source of light in the Co/PMS/UV processes have been reported.

(32) The aims of this work are to test the potential of the Co/PMS and solar-driven Co/PMS/UV processes to degrade 2,4-dichlorophenoxyacetic acid (2,4-D), an important environmental organic pollutant, and compare them with the Fenton and photo-Fenton processes using the sun as the source of energy.

(33) Due to more stringent statutory regulations and increasing public concerns on harmful organic substances, there is an imperative need to develop efficient technologies for the complete removal of organic pollutants from wastewater effluents.

(34) Over the past years, sulfate radicals based-advanced oxidation technologies (SRs-AOTs) have emerged as a promising way leading to the total mineralization of most organic pollutants.

(35) It has been reported that the coupling of transition metal ions like $Fe^{2+}$, $Cu^{2+}$, $Mn^{2+}$ and $Co^{2+}$ to PMS leads to the accelerated generation of SRs and hence higher oxidation efficiencies.

(36) Therefore, the activation of PMS by heterogeneous cobalt sources has been given great attention recently.

(37) In particular, it has been reported that $TiO_2$ plays a significant role in promoting this step.

(38) Herein, we report cobalt oxide supported on MgO (Co/MgO) as an efficient heterogeneous catalyst for activation of PMS.

(39) The current catalyst system is therefore expected to be highly desirable for competing with the conventional nondestructive treatment processes.

(40) In the past decades, some techniques such as adsorption, membrane separation, coagulation, chemical oxidation, and electrochemical degradation have been investigated to treat dyeing wastewater.

(41) However, low removal efficiency or high cost in operation often limit their application.

(42) In the past years, Fenton-type oxidation ($Fe^{2+}/H_2O_2$) using hydroxyl radicals has been comprehensively investigated.

(43) In recent years, chemical oxidation using peroxymonosulfate (PMS) conjunction with cobalt ions to generate sulfate radicals has raised an interest and is found as a very promising technique.

(44) In the past a few years, some investigations have been reported in using $SO_4^{-}$ radicals for oxidation of organics in water.

(45) For dye containing wastewater, few investigations have been reported while focusing on the effect of photo-assisted decolourization.

(46) In this study, we will compare the effectiveness of $Co/H_2O_2$ and Co/PMS in oxidative decomposition of different dyes at neutral pH range and investigate the effects of several operation parameters on dye degradation efficiency.

(47) For chemical oxidation processes, the reactivities of chemical oxidants (hydrogen peroxide, ozone, permanganate, and persulfate) that are used for both ex situ and in situ chemical oxidation have been relatively evaluated. Moreover, combinations of these oxidants have been studied.

(48) Recently, chemical oxidation using either persulfate ($S_2O_8^{2-}$) or peroxymonosulfate (PMS, $HSO_5^{-}$) has been a topic of interest.

(49) The objective of this study was to evaluate the feasibility of an advanced oxidation process incorporating the generation of sulfate radicals ($SO_4^{-}$) from PMS mediated by transition metals (i.e. Co(Ⅱ) and Fe(Ⅱ)) for the treatment of soils contaminated with diesel fuel.

(50) Therefore, the main objective of this study is to analyze the feasibility of decolorization and degradation of RBB by the use of $Co^{2+}$/PMS oxidative processes with such extremely low dosages of $Co^{2+}$ ($[Co^{2+}] = 1 \sim 10$ μg/L) and PMS (0.08~0.32 mM).

(51) Furthermore, the influence of different operational parameters (pH, PMS dosage, $Co^{2+}$ dosage, and the anion effects) which affect the efficiency of $Co^{2+}$/PMS processes in RBB oxidation was also investigated.

(52) There have been a number of studies dealing with the mechanism under which sulfate radicals attack organics; however, compared to hydroxyl radicals, sulfate radicals have not been extensively studied.

(53) The catalytic decomposition of Oxone using cobalt, first reported in 1958, was shown to proceed via a radical mechanism.

(54) Since then, there have been a few studies exploring the transition metal activation of peroxymonosulfate for the oxidation of organic compounds, but to our knowledge, cobalt is the best catalyst-activator of this versatile inorganic peroxide.

(55) Previous investigations on the metal activation of peroxymonosulfate are limited to homogeneous systems.

(56) Following the proof of heterogeneous activation of Oxone discussed here, future research should now focus on the preparation of immobilized $Co_3O_4$ on solid surfaces.

(57) Driven by the need of overcoming the limitations of the Fenton Reagent and seeking for a process that introduces stronger oxidants than hydroxyl radicals, generated by almost all AOTs, this work explores an alternative oxidation technology.

(58) To our knowledge, the Co/PMS system has never been investigated before in the environmental field.

(59) This study explores the reactivity of Co/PMS at certain conditions and compares its efficiency with that of the Fenton Reagent for the destruction of three typical organic contaminants in water: 2,4-dichlorophenol, atrazine, and naphthalene.

(60) Specific investigations with respect to the catalytic behavior of cobalt and the primary radicals formed are also reported.

(61) Several investigators have measured PBDE levels in household dust samples and calculated potential exposure rates. However, most household studies have not included biomonitoring. Thus, it is difficult to make direct comparisons between PBDE levels in dust and biological samples.

(62) To identify household sources of PBDE exposure, we designed a follow-up study to evaluate residential exposure routes such as indoor air, household dust, and exposure to bromine containing household items.

(63) Several technologies are available to remediate soils that are contaminated by heavy metals. However, many of these technologies (for example, excavation of contaminated material and chemical/physical treatment) are extremely costly or do not achieve a long term solution. Phytoremediation is defined as the use of plants to remove or sequester hazardous contaminants from various media such as soil, water and air. This technique has become a tangible alternative to traditional methodologies.

(64) In addition, very few plants are known that can accumulate more than one metal. Therefore, researchers are searching new plant species that could potentially be used to remediate different metals from contaminated sites.

(65) Therefore, the present study was undertaken to determine the Pb, Cu, Ni and Zn accumulation capability of S.drummondii and to examine the possible interactions between these metals at roots and shoots level. In addition, this study also reports the involvement of various antioxidants (enzymatic and non-enzymatic) in the tolerance against Pb, Cu, Ni and Zn induced stress, individually and in combinations.

(66) The results of this study should be helpful to determine the potential application of S.drummondii to remediate different metals from contaminated sites and to elucidate the biochemical detoxification mechanisms against Pb, Cu, Ni and Zn-induced stress.

(67) A few recently published cohort studies have focused on women and have reported stronger effect estimates for fatal coronary heart disease (CHD) and all-cause mortality than previous studies.

(68) Questions remain regarding sex differences in the associations of chronic PM exposures with all-cause and cardiovascular mortality.

(69) Previous research has shown that hardwoods and softwoods as woody biomass can be degraded to carbon dioxide and methane under anaerobic conditions, although biodegradation of wood has most often been demonstrated under aerobic conditions.

(70) There has been limited work on the behavior of wood in landfills.

(71) The objective of this study was to characterize the anaerobic biodegradation of major types of natural lumber and engineered woods in a laboratory-scale simulated landfill.

(72) The mechanism of action through which MeHg increases the risk of CHD remains to be identified.

(73) We conducted a comprehensive health survey in the Inuit population of Nunavik (Québec, Canada) during the fall of 2004. In the course of this study, we investigated the relation between blood mercury concentrations and plasma PON1 activities in 896 Inuit adults living in Nunavik, while taking into account the potential protective role of selenium, which has been shown to counteract the toxicity of mercurials. We also investigated the possible confounding or modifying role of several factors, including PON1 gene variants.

(74) This goal of this preliminary investigation was to assess the release of CNMs into the laboratory atmosphere during handling and sonication into environmentally relevant matrices.

(75) We used controlled laboratory studies to determine if ER killifish are more or less sensitive to PAH-induced chronic hepatic toxicity than killifish from an uncontaminated site.

(76) We measured pestcide residues in 24-hr duplicate food samples collected from a group of 46 young children participating in the Children's Pesticide Exposure Study (CPES).

(77) The purpose of this work was to determine the effect of biological conditions on mixed liquor (ML) properties and their effects on the process and membrane performance.

(78) The goal of this research was to provide a fundamental understanding of SMBR performance at a range of F/M values so that economical design and operation, especially at high F/M, would be possible.

### 5.1.4 实例简析

实例1：来自于《Applied Catalysis B：Environmental》上已发表论文"Introduction"部分。论文题目为：Iron-cobalt Mixed Oxide Nanocatalysts：Heterogeneous Peroxymonosulfate Activation, Cobalt Leaching, and Ferromagnetic Properties for Environmental Applications. 下面在文中插入中文说明，进行简要剖析。阅读时，注意各要素的体现和时态与语态。

Introduction

In general, advanced oxidation technologies (AOTs) are based on the activation of oxidants such as hydrogen peroxide and ozone to generate highly oxidizing transient species such as hydroxyl radicals (HRs, ·OH) [1]. Semiconductor materials such as $TiO_2$ are also able to generate HRs when irradiated with high photon energy above their band gap, typically by ultraviolet (UV). Among various AOTs, Fenton reagent (e.g. $Fe^{2+}+H_2O_2$) and photocatalysis (e.g. $TiO_2$+UV) have been intensively investigated for environmental applications during the last decades because of their promising performances in the degradation of many pollutants in water and wastewater [2-5]. However, the Fenton reagent requires an acidic pH condition close to 3 while $TiO_2$ photocatalysis needs UV irradiation in most cases [2,3]. These requirements and some other issues remarkably limit their practical applications [6]. Compared to HRs with standard reduction potential of 1.8-2.7V, sulfate radicals (SRs, $SO_4^{-}$) also demonstrate high reduction potential of 2.5-3.1V at neutral pH and are more selective for oxidation [7]. Recently, many research studies have been conducted to find an alternative and practical approach employing SR-based technologies to the traditional AOTs

[8-15].（提供研究领域的一般概况与进展，即一般背景。这里的研究领域为：advanced oxidation technologies（AOTs））

In general, SRs are generated via homogeneous activation of peroxymonosulfate（PMS）with transition metal ions, among which cobalt is the best activator[14]. As expected, this homogeneous approach is highly efficient to degrade water pollutants. However, aesthetic aspects and health concerns associated with the adverse effect of dissolved cobalt species in water still need to be addressed. Anipsitakis et al. first demonstrated the heterogeneous PMS activation using commercially available $Co_3O_4$ particles [12]. More recently, in order to increase the reactivity of the catalyst, we immobilized and distributed nanosized $Co_3O_4$ particles onto various support materials, among which $TiO_2$ was proven the most efficient for limiting the cobalt leaching due to strong metal-support（Co-Ti）interactions [8, 9]. Moreover, in comparison with the commercially available bulk $Co_3O_4$, the $Co_3O_4/TiO_2$ system exhibited a much higher efficiency for the PMS activation due to its engineered properties at nanoscale [8]. However, the absence of practical and efficient approaches to recover the nanosized $Co_3O_4/TiO_2$ catalyst is a bottleneck for its environmental applications since nanosized materials, when discharged, might cause secondary environmental problems [16-18].（与本课题紧密相关的研究工作回顾。首先叙述 homogeneous activation of peroxymonosulfate（PMS）的功效与不足，然后交代 heterogeneous PMS activation 的研究进展和存在的不足）

Consequently, it is critical to develop a highly efficient nanoscale metal catalyst for the heterogeneous PMS activation, where the adverse impact of metal leaching on aqueous environments is notably minimized and the catalyst is readily recovered after its application.（提出需要解决的问题，即要克服 heterogeneous PMS activation 面临的挑战）Considering the environmental and practical aspects, Fe as a candidate for the catalyst might be more relevant than Co. However, the efficiency of Fe(Ⅱ) ions, as well as $Fe_2O_3$, which can be typically formed during heat treatment of Fe salts, were found to be inferior for the activation of PMS to degrade 2,4- dichlorophenol（2,4-DCP）[14]. Coupling of Fe with a suitable amount of Co（i.e.Fe-Co mixed oxide catalysts）might be an effective way to attain efficient catalysts for PMS activation. In addition, the cobalt leaching from the Fe-Co catalysts can be controlled since strong metal-metal interactions are typically observed during their heat treatment. Moreover, under appropriate conditions, the nanocomposite catalysts can be transformed to $CoFe_2O_4$ which can be easily recovered using magnetic-based separation due to its ferromagnetic properties [19].（提出解决问题的可能思路及依据）

In this study, we report the development of a novel approach of using Fe-Co mixed oxide nanocatalysts as a mediator for the heterogeneous PMS activation. We focus on the effect of several factors such as cobalt content, calcination temperature, and $TiO_2$ support for the Fe-Co catalysts on cobalt leaching, Fe-Co interaction, and heterogeneous activation of PMS. The catalyst performance was evaluated in terms of the degradation of highly toxic and poorly biodegradable 2,4-DCP, which mainly originates from the environmental transformation of herbicides such as 2,4-dichlorophenoxyacetic acid [13]. Moreover, we discuss the environmentally friendly aspects of the Fe-Co nanocatalysts, including reduced cobalt leaching and ferromagnetic-based separation properties.（阐明本研究的主题和具体内容）

该引言通过研究背景的介绍，指出目前研究中存在的不足，引出值得研究的问题，最后点明本研究的主题和研究内容。写作中条理清楚，过渡自然，背景介绍详略恰当，与本课题紧密相关的研究工作回顾简洁、全面，从形式和内容来看都是一篇好的引言。

实例2：来自于《Environ. Sci. Technol.》上已发表论文"Introduction"部分。论文题目为：Lead Contamination of Potable Water Due to Nitrification. 下面在文中插入中文说明，进行简要剖析。阅读时，注意各要素的体现和时态与语态。

Introduction

As United States utilities switch to chloramines for residual disinfection of potable water (1, 2) there is concern about potential costs and health implications of corrosion induced by nitrification (1). Nitrification, conversion of ammonia to nitrite ($NO_2^-$) and then nitrate ($NO_3^-$) by nitrifying bacteria, could impact corrosion by decreasing pH, alkalinity, and dissolved oxygen (1). Production of organic carbon and accelerated disinfectant decay might also stimulate growth of corrosion-influencing microbes (1,3). A 1991 survey indicated that two-thirds of the medium and large utilities that use chloramines report nitrification problems in water mains (3), and it is very likely that even a greater percentage would have nitrification issues if premise plumbing is considered (1, 4).（作为引言部分的第一自然段，作者介绍研究领域的一般概况与进展，即一般背景。这里叙述焦点是"corrosion induced by nitrification"）

There is some limited anecdotal evidence of corrosion problems triggered by chloramines and nitrification in at least some circumstances. For example, recent work in Pinellas County, FL, highlighted some concerns related to iron corrosion control and red water (5). Likewise, elevated copper levels at the tap were suspected to be linked to action of nitrifying bacteria in Willmar, MN, homes (6). Nitrification also co-occurred with higher lead leaching in Ottawa (7), Washington DC, and Durham and Greenville, NC, homes (8-10). However, any link between nitrification and increased lead contamination was not definitive, nor were mechanisms postulated except for the case of Ottawa for which it was proposed that the higher lead resulted from decreased pH due to nitrification (7).（围绕"corrosion problems triggered by chloramines and nitrification"叙述已有研究工作，交代了取得的结果，指出了研究的不足）

Given the high costs and health implications of corrosion to utilities and consumers (8, 10) and further considering that prior research emphasized nitrification problems occurring in the main distribution system whereas virtually all lead and copper plumbing materials are located within premise plumbing (11), it is important to better understand effects of nitrification on corrosion and metal release.（指明了有待研究的问题及其缘由）The objective of this study was to scientifically verify anecdotal links established between nitrification and increased lead leaching through a well-controlled laboratory study, solubility modeling, and field work at water utilities.（表明本研究的目的(主题)和主要研究内容）

本引言篇幅适当，关于研究背景叙述简明而完整，较好地体现了引言各要素。英文表达规范、地道。

实例3：来自于《Environ. Sci. Technol.》上已发表论文"Introduction"部分。论文题目为：Pharmaceuticals and Endocrine Disrupting Compounds in U. S. Drinking Water. 下面在文中插入中文说明，进行简要剖析。阅读时，注意各要素的体现和时态与语态。

Introduction

Pharmaceuticals and endocrine disrupting compounds (EDCs) are subclasses of organic contaminants that have been detected in wastewater and surface waters throughout the world (1-4). (给出 Pharmaceuticals and endocrine disrupting compounds (EDCs) 的定义。) Their occurrence is most often a result of municipal wastewater discharge, as these compounds are not completely removed during treatment (4). Other sources of pharmaceuticals and EDCs in water include runoff from agricultural fields, concentrated animal feeding operations, landfill leachates, and urban runoff (5-7). (水中 EDCs 的来源) Scientists and regulators are concerned about what level of risk may be associated with the presence of pharmaceuticals and EDCs in drinking water, as many drinking water treatment plants (DWTPs) use source water impacted by wastewater. (饮用水中出现 EDCs 的原因和产生风险的程度受到有关人员关注) While some researchers have postulated that the long-term risk to humans from any single pharmaceutical at sub-μg/L levels is negligible (8), it is not clear what toxicological implications chronic exposure to suites of trace contaminants may pose (9, 10). (关于人受到的相关风险的研究结果还很少) The degree to which this issue has drawn interest across disciplines is illustrated by the voices of concern stemming from medical professionals, environmental scientists, drinking water municipalities, government agencies, and the general media (9, 11-13). (人受到的相关风险成为各界人士的关注点) However, if risk assessors and epidemiologists are to link any potential health outcomes with pharmaceutical and EDC exposure, a better understanding of their occurrence in drinking water is critical. (引出值得研究的工作)(本段作为引言的开头部分，通过叙述有关研究背景，得出有待研究的领域)

There is relatively sparse information regarding pharmaceutical and EDC occurrence in drinking water. Researchers in Germany measured ng/L concentrations of clofibric acid in Berlin tap water (14), a case which remains a strong illustration of the sometimes close wastewater to drinking water coupling of unintended water reuse. The elimination of pharmaceuticals at German DWTPs was attributed to ozone oxidation or adsorption to granular activated carbon (15): finished drinking water concentrations of five compounds were < 10 ng/L. The occurrence of 106 organic wastewater contaminants, including some pharmaceuticals and potential EDCs, at different stages of a U. S. DWTP was documented by Stackelberg et al. (16): 18 compounds were measured in finished drinking water at concentrations up to 258 ng/L. Bruchet et al. (17) investigated the occurrence of 21 antibiotics and X-ray contrast agents in the Seine River, through groundwater recharge, and in finished drinking water. In this case, only four X-ray contrast agents persisted into finished drinking water at concentrations up to 60 ng/L. Selected antibiotics were measured in the finished water of three DWTPs in the U. S. at concentrations up to 5 ng/L (18). As a precursor to the data presented in this study, concentrations of a more limited set of pharmaceuticals and EDCs were measured in source and finished drinking water from 20 utilities in the U. S. (19) (see Supporting Information Table S1). (本段对与本课题紧密相关的研究工作进行了回顾，表明已有结果是较有限的)

This paper describes results from a comprehensive survey of 20 pharmaceuticals, 25 known or potential EDCs, and 6 other wastewater contaminants in source water, finished drinking water, and distribution system (tap) water from 19 U. S. DWTPs sampled during 2006-2007. (交代本研究目的(论文导向)) The results provide an assessment of the actual concentrations to which people are exposed from drinking water. Occurrence data were used to propose a set of indicator compounds that can predict the presence of other pharmaceuticals and EDCs as well as monitor the efficacy of treatment processes.（指出本研究的意义）

本论文引言篇幅适当，通过研究背景叙述道出值得研究的领域，再对属于上述领域的、与本课题紧密相关的研究工作进行了回顾，简洁而全面，最后引出本研究目的和意义。该引言行文流畅，连接自然，较充分地反映了引言的要素组合。英文表达规范、地道。

实例4：来自于《Water Research》上已发表论文"Introduction"部分。论文题目为：Dissolved Organic Nitrogen Removal During Water Treatment by Aluminum Sulfate and Cationic Polymer Coagulation. 下面在文中，插入中文说明，进行简要剖析。阅读时，注意各要素的体现和时态与语态。

Introduction

Research in the drinking water field over the past quarter of a century has focused on dissolved organic carbon (DOC), thus avoiding potential issues associated with dissolved organic nitrogen (DON).（指出对DON问题关注的缺失）Natural organic matter (NOM) contains roughly 40%-60% carbon by weight and 1-5% nitrogen by weight (International Humic Substances Society). The authors' previous analysis of total Kjehldahl nitrogen and ammonia data for approximately 23,000 samples in the United States Geologic Survey (USGS) National Aquatic Water Quality Assessment (NAWQA) databases yielded a median DON concentration of 0.37 mg/L of N in surface water and DON concentrations of 0.24 and 0.18 mg/L of N in shallow and deep groundwater, respectively (Westerhoff and Mash, 2002). In the authors' DON occurrence sampling campaigns from 28 US water treatment plants (WTPs), the average DON concentration of raw waters was 0.19 mg/L of N, with the dialysis based pretreatment and nitrogen species analysis (Lee et al., 2006). DOC/DON ratios of the 28 raw waters averaged 18 mg DOC per mg DON.（通过提供DON在水环境（地表水和地下水）和水厂原水中的背景数据（本底含量），表明DON在环境中的客观存在）

During water treatment, nitrogenous moieties of NOM can react with disinfectants to form carcinogenic, nitrogenous disinfection by-products (e.g., haloacetonitriles, halonitromethanes, N-nitroso-dimethylamine) and can affect the speciation of regulated disinfection by-products such as trihalomethanes and haloacetic acids (Peters et al., 1990; Richardson, 2003). DON also reacts with free and combined chlorines to form organic chloramines that have little or no bactericidal activity (Isaac and Morris, 1983; Donnermair and Blatchley Ⅲ, 2003). Furthermore, proteinaceous materials have been implicated with polysaccharides as an important component of organic foulant on membranes during water treatment and wastewater reclamation (Lee et al., 2003; Her et al., 2004).（指出在水处理过程中存在的DON会产生的危害）A need thus exists to understand and quantify the removal of DON prior to chlorination, chloramination, or membrane treatment.（基于前面所述背景，引出值得研究的课题（研究领域），即"to understand and quantify the removal of DON during water treatment"）This paper focuses on DON removal during coagulation.（点明本研究主题属于该需要研究的课题）

During coagulation, NOM is removed through charge neutralization, entrapment, and sorption onto floc surfaces; the particulate phase is subsequently removed during solid-liquid separation

(Letterman et al., 1999). Previous laboratory jar tests demonstrated that Al and Fe salts do not efficiently coagulate organic nitrogen compounds (e.g., dimethylamine, cyclohexylamine, pyrrolidine, piperidine, morpholine, piperazine) and proteins (Vilge-Ritter et al., 1999; Pietsch et al., 2001). Widrig et al. indirectly evaluated DON removal during coagulation of extracellular organic matter from green algae by using pyrolysis-gas chromatographymass spectrometry (Py-GC-MS); algal-derived DOC (nitrogen enriched) was difficult to remove (Widrig et al., 1996).(概述有关"DON removal during coagulation"的已有研究结果)Except for these limited studies using model compounds (e.g., organic amines, proteins) or analytical groups examined by Py-GC-MS, the removal efficiency of bulk DON during water treatment has not been previously reported in the literature.(指出已有研究的不足)

In the authors' DON occurrence sampling campaigns from 28 US WTPs, the DON concentrations of individual waters decreased by an average of 20% across the full-scale WTPs (Lee et al., 2006). The seven utilities that did not use polymers averaged only 9% DON removal, while the 21 utilities using polymers averaged 23% removal. Eighteen of these 21 utilities used cationic polymers with dosages ranging from 0.5 to 3.5mg/L.(提供水厂实际运行中 DON 去除的情况)It is unclear, however, if the use of cationic polymers was the direct cause of the higher DON removal or simply coincidental.(指出对有关处理机理认识不足)The objective of the present study is to investigate DON removal in the laboratory during coagulation (aluminum salt and/or cationic polymer) of raw waters.(交代本研究的目的(主题))We hypothesize that cationic polymer (polyelectrolyte) addition improves the coagulation efficiency of aluminum to remove DON even though cationic polymers contain organic nitrogen.(提出对本研究结果的预期)Jar tests were performed on three surface waters to determine the effect of cationic polymer on DON removal during aluminum coagulation. Removal efficiencies of DON and DOC, ultraviolet absorbance at 254nm (UVA254), and turbidity were monitored. Molecular weight fractionations of DOC and DON were also performed to assess the effect of polymer on the removal of each molecular weight fraction of NOM.(交代具体实验研究内容)

本论文引言篇幅适中，通过研究背景叙述道出值得研究的课题(研究领域)，再对属于上述领域的、与本课题紧密相关的研究工作进行了回顾，简洁而全面，最后引出本研究目的(主题)、研究的期望和实验研究的主要内容。该引言表述条理清楚，行文流畅，充分地体现了引言的各要素组合。英文表达规范、地道。

## 5.2 材料与方法部分

材料与方法部分是英语科技论文正文的第二大组成部分。科学研究的基本要求是研究结果的真实性和可靠性，而研究结果能够被重复是重要标志，决定结果能否被重复的依据就是论文所描述的"材料与方法"。根据不同的研究领域、期刊种类，"材料与方法"的写作方式是有差异的，但一般学术期刊都有很详细的、约定俗成的规则，论文作者应给予足够关注，这样才能事半功倍。

### 5.2.1 功用

用准确与简洁的语言叙述实验用原料与材料、实验技术和方法等，为他人再现论文中的研究结果提供必要条件，保证本研究设计和研究结果的可靠性和可信度。

### 5.2.2 组成要素

英语科技论文中“材料与方法”部分通常包括以下几方面的信息，即组成要素。

1）实验用原料与材料

原料与材料的技术规格、数量、来源及制备方法、主要的物理和化学性质。

在说明来源和纯度时，若还需自行提纯，则要说明提纯方法和提纯后的纯度，有时要提供一些物理性质。一般不采用原料与材料的商品名称，而采用它们的属名或化学名称。

2）实验技术和方法

在任何实验里，可变的因素大致分为两类，即被研究的各种可变因素和实验条件中的各种可变因素。前一类是观察项目，应该写在实验结果节里，后一类是需要控制的实验条件，应该写在实验方法节里。后一类的可变因素中，有些是容易控制的，有些是不容易控制的。但必须尽力控制，使实验条件稳定，前后一致。应该详细说明控制这些可变条件的方法，以显示结果的可靠性和准确性。

提供的实验技术和方法要能达到用最少的实验次数获得最优的实验结果。

a. 实验设备和仪器

（1）对通用的、标准的、常见的设备和仪器，只提供型号、规格、主要性能指标；

（2）对前人用过的专门设备和仪器，只需要给出文献；

（3）对自行设计的设备和仪器，需较详细地说明其特性；

（4）必要的设备和仪器的操作条件和过程。

b. 实验过程和方法

（1）对一般的实验过程，不必详细叙述，对特殊的实验过程，要专门说明；

（2）对公知公用的方法写明其方法名称即可；

（3）引用他人的方法、标准，属于已有应用而尚未为人们熟悉的新方法，应注明文献出处，可作简要介绍；

（4）对自行改进或创新的实验方法应详细介绍。

描述实验技术和方法，通常按研究步骤的时间顺序或空间层次进行。

### 5.2.3 时态、语态与常用句型

在英语科技论文的“材料与方法”部分通常采用被动语态、一般过去时态。因为在这部分描写时，句子中的主题或中心是实验原料与材料、场地和方法本身，表达“做了什么”、“怎么做的”之意，而不在意“谁做了什么”。如果涉及表达作者的观点或看法，则可用主动语态或动词不定式结构。若描述的内容为不受时间影响的事实，采用一般现在时。常用的句型有：

（1）… was（were）done to do….

（2）… was（were）removed… and then weighed.

（3）… was（were）evaluated….

（4）… was（were）selected… and asked to do….

（5）… was（were）measured through….

（6）The work was carried out on A，which has been described in detail elsewhere.

（7）The samples were immersed in an ultrasonic bath for 3 minutes in A followed by 10 minutes in distilled water.

（8）For the second trial，the apparatus was covered by A. We believed this modification would reduce B.

(9) For the second trial, the apparatus was covered by A to reduce B.

其他实际典型例句(单句或连续语句)如下所示:

(1) This suspension was stirred for 24 h, and then dried under an infrared lamp at around 50℃ to remove water.

(2) Finally, the catalysts were calcined in a furnace (Paragon model HT-22D, Thermcraft) at 500℃ (if not specified) for 4 h with a ramp rate of 10℃/min.

(3) Based on this methodology, the various cobalt catalysts were also prepared using different supports, $SiO_2$(99.5%, Aldrich) and $\gamma$-$Al_2O_3$(PURALOX TH 100/150, 98%, Sasol), and different cobalt precursors, $CoCl_2 \cdot 6H_2O$ (98%, Aldrich) and $CoSO_4 \cdot xH_2O$ (Aldrich).

(4) The resulting catalysts were ground thoroughly and labeled as XCo/support, where X stands for the molar ratio of Co support.

(5) The crystallographic structure of the supported cobalt catalyst was investigated with X-ray diffraction (XRD) analysis using a Kristalloflex D500 diffractometer (Siemens) with Cu Kα ($\lambda$ = 1.5406Å) radiation.

(6) Temperature programmed reduction (TPR), as a powerful tool to differentiate various cobalt species possessing different interaction strengths with the supports, was conducted on a Micromeritics AutoChem 2910 TPD/TPR instrument.

(7) A Tristar 3000 (Micromeritics) surface area and porosimetry analyzer was used to determine Brunauer, Emmett, and Teller (BET) surface area and pore size of the cobalt catalysts.

(8) A quartz reactor (base: 10 cm × 10 cm; height: 25 cm) containing 1 L of 50 mg/L (0.307mM) 2,4-DCP solution with adjusted pH of 7.0 using $K_2HPO_4$ and $KHSO_4$ was placed on a magnetic stirrer plate and equipped with a cooling fan.

(9) For the measurement of 2,4-DCP concentration during 2 h of reaction, 10 mL sample was withdrawn at specific time interval and quenched with 5 mL of 2.47 M methanol (Aldrich) to prevent further reaction.

(10) The sample was filtered with 0.1mm filter (Magna Nylon, Fischer) and analyzed using a high performance liquid chromatograph (HPLC, Agilent 1100 Series) with a photo-diode-array detector.

(11) An atomic absorption spectrometer (Perkin-Elmer AA-300) was used to study cobalt leaching to the solution.

(12) The support materials ZnO (Kanto Chemical, 99%) and P25 titania (Degussa, >99.5%) were obtained from commercial suppliers and used without further purification.

(13) $ZrO_2$ was obtained by calcination of zirconium oxyhydroxide which was prepared by precipitation of $ZrOCl_2$(Kanto Chemical, 99%) using ammonia (Kanto Chemical, 28%-30% in water) according to the procedures described in a previous report.

(14) SBA-15 was obtained from Dr.Y.H.Yang Research Lab, which was prepared according to their reported method in the literature.

(15) The catalysts of cobalt oxide supported on various metal oxides were prepared by incipient wetness impregnation with an aqueous solution of $Co(NO_3)_2 \cdot 6H_2O$ (cobalt nitrate hexahydrate,

Fluka Chemika, >99%), followed by drying at 60℃ overnight and then calcination at 600℃ for 3 h in static air.

(16) The morphology and nanocrystal sizes were measured by transmission electron microscope (TEM) in JEOL 2010 with an accelerating voltage of 200 kV.

(17) X-ray photoelectron spectroscopy (XPS) analysis was conducted in an Axis Ultra Spectrometer (Kratos Analytical) using a monochromated Al Kα X-ray source (1486.7eV) operating at 15 kV.

(18) All chemicals used in this study were reagent grade.

(19) Double deionized water (DDW) with a resistivity above 18.2 MX cm was produced using a commercial deionization system (AquaMax System, Young-Lin Instrument Co., Korea).

(20) For pH adjustment, 0.1M sulfuric acid and/or 0.1M sodium hydroxide were used, and all the experiments were carried out in an air-conditioned room at 23(±2℃) in duplicates.

(21) The performance of spent cobalt oxide was also examined by collecting the used cobalt oxide after the tests and filtered using 0.45μm filter.

(22) The spent oxide was washed by DDW and then dried in a desiccator for later use.

(23) All the experiments were conducted in duplicate and the experiment errors were less than 3%.

(24) The degradation experiments were conducted at ambient temperature in 0.5 and 2L batch-reactors.

(25) It was decided not to cover the vessel, with aluminum foil for example, to be able to watch the color development in the reactor, which is indicative of the progress of the reaction (catalyst speciation, complexation, etc.).

(26) The initial concentrations of the contaminants were 50 mg/L (0.307 mM) for 2,4-DCP, 8 mg/L (0.037 mM) for atrazine, and 5 mg/L (0.039 mM) for naphthalene.

(27) The following pairs, $NaOH/K_2HPO_4$, $H_2SO_4/NaHCO_3$, $Na_2CO_3/NaHCO_3$, and $KH_2PO_4/K_2HPO_4$ at various molar ratios, were the buffering species introduced in solution for pH adjustment in the range of 5.8-9.0.

(28) The degradation experiments were carried out in erlenmeyer flasks.

(29) Dye solution with 200 mL was filled into the flasks and reacted at room temperature (26℃, with water bath) and 700 rpm stirring speed.

(30) The initial concentrations of basic blue 9 and acid red 183 were kept at 7 and 165 mg/L, respectively, in most of tests, unless indicated.

(31) For some tests, total organic content (TOC) was also determined using a Shimadzu TOC-5000 CE analyser.

(32) Reactive dye-Black B (Fig. 1) was purchased from Aldrich.

(33) Other chemicals used herein, including sulfuric acid, sodium hydroxide and the phosphate buffer solution (obtained from Scharlau), were of reagent grade and were used to adjust pH.

(34) All sample solutions were prepared using deionized water from the Millipore Milli-Q system.

(35) The absorption of RBB is maximum at $\lambda_{max} = 591nm$, and the color removal relating to its

decomposition was determined using a UV-visible spectrophotometer.

(36) All RBB samples in the oxidation processes were analyzed immediately after sampling so as to prevent further reactions.

(37) The following chemicals were used as received with no further purification:...

(38) Stock solutions of all chemicals were prepared in advance.

(39) Oxone ($2KHSO_5 \cdot KHSO_4 \cdot K_2SO_4$) was also tested at several doses from 0.5 : 1 to 64 : 1 as the mole ratio of peroxymonosulfate versus 2,4-DCP.

(40) Oxone powder was dissolved directly in the reaction vessel at amounts very close to its solubility limit (256 g/L) : 12.41 g in 50 mL, obtaining a concentration of 383.5 mM (or 235.8 g/L) as Oxone.

(41) All experiments were performed at room temperature and in deionized water without the pH being controlled.

(42) The following chemicals with purity ranging from 97% to 99% were obtained from Aldrich, USA and used as received....

(43) Quantitative analysis for chlorophenols was performed with a high performance liquid chromatography instrument (HPLC, Agilent 1100 series).

(44) Total organic carbon (TOC) was monitored using a Shimadzu TOC analyzer (Model: 5050).

(45) All the degradation experiments were conducted at neutral pH ($pH = 7.0$) which was maintained using 50 mM phosphate/NaOH buffer.

(46) Initial and final pH was monitored in all degradation experiments.

(47) Samples were taken periodically to measure the 4-CP, oxidant, and soluble iron concentration profiles with time.

(48) Selected experiments were performed in triplicate to assure accurate data acquisition and error bars in figures represent the standard deviation.

(49) Reactors were operated under conditions designed to maximize the rate and extent of decomposition.

(50) All reactors were monitored until they were no longer producing measurable methane.

(51) The methods employed for analysis of gas concentrations and volume, and leachate COD have been presented previously.

(52) This health survey was conducted in Nunavik, a northern region of Québec where approximately 9,500 Inuit live in 14 communities along the coasts of Hudson Bay, Hudson Strait, and Ungava Bay.

(53) Data on the consumption of traditional foods were obtained from a food frequency questionnaire, which was designed to measure season-specific consumption of food items derived from fishing and hunting during the year before the survey.

(54) Parents were instructed to collect 24-hr duplicate food samples of all conventional fruits, vegetables, and fruit juices equal to the quantity consumed by their children, similarly prewashed/prepared, and from the same source or batch. Individual or composite food items were analyzed for organophosphate (OP) and pyrethroid insecticide residues.

(55) At various time points, larvae were analyzed for CYP1A activity, BaP concentrations, nuclear and mitochondrial DNA damage, and liver pathology.

(56) A pilot-scale SMBR (Fig. 1), designed to operate at a range of hydraulic residence times ($\theta_H$) while maintaining a constant membrane flux of 30.6L/m$^2$h (LMH), was custom-built by ZENON Environmental, Inc. (Oakville, Ontario, Canada).

(57) The membrane was operated at a constant, but higher, flow rate (flux) than required for maintaining the desired $\theta_H$.

(58) The aeration tank was equipped with fine-bubble air diffusers. Aeration air was supplemented with pure oxygen to maintain adequate dissolved oxygen (DO) concentrations at the higher organic loading rates.

(59) The pilot-scale SMBR was fed with primary effluent from the Southeast Water Pollution Control Plant (SEP), San Francisco, CA (Table 1).

(60) The reactor was operated for three MCRTs prior to steady state data collection at each MCRT tested. ML DO concentration was ≥2 mg/L and $Na_2CO_3$ was added to the feed wastewater to control the aeration basin pH to ≥6.5 during nitrification. Sludge wasting was performed on a continuous basis by pumping from the upper portion of the membrane tank.

### 5.2.4 实例简析

实例1：来自于《Water Research》上已发表论文"Materials and Methods"部分。论文题目为：Dissolved Organic Nitrogen Removal During Water Treatment by Aluminum Sulfate and Cationic Polymer Coagulation。下面在文中插入中文说明，进行简要剖析。阅读时，注意各要素的体现和时态与语态。

2. Materials and methods

2.1. Raw water

Three surface samples of raw water for full-scale WTPs were collected from Harwood Reservoir (Yorktown, VA) on May 24, 2004; the Huron River (Ann Arbor, MI) on July 13, 2004; and the Salt River (Mesa, AZ) on August 15, 2004. Water samples were collected in four 20L Nalgene high-density polyethylene containers and delivered to the laboratory via overnight delivery (Harwood Reservoir and Huron River waters) or directly (Salt River water). The waters were selected to represent a range of specific UVA at 254 nm (SUVA) and alkalinity levels (Table 1). Harwood Reservoir water had relatively higher SUVA and lower alkalinity; Huron River water had higher SUVA and higher alkalinity; Salt River water had lower SUVA and moderate alkalinity. DON concentrations in the three waters were similar (0.25-0.35mg/L of N). In addition, the DON to total dissolved nitrogen (TDN) ratios of the waters were>40%, which can reduce interference from dissolved inorganic nitrogen (DIN) in the quantification of DON (Lee and Westerhoff, 2005). (Raw water 是本实验研究的主要原材料，这里单列详细说明，包括来源、水质特性等。在水处理类型的研究论文中，原水水样是重要原料，一般需要专门介绍)

2.2. Jar tests

Coagulation and flocculation experiments were performed in a jar test apparatus (Stirrer Model 7790-400, Phipps and Bird, VA). Each sample (1.5 L) was rapidly mixed at 100 revolutions per minute (rpm) for 2 min, slowly mixed at 30 rpm for 20 min, and then settled (0 rpm) for

60 min. During the rapid mixing, aluminum sulfate ($Al_2(SO_4)_3 \cdot 18H_2O$) and/or a cationic polymer (polydiallyldimethyl-ammonium chloride (polyDADMAC), Polydyne Inc., GA) were added. In aluminum sulfate/polymer jar tests, aluminum sulfate was added first, followed by polymer (after 30s) during the rapid mix. Coagulants were added using an automatic pipette with the tip submerged to deliver the coagulant into the most turbulent zone (directly above and to the side of the stirring paddles) (Bolto et al., 1998). Aluminum sulfate dosages ranged from 0 to 110mg/L. Cationic polymer dosages were 0 to 3 mg/L. Two duplicate jars were used for each condition. At the end of each jar test, the turbidity of the sample was measured. The samples were then filtered through pre-ashed (at 500℃) glass fiber filters (GF/F; pore size = 0.7μm; Millipore, MA) and analyzed for other parameters.

PolyDADMAC is made up of a long chain of $C_8H_{16}NCl$ (Fig. 1) and has a molecular weight of around 300, 000 g/mol (provided by manufacturer). The polymer has an intrinsic viscosity of 80-190 centipoise (provided by manufacturer) and a high cationic charge density of 6.4 meq/g (based upon titration). PolyDADMAC contains 8% nitrogen by weight in the active polymer, so a polymer dose of 1mg/L actually adds 80 mg/L of DON during water treatment. The polymer was provided as a 20% active aqueous solution but was diluted to give a stock solution of 1%. (Jar test 即混凝试验或搅拌试验，是常用的实验方法。但其中的操作参数，如时间、体积、混凝剂等各不相同，这里省略了常用混凝剂的描述，但对新混凝剂作了适当介绍)

2.3. Molecular weight fractionation

Raw waters and select jar-test samples were fractionated using two 400 mL commercial stirred cell units (Amicon 8400; Millipore Corp., MA) in parallel. Two types of regenerated cellulose membranes were used: (1) YM 1, which has a nominal molecular weight cut-off 1000 (1k) Da, and (2) YM 10, which has a nominal molecular weight cut-off 10,000 (10k) Da. The initial sample volume was 320 mL. After 250 mL of sample volume permeated the membrane, the remaining 70 mL was collected as the retentate.

In all the samples tested, mass balances for DOC, DON, and $UVA_{254}$ before and after molecular weight fractionation were within ±15%. (这里描述了一种比较专门的实验方法，提到两种装置 (stirred cell units 和 regenerated cellulose membranes)，介绍了三个计算式)

2.4. Analytical methods

A Shimadzu TOC-$V_{CSH}$ analyzer (high temperature combustion at 720℃; non-dispersive infrared detection) with a TNM-1 TN unit (chemiluminescence detection) (Shimadzu Corp., Japan) was used to measure DOC and TDN simultaneously. Nitrate and nitrite were measured using a Dionex DX-120 Ion Chromatography system (Dionex Corp., CA). Ammonia was measured by the automated phenate method (Standard Method 4500-$NH_3$ G) (APHA et al., 1998) using a TRAACS 800 autoanalyzer (Bran-Luebbe, Germany). UV spectra were measured using a MultiSpec-1501 spectrophotometer (Shimadzu Corp., Japan). SUVA ($UVA_{254}$/DOC) was calculated. A turbidity meter (HF Scientific Inc., FL) and a pH meter (Beckman Coulter Inc., CA) were calibrated prior to each use. Alkalinity was measured by a titration method (Standard Method 2320 B) (APHA et al., 1998). Soluble aluminum species and positive charge dose as a function of pH were calculated by MINEQL V.4.5 (Environmental Research Software, ME). Statistical analysis of experimental data was performed using SPSS V. 11. 0 (SPSS Inc., IL). (说明有关水质参数(物理、化学等的参数)的分析测定方法，指明了方法依据和仪器设备等，这是一个常见的"方法"描写片段，其具体内容随研究课题的不同而不同)

本论文属于饮用水处理类型的研究论文，"Materials and Methods"的描写，原料部分突出了实验用水样的介绍，其他则着重说明实验方法，其中涉及实验用药剂、仪器设备等。写作布局合理，表述详略适当，要素组合自然，英文表达规范、地道。

实例 2：来自于《Environ. Sci. Technol.》上已发表论文"Materials and Methods"部分。论文题目为：Comparison of Byproduct Formation in Waters Treated with Chlorine and Iodine: Relevance to Point-of -Use Treatment。下面在文中插入中文说明，进行简要剖析。阅读时，注意各要素的体现和时态与语态。

Materials and Methods

Natural Water Collection. Natural waters were collected in fluorinated high-density polyethylene containers. Except where noted, the waters were filtered through 0. 7-μm nominal pore size borosilicate microfiber filters (Environmental Express, Mt. Pleasant, SC) and stored at 4℃; the glass fiber filters had been baked at 400℃ to remove any organic contaminants prior to use. Dissolved organic carbon (DOC) analyses were conducted using a Shimadzu TOC-VCSH total organic carbon analyzer. Table 1 provides water quality characteristics of the waters examined in this study. (原水水样是重要原料，在这里单列专门介绍，涉及来源、预处理和水质特性等)

Materials. Free chlorine stock solutions (20 mM) were standardized by UV absorbance at 292 nm (21). Preformed monochloramine stock solutions (20 mM) were constituted by mixing free chlorine and ammonium chloride at a 1∶1. 2 molar ratio and standardized by UV absorbance at 245 and 295 nm, as described previously (21). A tincture of iodine stock solution was formulated as a mixture of 72 mM elemental iodine (Acros resublimed, Fair Lawn, NJ) and 172 mM potassium iodide (Acros, Greel, Belgium) in 50% deionized water and 50% methanol, and was standardized by titration against sodium thiosulfate. A Lifestraw Personal POU treatment unit was obtained from Vestergaard Frandsen (Lausanne, Switzerland); note that Vestergaard Frandsen has recently changed the design of the Lifestraw from an iodinated resin to a hollow-fiber filter system. Fisher Scientific iodoform (99%) and a 0. 2 mg/mL standard mix of the four regulated THMs (THM4; chloroform, bromodichloromethane, dibromochloromethane, and bromoform) in methanol (AccuStandard, New Haven, CT) were employed as standards. The remaining I-THMs and iodoacid standards were purchased at the highest level of purity from Orchid Cellmark (New Westminster, BC, Canada), CanSyn Chem. Corp. (Toronto, ON, Canada), and Sigma-Aldrich. (介绍实验用的药剂，包括来源及制备方法等)

Reactions. Reactions for analysis of THM4 and iodoform were performed in duplicate in 25 mL headspace-free vials with PTFE-lined septa, and were initiated by injection of 36 μM oxidant of tincture of iodine or 200 μM hypochlorite stock solutions via syringe injection through the septa. Triplicate reactions were conducted similarly in 100 mL headspacefree vials for the analysis of 6 I-THMs (dichloroiodomethane, bromochloroiodomethane, dibromoiodomethane, chlorodiiodomethane, bromodiiodomethane, and iodoform) and 6 iodo-acids (iodoacetic acid, bromoiodoacetic acid, diiodoacetic acid, (Z)-3-bromo-3-iodopropenoic acid, (E)-3-bromo-3-iodopropenoic acid, and

(E)-2-iodo-3-methylbutenedioic acid). At the 36 μM total oxidant dose of iodine tincture employed, calculations indicate that the initial iodine speciation was 32.7 μM $I_2$ and 3.3 μM $I_3^-$(22). Reactions using 6 g iodine tablets (Potable Aqua Purification Technology; 16.7 % tetraglycine hydroperiodide) employed two tablets per liter of solution in 1L amber jars. The initial total residual iodine concentration measured after dissolution of the tablets into deionized water was 3.4 mg/L as $Cl_2$. After injection of the disinfectant, reaction vials were shaken for 1 min and then stored in the dark. For treatments using the Lifestraw Personal, water was pumped from the Teflon sample jar through Teflon tubing and through the Lifestraw via a peristaltic pump at 1.4 mL/sec. The Lifestraw contains an iodinated resin followed by activated carbon post-treatment, all contained within a~170 mL cylinder (empty bed contact time=2 min). Samples were collected into either 500 mL volumetric flasks for nitrosamine analysis, or in 25 mL headspace-free vials for trihalomethane analysis. A more limited number of samples were analyzed for total organic chlorine (TOCl), total organic bromine (TOBr), and total organic iodine (TOI). Samples were treated with oxidants in 500 mL amber glass bottles under headspace-free conditions.（本节介绍对水样实施处理的方法，也是本实验研究的主体方法，包括实验条件、目的、反应装置、供分析的样品的收集等）

Analyses. One sample aliquot was analyzed for total oxidant residual by the DPD colorimetric method (23). Oxidants in other aliquots were quenched after 24 h with either 200 μM ascorbic acid (for THM4, iodoform, and NDMA measurements), 220 μM sodium sulfite (for 6 I-THMs and 6 iodo-acids), or a stoichiometric concentration of sodium sulfite for total organic halogen measurements. THMs and iodo-acids were extracted by liquid-liquid extraction with methyl *tert*-butyl ether (MTBE) within 5 min after quenching the disinfectant residual. THMs (including I-THMs) were analyzed by gas chromatography (GC) with electron capture detection or electron ionization-mass spectrometry (MS) against standards spiked into 20 mM phosphate buffered water and extracted and analyzed as for the samples. Iodoacid measurements were carried out using diazomethane derivatization, and detection by GC/negative chemical ionization (NCI)-MS, also against standards spiked into buffered water and extracted as for the samples. Samples (500 mL) were analyzed for nitrosamines by EPA Method 521. TOCl, TOBr, and TOI analyses were conducted in triplicate using a previously published procedure (24, 25), with minor modification. A Mitsubishi AQF-100 precombustion station (Cosa Instruments, Norwood, NJ) was interfaced to an ion chromatography (IC) system (LC30 chromatography oven, AD25 absorbance detector, Dionex), which was used to separate and detect the halide ions. The relative standard deviation of replicate analyses was generally<25%. Further details on these methods are available in the Supporting Information (SI).（介绍本实验研究中所用的全部分析测定方法，包括分析的具体项目、方法的依据、仪器设备、测定误差等）

本“Materials and Methods”部分，根据实际需要，“材料”和“方法”描写并重，重点部分叙述充分(如水质列表说明等)，注重言之有据，避免简单的重复介绍，信息组织有序、有理，英文表达规范、地道。

实例 3：来自于《Environ. Sci. Technol.》上已发表论文“Experimental Section”部分。论文题目为：Lead Contamination of Potable Water Due to Nitrification. 下面在文中插入中文说明，进行简要剖析。阅读时，注意各要素的体现和时态与语态。

Experimental Section

Water Chemistry. Lead pipes (1.9 cm×30 cm) were aged by exposure to a synthesized water for 1 year without nitrification and then exposed to water with 2 mg/L-N ammonia (and resulting nitrification) for 15 months as described elsewhere (4). No disinfectant had ever been added to the pipes. Thirty pipes were exposed at 5, 60, and 1000 ppb orthophosphate-P (10 at each phosphate level), representing low, moderate, and high levels of phosphate typically encountered in potable water distribution systems. The 10 replicate pipes were separated into three groups. The first group continued as a control (4 pipes), and the second group was modified by addition of free chlorine to a final concentration of 10 mg/L total chlorine (3 pipes). The added chlorine reacted with the ammonia to form almost exclusively monochloramine (residual free chlorine and free ammonia undetectable). The third group of pipes was modified by addition of 1mg/L chlorite (3 pipes). The high chloramines (12) and chlorite (13, 14) levels were added to inhibit nitrification, which was allowed to proceed unimpeded in the control. The pH of each type of water was adjusted to 8 before filling up the pipe. The alkalinity of the water was dropped stepwise from 100 mg/L to 30, 15, and then 0 mg/L alkalinity by decreasing the amount of $NaHCO_3$ added. Each alkalinity level was maintained for sufficient time for nitrification, final pH, and lead leaching to stabilize. Water in the pipes was changed twice a week, and pipes were maintained at room temperature. (本段冠名为:"Water Chemistry",实际为基本实验过程描述,其中指明有关物料,该实验过程也有一定特性)

Analytical Methods. Nitrifier activity was measured by loss of ammonia, production of nitrite and nitrate, and reduction of pH. pH was monitored using a pH electrode according to Standard Method 4500—$H^+$B(15). Total ammonia ($=NH_3+NH_4^+$) was measured using a salicylate method with a HACHDR/2400 spectrophotometer according to Standard Method 4500-$NH_3$(15). $NO_2$—N and $NO_3$—N were measured using DIONEX, DX-120 ion chromatography, according to Standard Method 4110 (15). Dissolved lead was operationally defined as that which passed through a 0.45 μm pore size syringe filter. Total metal release was quantified by digesting samples with 2% nitric acid for 24 h in a 90℃ oven. Metal concentrations were quantified using an inductively coupled plasma mass spectrophotometer (ICP-MS) according to Standard Method 3125-B (15). (本实验研究中涉及有关生物、化学变量的测定,本节具体介绍各相关变量的测定方法,包括所用的仪器和装置)

Case Studies. Two types of case studies were conducted. (本研究的特性研究方法——案例研究)

*Water Utility Studies.* Five participating utilities coordinated the collection of samples at three consumer homes before and after stagnation. Samples were analyzed for ammonia, nitrite, nitrate, lead, and copper release and other basic water quality parameters (pH, chlorine, alkalinity, temperature, HPC, etc). More details of the sampling procedures are included in the Supporting Information. (特性研究方法 1——实际案例调查)

Montana Bench Test. A plumbing rig was constructed to directly test the effect of pipe material (PVC versus copper) on nitrification and resulting lead contamination of water by leaded brass. A brass rod (0.64 cm diameter×10 cm length, C35300 alloy with 2% lead) was machined and placed inside a PVC or copper pipe (1.3 cm diameter×61 cm length) to simulate the situation in homes with PVC/copper plumbing and leaded brass faucets. The brass rod was not in electrical contact with the PVC or copper pipe. Each experiment was run in triplicate using synthesized potable water containing nitrifying bacteria. The synthesized water contained $(NH_4)_2SO_4$(2.13 mg/L-N), initial pH of 8.15, $Na_2HPO_4$(1mg/L-P), $NaHCO_3$(35 mg/L as $CaCO_3$), Elliot Humics (4mg/L as C), and other salts described elsewhere (16). Water in the pipes was changed every Monday, Wednesday, and Friday, and samples were analyzed for ammonia, pH, lead, and zinc release as described above.(特性研究方法 2，该方法的目的是测试"the effect of pipe material (PVC vs copper) on nitrification and resulting lead contamination of water by leaded brass")

本论文的"Materials and Methods (Experimental Section)"主要针对课题特点，介绍本研究中采用的各种方法，大多为特性研究方法，在方法叙述中自然地介绍研究所用材料、仪器和设备。本部分行文篇幅紧凑，紧扣要点，详略得当。

实例 4：来自于《Applied Catalysis B：Environmental》上已发表论文"Experimental"部分。论文题目为：Supported Cobalt Oxide on MgO：Highly Efficient Catalysts for Degradation of Organic Dyes in Dilute Solutions。下面在文中，插入中文说明，进行简要剖析。阅读时，注意各要素的体现和时态与语态。

Experimental

2.1. Preparation of catalysts

The support materials ZnO (Kanto Chemical, 99%) and P25 titania (Degussa, >99.5%) were obtained from commercial suppliers and used without further purification. MgO and $Al_2O_3$ were obtained after calcination of commercial $Mg(OH)_2$ (Fluka, >99%) and $Al(OH)_3$ (Riedei-de Haen, Al content 63%-67%), respectively in static air at 400℃ for 3 h. $ZrO_2$ was obtained by calcination of zirconium oxyhydroxide which was prepared by precipitation of $ZrOCl_2$ (Kanto Chemical, 99%) using ammonia (Kanto Chemical, 28%-30% in water) according to the procedures described in a previous report [24] in static air at 400℃ for 3 h. SBA-15 was obtained from Dr. Y. H.Yang Research Lab, which was prepared according to their reported method in the literature [25]. The catalysts of cobalt oxide supported on various metal oxides were prepared by incipient wetness impregnation with an aqueous solution of $Co(NO_3)_2 \cdot 6H_2O$ (cobalt nitrate hexahydrate, Fluka Chemika, >99%), followed by drying at 60℃ overnight and then calcination at 600℃ for 3 h in static air. To study the effect of calcination temperature, MgO supported cobalt catalyst was also calcined at 400 and 800℃. The unsupported $Co_3O_4$ catalyst was prepared by calcination of $Co(NO_3)_2 \cdot 6H_2O$ at 600℃ for 3h in static air.(描述本研究所采用的主要实验方法之一——催化剂的制备，在叙述过程中自然对有关试剂、有关制备方法依据进行了交代)

2.2. Catalyst characterization

The powder X-ray diffraction (XRD) patterns of as-prepared samples were recorded on a Bruker AXS D8 X-ray diffractometer with Cu Ka (l=1.5406Å) radiation at 40 kV and 20 mA. The BET surface areas were measured in Autosorb-6B (Quantachrome Instruments) using the liquid nitrogen adsorption method. The leached $Co^{2+}$ ion concentration was measured by inductively coupled plasma (ICP) optical emission spectroscopy on a Prodigy High Dispersion ICP (Leeman Teledyne). Solutions were taken during the reactions, filtered and mixed with 2% nitric acid aqueous solution before analysis. The morphology and nanocrystal sizes were measured by transmission electron microscope (TEM) in JEOL 2010 with an accelerating voltage of 200 kV. X-ray photoelectron spectroscopy (XPS) analysis was conducted in an Axis Ultra Spectrometer (Kratos Analytical) using a monochromated Al Ka X-ray source (1486.7 eV) operating at 15 kV. The binding energies were calibrated with the position of C 1s peak at 284.6 eV arising from the adventitious hydrocarbon.（描述本研究所采用的主要实验方法之二——催化剂的表征，在叙述过程中自然对采用的表征方法和仪器设备进行了交代）

2.3. Evaluation of catalytic activity

In a typical reaction of methylene blue (MB, Alfa Aesar, high purity) degradation, 10 mg of catalyst and 0.1 mmol of Oxone (0.5 mM, $2KHSO_5 \cdot KHSO_4 \cdot K_2SO_4$, Alfa Aesar, 4.7% active oxygen) were added in 200 mL of MB solution. The starting concentration of MB was 40 ppm (~15 mg/L, 1 ppm = 1 μM). The reaction was conducted in a beaker with constant stirring at around 300 rpm. The beaker was covered and wrapped with aluminum foil to block the incidence of indoor light to the reaction mixture. At certain reaction time intervals, liquids of about 1mL were withdrawn from the suspension and then filtered. The measurement of MB concentration in the filtrate was carried out in a UV-vis spectrophotometer (Shimadzu 2450). The area of the absorption bands integrated in the range of 500-750 nm was used to monitor the reaction progress. The activity of the Co/MgO catalyst was also evaluated for the degradation of orange Ⅱ and malachite green under the same experimental conditions. For the recycle runs of MB degradation, the used catalyst was collected by centrifugation, washed thoroughly and dried at 60℃ overnight before the next run. Due to the small particle sizes, certain catalyst loss was unavoidable during the washing and drying process. Therefore, several parallel reactions were conducted in the first and second runs to ensure that the recycled catalyst amount was enough for the next run. Catalyst dose and other reaction conditions remained the same for the subsequent runs. Altogether three runs conducted for this type of recycle study. To further evaluate the stability of the catalyst in a greater number of recycle runs, a higher catalyst dose (0.2 g) was used for the reaction while keeping other conditions the same. After the complete decolorization of the MB solution, the catalyst was collected by centrifugation and washed with deionized water three times before the next run. The reactions were repeated for 10 times. The final catalyst was collected by centrifugation, washed thoroughly and dried at 60℃ overnight. Then 10 mg of the dried catalyst was used for the last reaction run under the same conditions.（描述本研究所采用的主要实验方法之三——催化剂的催化活性评价（有机物降解试验）方法，在叙述过程中对实验条件、方式等进行了说明）

本论文的“Experimental”部分的安排具有同类论文的共同特征，即制备-表征-应用性能，以叙述方法为主线，依次进行，在方法叙述中有机进行材料、仪器和装置等的说明。英语方面，可见动词时态大多为一般过去时，语态以被动语态为多。表述规范。

## 5.3 结果部分

英语科技论文的结果（研究结果）“Results”部分是正文的第三大组成部分。有些期刊论文将结果与讨论（“Discussion”）归并在一节中，冠名为“Results and discussion”。

### 5.3.1 功用

结果(研究结果)部分是对研究中所发现的重要现象的展示与归纳，论文的讨论由此引发，对问题的判断和推理由此产生，全文的一切结论由此推出，是论文的核心。其功用是对研究(实验研究)结果的展示、描述及解释，也就是说，以陈述客观数据为主，在数据之后对结果作必要的解释和说明，但不作进一步的讨论。

### 5.3.2 组成要素

结果部分一般由下述要素构成。要避免按照先后顺序描述结果，好的组织方式是按重要性由高到低组织，或按主题的不同展现结果。

(1) 必要的实验目的(缘由)、实验条件和方式的简述

(2) 实验数据的展现(数据包括事实、数字和实验或观察的细节)

首先运用有效方法对实验数据进行整理、分析、归纳和统计处理，产生直观、明了、有逻辑性和规律性的数据，并带有数据误差说明。然后，采取图、表、公式和文字叙述等相结合的方式展示实验数据。一般用图、表、公式等表达实验数据，列入文中的图和表必须是精选的。由于图便于显示变化的规律性和对不同变化条件进行对比，而且很直观，所以在表达实验数据时尽量用图，但对从图中读取数据时产生的误差较大、必须列出具体的准确数据、数据不够多不便于绘图，或数据变化复杂不易用图表示等情况，可列表表示。如实验数据很少，一般在文中可直接用文字叙述表示。另外，为了体现理论性和可用性，应尽可能实现数据数学模型化。在用图、表和公式等表示实验数据时应有文字叙述的引领，指出实验数据在哪些图、表和公式中给出。

(3) 实验结果的展现(结果是对数据的一般解说，与实验数据有区别)

在采用图、表、公式和文字说明相配合的方式展现实验数据后，应对实验数据进行解说，指出实验数据出现何种趋势，隐藏何种规律和特性，这样的文字叙述才是实验研究的结果。不能在图、表中列出一堆数据而由读者自己来解读这些数据。

(4) 对实验结果的基本解释、说明和评论

在实验结果展现后，应对其进行解释、说明、与模型或他人结果比较等(需要引用文献)。说明实验结果有何重要特性、意义和可以得出的推论和结论。有些论文结构中将结果与讨论合并，此部分内容有时也包含在“讨论”的篇幅中。

### 5.3.3 时态、语态与常用句型

英语科技论文中结果部分的动词时态以一般现在时为多，动词的语态较灵活，根据需要，主动态和被动态兼用。常用句型可根据不同功用分述如下：

1) 对结果介绍的常用句型

即指出什么结果在哪些图表中列出。

(1) “Table+序号”(“Figure(Fig.)+序号”) shows (provides/gives/presents/summarizes/illustrates/reveals/displays/indicates/suggests)....

(2) From Fig. 1 we learn....

(3) As shown in Table 3,....

(4) As can be seen from the data in Table 1,....

(5) As shown by the data in Table 1,....

(6) As described on page 20,....

(7) The previous(preceeding/forgoing) example shows....

（8）现在有趋势避免使用冗长的词汇或句子来介绍或解释图表。为简洁、清楚起见，避免把图表的序号作为段落的主题句，而在句子中指出图表所揭示的结论，把图表的序号放入括号中。A was significantly higher than B at all time points checked（Fig. 1）.（Fig. 1 shows the relationship between A and B.）

（9）表达“比较”时，避免使用“compared with”，应直接明确指出比较的结果。

A was significantly higher than B（Fig. 1）.

（10）The variation in the temperature of the samples over time is shown in Figure 2.

（11）Figure 2 shows the variation in the temperature of the samples over time.

（12）As Figure 2 shows，the temperature increased rapidly.

（13）The temperature increased rapidly，as shown in Figure 2.

（14）The temperature increased rapidly（see Figure 2）.

（15）对结果描述的常用句型

对结果描述往往是对已发生的事实的叙述和总结，所以通常采用过去时。

（1）After flights of less than two hours，20% of the army pilots and 30% of the civilian pilots reported A.

（2）Female listeners found loud music more irritating than male listeners did.

2）理论分析的常用句型

在英语科技论文的实验结果（实验结果和讨论）部分，常常包含一定篇幅的理论分析，用于介绍所采用或开发的理论模型等。在作假设、理论分析、数学模型的建立以及计算过程中，常常要推导和描述公式，表达数字与逻辑关系，由于这类表述是不受时间影响的普遍事实，所以常用一般现在时态。常用句型如下：

（1）Substituting M in（into）N，we obtain（have/get）….

（2）Substituting M（Substitution of M）in（into）N gives（yields/produces/results in/leads to）….

（3）Let us now consider the case….

（4）Suppose that X is a solution of….

（5）Let d and v be scalings for….

（6）If m = 1.0，then we have the following equations：

（7）Given that m = 1.0，we obtain….

（8）The equation A can be written（expressed）as….

（9）The relationship between A and B is as follows：

（10）A is inversely proportional to B，as shown below.

（11）If the same material is used，the above equation reduce to….

（12）We will now reduce Eq.（3）to a simpler form.

（13）On substituting this equation into Eq.（5）and solving for A….

（14）A is defined as（given by）….

（15）We will now integrate Equation（3）in order to do….

（16）We can now derive the solution to Equation（3）.

（17）The theory comes（stems/emerges/originates）from….

（18）The theory is obtained（provided/furnished）from….

3）对实验（研究）结果说明和解释的常用句型

在对实验（研究）结果说明和评论的常用句型中，动词一般用一般现在时态。

a. 根据本人的研究结果作出推论。

（1）The results suggest that….

（2）Typical values range（vary）from A to B.

（3）This equation indicates（implies/shows/suggests/demonstrates/reveals/establishes/bears out/confirms/supports/favors/provides）….

（4）There is evidence that….

（5）There is evidence to show（indicate/suggest）that….

b. 作者解释研究结果或说明产生研究结果的原因。

（1）These findings are understandable because….

（2）A possible explanation for this is that all of the oxygen was used up in the early stages of the reaction.

（3）The higher incidence of A may be due to B.

（4）It appears that because of A，B（复数）are subjected to higher incidence of C.

（5）This is due to….

（6）This is caused by…

（7）This results from（arises from）…

（8）… is the result of….

（9）A is an indication of（gives an indication of/indicates/is indicative of）….

（10）The theory holds（maintains/claims/implies）that….

c. 作者对本研究结果与其他已有结果作比较，指出相异或相同。

（1）These results agree with….

（2）This has nothing（little/something/much）in common with….

d. 作者对本研究方法或技术的性能与其他研究者的方法或技术的性能进行比较。

（1）They put forward（advanced/developed/proposed/suggested/created/constructed/formulated/eleborated）this theory，which is based on（rests on/proceeds from）….

（2）A is significantly higher than….

e. 作者指出自己的理论模型是否与实验数据相符合，若此类句子中出现以 that 开头的从句，则从句中也使用一般现在时态。

（1）The data confirm closely A.

（2）A（复数）are all highly consistent with….

（3）The theoretical model fits（agrees well with）the experimental data well.

（4）The experimental measurements are very close to the predicted values.

（5）There is a high level of agreement between the theoretical predictions and the experimental data.

f. 当对研究结果作可能的证明时，句子的主要动词之前通常加上 may 或 can 等一般现在时态的情态动词。

（1）A may be the most suitable for C.

（2）The results can be explained by C.

(3) One reason of this advantage may be that....

(4) A possible explanation for this is that....

(5) This may have occurred because....

(6) These results agree well with the findings of....

其他实际典型例句(单句或连续语句)如下所示：

(1) This finding agrees with the studies of Ball and Edwards who first suggested that cobalt is the best catalyst for the decomposition of peroxymonosulfate.

(2) Figure 2a shows the limitations of the Fenton Reagent with respect to pH for the treatment of 2,4-DCP. Significant degradation efficiencies of 2,4-DCP were achieved only at acidic pH (6.0 and below).

(3) It is shown that, allowing sufficient reaction time, the Co/PMS process gives higher degradation efficiencies even at that particular pH.

(4) When atrazine was the model contaminant tested, similar results were obtained as shown in Figure 3.

(5) In this case, the Fenton Reagent showed better results than the Co/PMS process.

(6) Data obtained by varying the molar ratio of the compared oxidizing reagents versus the contaminant are depicted in Figure 5.

(7) Figure 6 shows the results from the treatment of 2,4-DCP with Co/PMS, at buffered solutions with phosphates in the pH range of 6-8.

(8) Throughout this pH range and under certain loading of oxidizing system, 2,4-DCP was completely transformed and a 20%-30% TOC removal was achieved.

(9) It seems that the sulfate radicals generated by the Co/PMS process are not affected by carbonate species, since the removal efficiencies of chlorophenol and TOC were still at high values.

(10) The higher TOC removal at pH 9.0 as compared to pH 6.5 with the Co/PMS reagent is attributed to different kinetics at these particular conditions.

(11) It must be also underlined that during the experiments at pH 6.5 and 6.7 the pH increased instead of decreasing, as in the cases of most AOTs due to the formation of organic acids from the degradation of the contaminants and the proton release from the hydrolysis of the transition metals. This is because the system was initially close to the $pK_a$ value of the first dissociation of carbonic acid ($pK_{a_1} = 6.35$) and carbon dioxide was generated that escaped the solution (bubble formation observed). Consequently, hydrogen and bicarbonate ions were consumed to maintain equilibrium leading to an increase of the pH.

(12) The degradation trends of 4-CP at buffered neutral pH employing three different oxidation systems ($Fe(II)/H_2O_2$, Fe(Ⅱ)/PMS, Fe(Ⅱ)/PS) are shown in Fig. 1(a).

(13) Among the three oxidation systems, Fe(Ⅱ)/PMS system showed relatively higher reactivity compared to the other two systems and led to a maximum of 65% 4-CP removal after 4 h.

(14) This observation indicates that the PMS might be activated by iron precipitates resulting in 4-CP removal.

(15) Table 1summarizes the results of a 7-day degradation experiments conducted to evaluate the effectiveness and longevity of each oxidative system.

(16) Only 1% of iron was available in the soluble form at the end of 7 day, indicating the unavailability of activating agent, which is the most likely reason for limited 4-CP removal.

(17) Fig. 3 presents the effect of $H_2O_2/Co^{2+}$ ratio on dye degradation.

(18) As shown, there is an optimum ratio ($H_2O_2/Co^{2+}=6$) for achieving the highest degradation rate.

(19) Fig. 7 illustrates the degradation of acid red 183 at varying $Co^{2+}$ concentration with PMS concentration of $4\times10^{-4}$ M and the relationship between degradation efficiency and PMS/$Co^{2+}$.

(20) As shown, the second-order kinetics exhibits better fitting results in terms of initial concentration and regression coefficient, suggesting the dye degradation may follow the second-order kinetics.

(21) It is determined that the order (n) respective to dye concentration is 1.99 and rate constant (k) is 156 $ppm^{-3}\ min^{-1}$, thus confirming the second order kinetics.

(22) From the kinetic results at different temperatures, an activation energy value of 34.3 kJ/mol was determined for the process using the Arrhenius equation.

(23) The effect of pH on the decolorization of RBB by the $Co^{2+}$/PMS oxidative process is shown in Fig. 2.

(24) In the acidic range, the decolorization rate increases sharply as the solution pH increases from 3.5 to 5 in the $Co^{2+}$/PMS process; on the contrary, it decreases sharply as the pH of the solution increases from 6 to 8.4.

(25) Further negligible iron precipitation was observed mainly due to stable acidic pH conditions, which indicates that the degradation process was not limited by iron concentration.

(26) As it can be seen from Fig. 1(b and c), once the Fe(Ⅱ) and oxidant molar ratio is increased over 1 : 1, the 2-CB degradation curve reaches a plateau, which indicates that most of the generated sulfate radical are getting utilized in side.

(27) We collected a total of 239 24-hr duplicate food samples collected from the 46 CPES children.

(28) We measured a total of 11 OP insecticides, at levels ranging from 1 to 387 ng/g, and three pyrethroid insecticides, at levels ranging from 2 to 1,133 ng/g, in children's food samples.

(29) We found that many food items consumed by the CPES chidren were also on the list of the most contaminated food commodities reported by the Environmental Working Group.

(30) CYP1A activity was induced by BaP in KC but not ER larvae, and KC larvae demonstrated a greater reduction in whole-body concentrations of BaP over time.

(31) The median MLSS concentrations for the pilot-scale SMBR (Table 3) ranged from 6.9 to 8.6 g/L and the median percent volatile fraction was 85.5% at all conditions tested.

(32) Complete nitrification ($NH_4^+$<1mg-N/L) occurred at all conditions except at 2-d MCRT (Table 3) where the effluent $NH_4^+$ concentration was 7.0 mg/L.

(33) Fig. 3 shows the SMBR membrane performance at a 10-d MCRT (F/M = 0.34 gCOD/gVSS d).

(34) The 90-d start-up period allowed adequate time for conditioning the new membrane and establishing stable SMBR pilot operation.

(35) Fig. 4 presents membrane performance results at the 4-d MCRT ( *F/M* = 0. 73 gCOD/gVSS · d).

### 5.3.4 实例简析

实例 1：来自于《J. Phys. Chem. B》上已发表论文"Results"部分。该论文正文包括：Introduction、Materials and Methods、Results 和 Discussion 四部分。即，将 Results、Discussion 分开，不设"Conclusion"部分。论文题目为：Heterogeneous Activation of Oxone Using $Co_3O_4$。下面在文中，插入中文说明，进行简要剖析。阅读时，注意时态、语态和句型。

Results

The first two rows of Table 1 show the results from the screening experiments, and in particular those from the transformation of 2,4-DCP with CoO/Oxone and $Co_3O_4$/Oxone as well as the dissolution of cobalt from the two oxides. Although with CoO almost complete destruction of 2,4-DCP took place within 30 min of reaction while the respective 2,4-DCP transformation using $Co_3O_4$ was 74%, significantly higher amounts of cobalt from CoO leached in solution compared to $Co_3O_4$; 3. 01 mg/L versus 0. 59 mg/L, respectively. (上述语句都为对实验结果的介绍) The extent of mineralization of the organic material with the homogeneous Co/Oxone was investigated in our earlier publications. Intermediate formation from the sulfate radical attack on phenolic compounds has been the subject of a recently submitted publication to Environ. Sci. Technol. (提供有关背景) CoO at acidic pH leads to significant dissolution of Co, and the CoO/Oxone system is homogeneous. As with pure CoO, the cobalt leached from $Co_3O_4$ must have resulted from the CoO component of $Co_3O_4$($CoO \cdot Co_2O_3$). (以上语句(含有关背景)是对实验结果的解释和说明)

Given these comparative results and the fact that CoO is the most soluble cobalt oxide form, it was decided to further explore the use of $Co_3O_4$ under both acidic (Oxone addition) and neutral (0. 1M $NaHCO_3$, $pH_0 = 7.0$) conditions. (交代实验内容) Our objective was to determine whether heterogeneous activation of Oxone could take place and, at the same time, achieve as limited as possible cobalt dissolution from $Co_3O_4$. (交代实验目的) For this, $Co_3O_4$ at 157 mg-Co/L was tested and the induced transformation of 2,4-DCP as well as the dissolution of cobalt both in buffered and unbuffered solution were monitored. (交代实验方式) Figure 1 shows the transformation of 2,4-DCP with the $Co_3O_4$/Oxone reagent under these two conditions as well as the soluble Co evolution versus the reaction time. It is shown that at acidic conditions, Co from $Co_3O_4$ is slowly dissolved in solution, reaching a value of 0. 73 mg/L after 2 h of reaction. At neutral pH, dissolved Co was always below 0. 07 mg/L. (介绍实验结果) The initial slight increase in Co concentration followed by a gradual drop is most probably due to the increase of the pH. From an initial value of 7. 0, the pH increased to 8. 0 after 2 h of reaction, and some coadsorption of $Co^{2+}$ ions might have taken place. (对实验结果的解释和说明) In addition, the transformation of 2,4-DCP at neutral pH was faster compared to that at acidic pH. (再介绍实验结果)

Based on the results from the dissolution of Co from $Co_3O_4$ and using instead $Co(NO_3)_2 \cdot 6H_2O$, homogeneous experiments at dissolved cobalt concentrations equal to those leached from $Co_3O_4$ were performed. (交代实验内容) Our objective was to compare the homogeneous Co/Oxone reagent with the $Co_3O_4$/Oxone and explore whether heterogeneity with the latter is achieved. (交代实验目的) Figures 2 and 3 show the transformation of 2,4-DCP using the homogeneous Co/Oxone and the

$Co_3O_4$/Oxone reagents under acidic and neutral pH conditions, respectively. Figure 2 shows that, under acidic conditions, $Co_3O_4$/Oxone demonstrated some heterogeneity as compared to the homogeneous Co/Oxone reagent, but again the amount of Co leached in solution, most probably from the CoO component of $Co_3O_4$, was rather elevated at 0.75±0.06 mg/L. Figure 3 shows that, at neutral pH, the heterogeneous character of Oxone activation with $Co_3O_4$ was much more pronounced and the amount of Co leached in solution was approximately 10-fold less than the previous case, at 0.07±0.02 mg/L.（上述几句都为对实验结果的介绍）Error bars correspond to standard deviation from replicate experiments.（对实验结果误差的说明）Several control experiments were performed and showed that no significant adsorption of 2,4-DCP is taking place on the oxide surface and Oxone alone does not induce any significant transformation of 2,4-DCP within the 2 h of reaction time.（交代实验内容和实验结果）Some limited 2,4-DCP transformation was observed when Oxone alone (no cobalt catalyst) was used in buffered water due to the interaction with bicarbonate species (0.1M) and probably the generation of percarbonate, but still it was much less than the homogeneous cobalt-mediated activation of Oxone shown in Figure 3.（交代实验结果和做简单解释）

For comparison purposes and under the same conditions, CoO was also tested at 157 mg-Co/L and 66.7 mg-CoO/L, with the latter being equivalent to CoO contained in 157 mg Co/L $Co_3O_4$($Co_3O_4$ contains equimolar amounts of CoO and $Co_2O_3$).（交代实验内容、目的和条件）All results with respect to cobalt dissolution and the transformation of 2,4-DCP are summarized in Table 1. At acidic conditions, significant cobalt dissolution from CoO was observed. At neutral pH, as expected, the amount of cobalt dissolved was much less, but still significantly higher compared to $Co_3O_4$. In particular for the case that CoO was used at the same amount as it is contained in 157 mg Co/L $Co_3O_4$, the amount of cobalt dissolved was 1.01 mg/L from pure CoO and 0.75 mg/L from $Co_3O_4$.（交代实验结果）The values are close and rather elevated, but still the dissolution of CoO appears to be somewhat suppressed when it is contained in $Co_3O_4$ as opposed to its pure form.（对实验结果的解释和说明）At neutral pH, the overall cobalt dissolution either from CoO or $Co_3O_4$ was more limited. The respective values were 0.39 mg/L from pure CoO and 0.07 mg/L from $Co_3O_4$.（交代实验结果）Due to this relatively high solubility, there was no evidence of pure CoO participating heterogeneously in the activation of Oxone. We have previously reported almost the same percent transformation of 2,4-DCP in an unbuffered homogeneous system where $CoCl_2$ at 1.00 mg/L as $Co^{2+}$ was mixed with Oxone at 1.23 mM as $KHSO_5$. The slight difference in the 2,4-DCP as shown in Table 1 is due to the Oxone dose used; a 10∶1 molar ratio of $KHSO_5$ versus 2,4-DCP was used in the previous study while 11∶1 was used here. Even at neutral pH, dissolved cobalt concentrations as high as 0.39 mg/L, either coming from dissolution of CoO or a simple cobalt salt would induce the same 2,4-DCP transformation and pure CoO appears to react homogeneously with Oxone.（以上几句就“No pure CoO participating heterogeneously in the activation of Oxone（从实验结果引伸出来）”进行解释和说明）

以上四个自然段构成了一篇论文的“Results”部分，表达规范，句子结构地道，叙述条理清楚，逻辑性强，是“Results”部分组成要素的有机结合，注重与“Discussion”部分的分工，体现了论文“Results”部分的功能特点。

实例2：来自在《Water Research》上一篇已发表论文的"Results"部分。该论文正文包括：Introduction、Materials and methods、Results、Discussion 和 Conclusions 五部分。"Results "下又分若干小部分，都有各自的标题。论文题目为：Hydrogen and Methane Production from Household Solid Waste in the Two-stage Fermentation Process。下面在文中插入中文说明，进行简要剖析。阅读时，注意要素、时态、语态、句型和动词选用。

Results

3.1. Two-stage process performance

The two-stage process was operated in the lab for around 3 months.（简述"two-stage process"的实验条件）For hydrogen production stage in stable conditions, total gas production was 1530 mL/d and 42% of this (640 mL/d) was hydrogen. No $CH_4$ and $H_2S$ were detected in this stage. The VS was destructed from 7.5±0.31% (influent) to 6.1±0.42% (effluent) with 18.7% VS removal efficiency.（介绍"hydrogen production stage in stable conditions"中的实验研究结果）. Fig. 2 shows the monitoring profiles of hydrogen, pH and VFA in R1. Hydrogen was produced immediately from day 1, then its production fluctuated between 400 and 700 mL/d after day 17. It decreased to low level from day 28 and increased again at day 55. During this period, the hydrogen production was approximately 440 mL/d. Except this period, the hydrogen production rate was approximately 640 mL/d (or 1.6 $m^3/m^3/d$). At day 74, R1 was sparged by treated biogas (methane). Gas sparging appeared to increase hydrogen production from 640 to 1200 mL/d and $H_2$ percentage dropped from 42% to 17%. The step of 900 mL/d (day 75-78) was caused by the moving of pump tube.（介绍"hydrogen production"阶段的产氢量随时间的变化(实验研究结果)）

pH started from 5.2, dropped to 4.8 at day 30 and increased back to 5.2 after day 45 (Fig. 2). This pH change was mainly due to the pH variations in feedstock. Total VFAs concentration was 80-90 mM in the whole period. Acetate and butyrate were the main VFA species in this experiment. When pH was 5.2, 90% of total VFAs was acetate. When pH was 4.8 at day 30, acetate was 53% and butyrate was 20% of total VFAs on a molar basis.（介绍"hydrogen production"阶段的 pH、VFAs 随时间的变化(实验研究结果)）

Biogas production from R2 was an average of 11500 mL/d with a 65% V/V methane concentration. The reactor VS was removed from 6.1±0.42% (influent) to 1.05±0.46% (effluent), resulting in 82.8% VS destruction for this stage or 86% for overall two-stage process. Profiles of methane production, pH and VFAs are shown in Fig. 3. Methane production rate stayed at 7500 mL/d (or 2.5 $m^3/m^3/d$) after day 15 and pH was stable at 7.5. Total VFA concentration was at very low concentration of 1.6-2 mM. Acetate was the main organic acid detected.（介绍"Biogas production"阶段的实验结果）

3.2. One-stage process

The one-stage reactor produced 9800 mL/day biogas with a methane production rate of 6200 mL/d (or 2.1 $m^3/m^3/d$) found. The VS in the reactor was 1.55±0.58%, resulting in 79.3% VS removal efficiency. Fig. 4 shows the monitoring profiles of methane production, pH and VFAs in R3. Methane production and pH (7.5) were stable. Total VFAs were maintained between 3.4 and 3.7 mM and only acetate was found.（介绍"One-stage process"的实验结果）

3.3. pH effect on hydrogen production in batch experiments

Fig. 2 illustrates that hydrogen production fluctuated with pH change. pH batch experiments aim to evaluate the short-term effect of pH on hydrogen generation.（交代实验目的）As shown in Table 2, hydrogen was produced from pH 5 to 8.5 within the first 12 h but only at pH 5 and 5.5, hydrogen continued to be generated within 90 h. The highest hydrogen production was always at pH 5.5 in the whole experimental period, but after 60 h, pH 5 had a very similar hydrogen value as the one at pH 5.5, indicating the optimum pH should be around 5-5.5. For pH from 6 to 8.5, hydrogen productions decreased after 30 h and went down to zero at 90 h.（介绍"effect of pH on hydrogen generation"的实验结果。）At the end of experiments, methane was found as 12.56, 15.53, 17.99, 22.84, 14.71, 12.51 mmol for pH 6, 6.5, 7, 7.5, 8 and 8.5, respectively. The highest methane production was at pH 7.5, which consisted with the long-term pH (7.5) in methane stage (Fig. 3). Neither hydrogen nor methane was found throughout the experiments at pH 3.5-4.5. This indicates that heating the inoculum at 100℃ for 1h did not inhibit methanogenesis successfully. pH was the most critical factor for inhibition of methanogenesis in the HSW experiment.（另外提及 pH 对 methane production 影响，然后得到重要结论"pH was the most critical factor for inhibition of methanogenesis in the HSW experiment"）

以上材料分三个专题介绍本论文的"研究结果"（正文中"讨论"部分单独设置），每一专题主要为客观性的实验结果描述，具有典型的论文"Results"部分的特征。英语表达地道、规范。

实例3：来自在《Journal of Hazardous Materials》上一篇已发表论文的"Results"部分。该论文正文包括：Introduction、Experimental（Materials and methods）、Results、Discussion 和 Conclusion 五部分。"Results"下又分若干主题。论文题目为：Interactive Effects of Lead, Copper, Nickel, and Zinc on Growth, Metal Uptake and Antioxidative Metabolism of Sesbania Drummondii。下面在文中，插入中文说明，进行简要剖析。阅读时，注意要素、时态、语态、句型和动词选用。

3. Results

3.1. Metal concentrations in plant tissues

The concentrations of metals (Pb, Cu, Ni and Zn) in the roots and shoots of S.drummondii seedlings grown at different treatments (Pb, Cu, Ni, Zn, Pb+Cu, Pb+Ni, Pb+Zn, Cu+Ni, Cu+Zn, Zn+Ni and Pb+Cu+Ni+Zn) are shown in Table 1.（展现实验研究结果）The results show that the metal contents in the plant tissues varied among metals in different combinations. Accumulation of all the metals was substantially higher in roots than in shoots. S. drummondii accumulated significantly ($P<0.05$) higher Pb in its roots as well as in shoots compared to other metals (Cu, Ni and Zn). The metal concentrations followed the order Pb>Cu>Zn>Ni in roots and Pb>Zn>Cu>Ni in shoots. For all different combinations of metal accumulation studied with S. drummondii seedlings, bioaccumulation of a single metal in the roots as well in the shoots was affected by the presence of a second metal, resulting in the inhibition or increase in the bioaccumulation of one metal over other (Table 1).（介绍实验研究结果）

3.2. Effect of different metal treatments on plant growth

Plant biomass is a good indicator for the overall health of S.drummondii growing in the presence of heavy metals.（实验研究的方法）Growth of S.drummondii seedlings was significantly（$P<0.05$）inhibited with metal treatments（Fig. 1).（展现实验研究结果）The level of inhibition depended on metal types and their combinations. However, we have not noticed any toxicity symptoms like necrosis in the plant. Among the four metals tested in this study, Cu was the most toxic followed by Ni>Zn>Pb. Among the binary mixtures of metals, Ni+Zn were most toxic. Furthermore, maximum inhibition in S.drummondii growth was noticed in the mixture of all metals（Pb+Cu+Ni+Zn). Compared to the control（no metal), seedling growth significantly（$P<0.05$）reduced by 21.0%, 46.3%, 31.5%, 25.2%, 45.7%, 41.0%, 36.3%, 42.1%, 41.5%, 47.3% and 59.0% at Pb, Cu, Ni, Zn, Pb+Cu, Pb+Ni, Pb+Zn, Cu+Ni, Cu+Zn, Zn+Ni and Pb+Cu+Ni+Zn treatments, respectively.（介绍实验研究结果）

3.3. Effect of different metal treatments on photosynthetic activities

The photosynthetic efficiency of S.drummondii in the presence of different metals was assessed by measuring Fv/Fm and Fv/Fo ratios. The level of response depended on metal types and their combinations.（介绍实验研究内容和方法）In the present study, Fv/Fm ratios were higher than 0.80 in the Pb, Cu, Zn, Pb+Cu, Pb+Zn, Cu+Ni, Cu+Zn and Ni+Zn treatments. Though, Fv/Fm ratios were lower than 0.80 in Ni, Pb+Ni and Pb+Cu+Ni+Zn treatments. A similar trend was also exhibited in Fv/Fo values. Fv/Fo ratios were higher than 4.0 in the Pb, Cu, Zn, Pb+Cu, Pb+Zn, Cu+Ni, Cu+Zn and Ni+Zn treatments(Fig. 2). Though, Fv/Fo ratios were lower than 4.0 in Ni, Pb+Ni and Pb+Cu+Ni+Zn treatments.（介绍实验研究结果）

3.4. Effect of different metal treatments on superoxide dismutase activity

The activity of SOD was significantly（$P<0.05$）increased in S. drummondii seedlings with metal treatments, when compared to the control（no metal). However, SOD activities differed with metals as well as with different combinations of metals. Among metals and binary combinations, Cu and Pb+Ni had maximum SOD activity while Pb, Zn and Pb+Zn had the least（Fig. 3). Though, the highest increase in SOD activity was noticed when all four metals were applied together in the mixture（Pb+Cu+Ni+Zn). The SOD activity at Pb+Cu+Ni+Zn treatment was 2.6 folds higher with respect to the control.（介绍实验研究结果。）

3.5. Effect of different metal treatments on ascorbate peroxidase activity

Compared to the control, metal treatments significantly（$P<0.05$）enhanced the activity of APX in S. drummondii seedlings. However, in general the APX activities were higher in the seedlings treated with a combination of metals as compare to those treated with a single metal. The APX activity showed maximum increase in Pb+Cu+Ni+Zn treatment which was 112% higher than the control. Similar to SOD activity, among metals and binary combinations, Cu and Pb+Ni had highest activation in APX activity, while Zn and Pb+Zn had the least（Fig. 4).（介绍实验研究结果）

3.6. Effect of different metal treatments on glutathione reductase activity

Fig. 5 shows the influence of different metal treatments on GR activity in S. drummondii seedlings.（展现研究结果）GR activity increased significantly（$P<0.05$）in all metal treatments but higher activity was noticed when all four metals were applied in combinations. However, no significant（$P<0.05$）change in GR activity was noticed among the different binary combinations of metal. The activity at Pb + Cu + Ni + Zn treatment increased by 80% with respect to the control (Fig. 5).（介绍研究结果）

3. 7. Effect of different metal treatments on glutathione level

Different metal treatments altered the levels of GSH and GSSG in S. drummondii seedlings (Table 2). The GSH content as well as GSH/GSSG ratio were found to be higher in all metal treatments when compared with the control (no metal). However, higher levels of GSH and GSSG contents were found in the seedlings treated with binary and quaternary mixture of metals (Pb+Cu, Pb+Ni, Pb+Zn, Cu+Ni, Cu+Zn, Zn+Ni, Pb+Cu+Ni+Zn) as compared to those treated with a single metal. However, the highest increment in GSH content was noticed at Pb+Cu+Ni+Zn treatment. In the Pb+Cu+Ni+Zn treatment, GSH level significantly ($P<0.05$) increased by 119. 7% with respect to the control.（介绍研究结果）

以上材料分七个专题介绍论文的"研究结果"(正文中单独设立"讨论"部分)，每一专题主要为客观性的实验结果描述，在着重正面结果叙述时，也有适当的负面结果描述，文风朴实，通俗易懂，具有典型的论文"Results"部分的特点。英语表达地道、规范。

实例4：来自在《Environmental Health Perspectives》上一篇已发表论文的"Results"部分。该论文正文包括：Introduction、Materials and methods、Results、Discussion 和 Conclusion 五部分。论文题目为：Particulate Matter Exposures, Mortality, and Cardiovascular Disease in the Health Professionals Follow-up Study. 下面在文中，插入中文说明，进行简要剖析。阅读时，注意要素、时态、语态、句型和动词选用。

Results

A total of 17,545 men were included at baseline, with a mean age of approximately 57 years. Most were never or former smokers at baseline, with current smokers decreasing to about 5% of the study population by the end of follow-up. The percentages of men who reported hypertension, hypercholesterolemia, or diabetes increased considerably from baseline (26. 5, 24. 3, and 4. 3%, respectively) to the end of follow-up (47. 5%, 56. 6%, and 9. 8%, respectively). About 33% of the men reported ≥ 27 MET hr/week of physical activity in 1989, with the percent of respondents in this category growing by the end of follow-up. About 23% of the study population lived in New York, with 15% in Pennsylvania, 14% in Michigan, 13% in Ohio, and 10% in New Jersey.（介绍与 "Study population" 有关的研究结果）At the beginning of the study, the average annual estimated exposure was 27. 9 $\mu g/m^3$ for $PM_{10}$, 17. 8 $\mu g/m^3$ for $PM_{2.5}$, and 10. 1 $\mu g/m^3$ for $PM_{10-2.5}$. In general, annual predicted PM exposures decreased over the follow-up period, with $PM_{10}$ showing the largest decline (Figure 1). We performed correlations among various time windows of average exposure (12, 24, 36, and 48 months) for each PM fraction. All time windows were highly correlated ($\rho>0.86$).（介绍与 "average exposure" 有关的研究结果）

There were 2,813 deaths, 746 cases of fatal CHD, 646 cases of nonfatal MI, 1,661 cases of total CVD, 230 ischemic strokes, and 70 hemorrhagic strokes.（介绍有关"mortality" 的研究结果）An interquartile range (4 $\mu g/m^3$) change in average $PM_{2.5}$ exposure in the 12 previous months was not associated with all-cause mortality (HR = 0. 96; 95% CI, 0. 90-1. 03) or ischemic strokes (HR = 0. 84; 95% CI, 0. 67-1. 06) in basic models adjusting for time and state of residence and stratified by age. The HR for fatal CHD was 1. 01 (95% CI, 0. 89-1. 15), 1. 02 for total CVD

(95% CI, 0.94-1.11), and 1.06 for nonfatal MI (95% CI, 0.92-1.22) in similar models. For comparison purposes with other studies, translating these relative risks (RRs) to a 10-$\mu g/m^3$ unit change in $PM_{2.5}$ exposure results in an HR for all-cause mortality of 0.90 (95% CI, 0.76-1.06), HR for fatal CHD of 1.02 (95% CI, 0.74-1.41), and HR for total CVD of 1.05 (95% CI, 0.85-1.30). The HR for an interquartile range change in average $PM_{2.5}$ exposure hemorrhagic strokes was highest, but the number of cases was small (HR = 1.14; 95% CI, 0.79-1.65). The HRs were very similar in fully adjusted models with additional adjustment for BMI, hypertension, hypercholesterolemia, diabetes, family history of MI, smoking (status and pack-years), physical activity, healthy diet, and alcohol consumption. For example, the fully adjusted HR for total CVD was 1.01 (95% CI, 0.93-1.10). Results were similar for interquartile range increases in $PM_{10}$(7 $\mu g/m^3$) and $PM_{10-2.5}$(4 $\mu g/m^3$) separately, as well as in copollutant models with $PM_{2.5}$ and $PM_{10-2.5}$(Table 3).(介绍有关"interquartile range increases in PM exposures 与 all-cause mortality 关系"的研究结果)

Effect modification was evident for $PM_{2.5}$ associations (Table 4). Men without a family history of MI were at significantly lower risk for all-cause mortality associated with an interquartile range increase in chronic fine particulate exposure compared with men with such a history. Analysis of $PM_{2.5}$ exposures and fatal CHD stratified by smoking status show an increased risk for fatal CHD associated with chronic fine particulate exposure among never smokers. Although the HR for current smokers was higher than that estimated for never smokers, there was only a small number of current smokers with fatal CHD, and there was no apparent relationship for former smokers.(介绍与"Effect modification"有关的研究结果)

Sensitivity analyses showed that model results were stable using different time windows of average exposure for each size fraction of PM, as well as excluding state of residence from the final models. Findings also did not change when participants residing outside of MSAs were excluded or when the addresses used for exposure assessment were limited to those we were able to confirm as residential (data not shown). Finally, we conducted sensitivity analyses excluding men with prior cancers (other than nonmelanoma skin cancer) as well as including men with prior stroke and CHD, and these results were also comparable (data not shown).(介绍 与"Sensitivity analyses"有关的研究结果)

以上材料不设分标题介绍论文的"研究结果",正文中单独设立"讨论"部分,正文的实验部分"Materials and Methods" 分专题介绍各研究方法,"Results"部分则相应地依次介绍各方法下的研究结果。研究结果包括数据、对数据的说明(即结果)和由结果得到的推论。本论文研究与一般的实验研究方法不同,故获得研究结果的方式也有其自身的特点。本材料英语表达地道、规范。

## 5.4 讨论部分

讨论(Discussion)部分是英语科技论文的又一重要组成部分,在全文中除摘要和结论部分外,最令人关注,是读者最感兴趣的部分之一。有些期刊论文将结果与讨论归并在一节中,冠名为"Results and Discussion"。

### 5.4.1 功用

英语科技论文的讨论部分反映了作者对事物的认识水平，能对读者产生启迪作用，在这部分要回答引言中所提的问题，评估研究结果所蕴涵的意义，对研究结果做有意义的深入分析，充分体现论文的价值。讨论部分的重点是对研究结果的深入的解释和推断。

### 5.4.2 组成要素

“讨论(Discussion)”可作为独立的部分置于结果(Results)之后、结论(Conclusion)之前，也可与结果部分合并在一起(Results and Discussion)。但作为讨论部分的篇幅，都有其基本组成要素，一般有以下几方面。但对于一篇具体论文而言，未必每一项都出现。

(1) 再次概述研究目的或假设，说明预期结果是否实现

(2) 简述最重要的研究结果

(3) 对结果进行深入分析、说明、比较和评价

用已知的理论或他人的结论论证有关观点，说明、解释结果和新发现，将研究结果与已有结果(文献结果)相比较，突出自己的创新点，分析与他人结果不一致的原因。

(4) 本研究的特色

(5) 由本研究结果所能得出的推论或结论

(6) 本研究方法或研究结果的不足及其原因

(7) 本研究结果的理论意义或实际应用价值

(8) 对未来研究工作的展望(今后的研究方向、设想和建议)

讨论部分的重点是要涉及论文内容的可靠性、外延性、创新性和可用性。这方面的叙述分散在以上各组成要素中。

可靠性指论文提供的实测值或计算值是否可靠，要通过重复性和误差分析来说明，还要尽可能与其他人的结果(已有的文献值)进行对比，以说明本论文数据的合理性。

外延性指通过本论文所提供的数据可供读者在更大范围内使用，因此要尽可能给出数据关联式，即数学模型。借助于该数学模型，在实验范围内能可靠地进行内插，在适当的条件下，可进行外延，以便在更大范围内应用实验数据所蕴涵的规律。

创新性要与引言部分一致。引言部分指出的是本论文总的创新性，而在讨论部分要把这一点具体化。

可用性有两层意思。第一层意思是与外延性一致。另一层意思是把数据变活，把不同条件的数据做对比，并尽可能做出优化选择，提出最优条件或最佳结果。

### 5.4.3 时态、语态与常用句型

在英语科技论文的讨论部分，作者通常对研究结果进行概述、分析和解释等，因为陈述的是作者的见解和结论，所以句子时态多用一般现在时，语态则较灵活，主动和被动态兼用。常用的句型有：

(1) 回顾研究目的

(a) This research investigated the effects of two different methods.

(b) In this study, the effects of two different methods were investigated.

(c) We originally assumed that worker who did… would be more satisfied with….

(d) We have performed (done/made/conducted) a number of experiments to test (verify/prove/check/) the theory.

(e) We have succeeded (have been successful) in doing…

(f) We have successfully done….

(2) 概述结果

(a) The results indicate(prove/show/suggest/reveal) that….

(b) These results provide substantial evidence for the original assumptions.

(c) These experimental results support the original hypothesis that….

(d) Our findings are in substantial agreement with those of….

(e) The present results are consistent with those reported in our earlier work.

(f) The experimental and theoretical values for A agree well.

(g) The experimental values are all lower(higher) than the theoretical predictions.

(h) These results contradict the original hypothesis.

(i) These results appear to refute the original assumptions.

(j) A seems to indicate….

(k) The results given in Fig. 3 validate (support) the hypothesis.

(l) This equation holds for(holds true for/is true for/is valid for/applyes to)….

(m) This causes(gives rise to/leads to/results in/brings about)….

(3) 阐述相关推论

(a) It is possible (may be/is likely) that adding water causes the reaction rate to increase.

(b) These results can be explained by assuming (This inconsistency indicates) that adding water caused the reaction rate to increase.

(c) The data reported here suggest (These findings support the hypothesis/Our data provide evidence) that the reaction rate may be determined by the amount of oxygen available.

(d) The reaction rate may be determined by the amount of oxygen available.

(e) The reaction rate is determined by the amount of oxygen available.

(4) 表示研究的不足

(a) It should be noted that this study has examined only….

(b) This analysis has concentrated on….

(c) The findings of this study are restricted to….

(d) This study has addressed only the question of….

(e) The limitations of this study are clear….

(f) We should (would) like to point out (indicate) that we have not….

(g) However, the findings do not imply….

(h) The result of the study cannot be taken as evidence for….

(i) Unfortunately, we are unable to determine from this data….

(j) Only two sets of conditions were tested.

(k) The method presented here is accurate, but can not be implemented in practical applications.

(l) Our findings may be only valid for A.

(m) We recognize that A may not fully reflect B.

(n) This experiment failed to show (demonstrate)….

(o) This experiment has not shown (demonstrated)….

（5）进一步研究的建议

（a）We suggest（recommend）that these experiments be repeated using a wider range of initial conditions.

（b）It would be interesting to learn why oxygen is depleted during A.

（c）Experiments similar to those reported here should be conducted using A.

（d）This should arouse（attract/gain/have/receive）our attention.

（e）This bears（deserves/demands/requires/calls for）further research（effort/study/work/investigation）.

（6）结果的理论意义或实际应用

（a）The results of this study may lead to the development of effective methods for....

（b）Our findings may be useful to A.

（c）The technique presented in this paper should be useful in reducing the amount of sludge in wastewater from semiconductor plants.

（7）其他

在"讨论"中应选择适当的词汇来区分推测与事实。

可选用"prove"、"demonstrate"等表示作者坚信观点的真实性；选用"show"、"indicate"、"found"等表示作者对问题的不确定性；选用"imply"、"suggest"等表示推测；选用情态动词"can"、"will"、"should"、"probably"、"may"、"could"、"possibly"等表示问题的确定性程度。

其他实际典型例句(单句或连续语句)如下所示：

（1）The high efficiency at elevated pH values of the process proposed herewith is a very important advantage for its future applicability in the large scale. This is because the pH of most contaminated natural waters is in the range of 6-8. It is also important to state that cobalt is needed only in small catalytic amounts, thus its potential toxicity may be overcome.

（2）The development of a heterogeneous system would be certainly more appropriate, especially for drinking water applications.

（3）With respect to the oxidants, hydrogen peroxide is more environmentally friendly than peroxymonosulfate. However, peroxymonosulfate, although it releases additional sulfate ions in the aqueous phase, is easier to handle（crystalline form）, and sulfate radicals are stronger oxidants than hydroxyl, especially at elevated pH.

（4）The cost of Co/PMS includes only the cost of PMS, since cobalt is needed in very small amounts and no pH adjustment or sludge treatment are required.

（5）Hence, it is believed that the Co/PMS reagent might be proven an attractive alternative remediation technology for practitioners in the field.

（6）Higher stability of PS and unavailability of sufficient amount of iron in soluble form could be the main reasons for the lower extent of 4-CP using PS at neutral pH conditions.

（7）Therefore, it could be postulated that the solution pH and associated dissociation of PS may have played a significant role in the performance of Fe(Ⅱ)/Citrate/PS oxidation system.

（8）The latter is most probably due to competition of citrate with 4-CP for the oxidizing radicals.

(9) Since PMS is consumed very quickly in the presence of iron/EDDS complex, there is a possibility of fast EDDS degradation and subsequent loss of chelating power.

(10) Effective use of generated reactive radicals due to high organic loading can be the main reason for the efficient chlorophenol degradation.

(11) Overall, it was demonstrated that the catalytic systems such as Fe(Ⅱ)/Pyrophosphate/PMS are capable of degrading a variety of chlorophenols and the oxidation system performs comparably with increasing chloro-functional groups in phenol.

(12) The decreasing rate of dye decolourisation is attributed to unreacted $H_2O_2$ acting as a scavenger of $\cdot OH$ and producing a less potent perhydroxyl radicals.

(13) Several investigations of wastewater decolourisation using Fenton oxidation have been conducted and the similar observation was reported.

(14) Therefore, it seems that kinetics of dye degradation in metal/PMS depends on dye concentration and metal ion.

(15) The second-order kinetics observed in this investigation is probably due to the higher concentration of dye and much lower PMS/AR183 ratio, which is at 1-4.

(16) These results are interesting and may provide insights on the role of solar radiation in driving this process.

(17) These results are comparable and showed evidence of the improvement in the degradation of organic pollutants with solar radiation using the Fenton reaction like in previous studies.

(18) Even a total aromatic decomposition may be achieved if there is a much longer period of contact or conjunction with other treatments of AOPs in this optimum situation.

(19) However the tendency of absorbance decline rate and the inflexion is not obvious in Fig. 8a; this is probably due to the fact that the participation of the initial $Co^{2+}$ concentration was relatively and significantly lower than that applied in Fig. 8b, leading to an insufficient amount of the Co(Ⅱ, Ⅲ)-catalyst (Eqs. (5)) to continuously activate the 100 ppm PMS efficiently via the chain oxidative pathways suggested in Section 3.6.

(20) Consequently, based on this work and a related literature review, the possible chain oxidative pathways for the degradation of RBB with $Co^{2+}$ and PMS can be simply generalized and suggested below to describe the oxidation of RBB: ....

(21) Taking into the account of low cobalt ion concentrations in the actual reaction (typically < 50 μg/L) in this study, it is therefore suggested that the activation of PMS by Co/MgO catalyst is through the heterogeneous pathway.

(22) Apparently, the dispersed $Co_3O_4$ nanoparticles on MgO support function as a stable cobalt source to accelerate the generation of sulfate radicals from PMS, and the presence of the MgO support leads to a much higher acceleration efficiency.

(23) Based on the reaction rate constant between sulfate radicals and Fe(Ⅱ), it can be said that quenching of sulfate radicals by Fe(Ⅱ) can be a major side reaction (Reaction(3)) especially at higher concentrations of Fe(Ⅱ).

(24) From these results it can be said that minimum amount of Fe(Ⅱ) is required to effectively activate PMS, while excessive Fe(Ⅱ) can be detrimental for process degradation efficiency.

(25) During redox reaction, the electron transfer from one species to another takes place by two mechanisms, namely the bridged mechanism and the outer sphere mechanism.

(26) It is believed that in Fe(Ⅲ)-peroxo complexes, unlike Fe(Ⅱ)-peroxo complex, no O-O bond cleavage takes place, but instead a Fe(Ⅲ) hydroperoxo intermediate forms as the first step via hydrolysis.

(27) In Fe(Ⅲ) based system, slow generation of sulfate radicals due to the rate limiting reduction of Fe(Ⅲ) can be a probable cause of the observed slower kinetics in these systems.

(28) The advantage of Fe(Ⅲ)/PMS system includes: (i) slower radical generation, which facilitate better utilization of generated radials, and (ii) virtually no radical quenching by Fe(Ⅲ), which was predominant in Fe(Ⅱ)/PMS system, especially at higher Fe(Ⅱ) concentrations.

(29) One possible reason for this observation could be stabilization of soluble iron in presence of PMS since addition of PMS reduces the system pH because of dissociation of $HSO_4^-$ present in Oxone salt.

(30) Since clean sediment contains relatively large amount of iron (14.0 g/kg), mainly in form of iron oxides, this degradation of 2-CB with only the oxidant can be attributed to the heterogeneous activation of these oxidants by the transition metals (mainly iron/iron oxides) present in the sediments.

(31) Our future studies will focus on fundamental understanding of the heterogeneous activation of these stable oxidants (PMS, PS) and application of these systems for treatment of PCB contaminated sediments.

(32) This implies that $TiO_2$ as a support material for Co might play a crucial role in the hetero-PMS-act.

(33) The changes in cobalt species might be responsible for the variation of catalytic activity and cobalt leaching described in Figs. 1-3.

(34) Furthermore, transformation of $Co_3O_4$ to $CoTiO_3$ can be a possible reason for the low activity of Co/$TiO_2$-700.

(35) The small 10 – 15 nm $Co_3O_4$ particles were well-incorporated into the edge of large 30–40 nm $TiO_2$ particles, forming a composite of $Co_3O_4$ and $TiO_2$ with heterojunction structure.

(36) The peaks at 782.1 and 797.5 eV resulted from a chemical shift of the main spin-orbit components because the Co cations on the nanocrystal surface are chemically interacted with surface hydroxyls which correspond to the shake-up satellites I at+5.6 eV from the Co 2p bands, suggesting a strong possibility of the presence of many hydroxyl groups on the surface of Co/$TiO_2$.

(37) The band at 600-740 nm is ascribed to Co-OH complexes and $Co_3O_4$, and thus the reduced intensity of this absorption band for spent 0.5Co/$TiO_2$ suggests that Co-OH complexes might be consumed during the PMS activation.

(38) The principle of heterogeneous cobalt-mediated PMS activation is supposed to be analogous to that of homogeneous system.

(39) In principle, a homogeneous cobalt-mediated PMS activation involves one electron transfer between Co(Ⅱ) and Co(Ⅲ) as a result of PMS decomposition, as expressed in Eqs. (1) and (2).

(40) However, it is expected that the hydroxyl groups at the surface of $Co_3O_4$ cannot be regenerated rapidly from the interaction of cobalt species with water molecules, especially at basic pH because the surface of $Co_3O_4$ is negatively charged due to its point of zero charge of around 7.5.

(41) This is mainly due to the surface hydroxyl groups generated from $TiO_2$ and uniform distribution of well defined 10–15 nm nanocrystalline $Co_3O_4$ particles on 30–40 nm $TiO_2$ nanoparticle surface, resulting in increase in catalytic surface area of Co–OH species to activate PMS.

(42) This prediction applies to average $CO_2$ emissions from large, multiplant refinery groups with diverse, well-mixed crude feeds and appears robust for that application. However, the method used here should be validated for other applications.

(43) Because it is assumed that most individuals spend many hours per day in close proximity to their pillow and bedding, it is not surprising that when PBDEs are present in these they would be a major contributor to exposure.

(44) This study demonstrates the usefulness of the XRF analyzer in identifying items that may contribute to human exposure to PBDEs.

(45) Differences between our study findings and previously published results may stem from differences in the populations examined, pollution measures, and covariates considered in analyses.

(46) Findings from the current study may not be generalizable to other populations of men.

(47) Findings have been somewhat inconsistent regarding the impact of socioeconomic status on relationships between mortality, CVD, and air pollution exposures; however, as Bateson and Schwartz suggest, inconsistencies may be attributable somewhat to differences in the geographic scale of these measures, with individual level measures showing higher risk for lower educational attainment.

(48) One explanation for the higher yield in the HW-red oak is that its lower lignin content and higher (C+H)/L relative to HW-eucalyptus and the two SW species, controlled the extent of degradation. However, (C+H)/L does not explain the methane yield difference between two SW species with similar lignin contents. A second explanation for the additional biodegradability of the HW-red oak is that the lignin polymers in HWs have a lower structural integrity and typically are chemically degraded more easily than softwood lignin.

(49) In general, the results reported previously are not comparable to the results measured here, suggesting that it may not be appropriate to apply a specific decomposition factor to wood products as a whole.

(50) Collectively, these findings have implications for national greenhouse gas emissions, since the IPCC allows the use of country-specific $DOC_f$ values, provided they are based on well documented research.

(51) A strong point of the present study is its sampling design that allows a proper representation of the Inuit population of Nunavik.

(52) To our knowledge, this is the first study to evaluate contamination between source and POU drinking water by sampling the same source water as collected by households in real time, and to follow its fate over the course of several days of storage within a household.

(53) The observed reductions in bacterial loads could be due to settling of organisms to the bottom of storage containers or die-off these organisms caused by predation by other microorganisms, lack of nutrients, or other factors contributing to inhospitable conditions in the container.

(54) Future studies could be enhanced by the collection of behavioral and other data related to potential recontamination events, such as frequency of water access, method of water access, location and height of containers, presence of a spigot, and exposure to sunlight. It would also be extremely useful to study the effects of turbidity of source waters on household water quality and natural attenuation.

### 5.4.4 实例简析

实例1：来自于《J. Phys. Chem. B》上已发表论文的"Discussion"部分。该论文正文包括：Introduction、Materials and Methods、Results 和 Discussion 四部分。即，将 Results、Discussion 分开，不设"Conclusion "部分。论文题目为：Heterogeneous Activation of Oxone Using $Co_3O_4$。下面在文中，插入中文说明，进行简要剖析。阅读时，注意时态、语态、句型和动词选用。

Discussion

In $Co_3O_4$, Co appears in two redox states; as Co(Ⅱ) in CoO and Co(Ⅲ) in $Co_2O_3$. As with the homogeneous system, it is believed that the mechanism of the heterogeneous catalysis shown in Scheme 1, involves a one-electron transfer process: the oxidation of $Co^{II}$ to $Co^{III}$ with peroxymonosulfate and the generation of sulfate radicals as well as the reduction of $Co^{III}$ to $Co^{II}$ and the generation of the peroxymonosulfate radical.（以上语句提供基本知识和一般认识）It was already demonstrated that the latter transient species are too weak compared to sulfate radicals to induce any transformation on 2,4-DCP.（提供已有研究的结论）On the basis of the results presented and the cobalt species tested here, there are two main hypotheses with respect to the species responsible for the heterogeneous catalysis observed with $Co_3O_4$.（得出结论——有两个假说）These hypotheses are associated with the two oxides bound together and contained in $Co_3O_4$($CoO \cdot Co_2O_3$).（交代结论的关联性）

The first suggests that cobalt redox cycling is taking place mainly in $Co_2O_3$(contained in $Co_3O_4$), which is an unstable oxide, given also the fact that pure CoO is soluble and did not appear to react heterogeneously. CoO is the most stable cobalt oxide and, presumably, Co from CoO needs to be in ionic form (dissolved) to participate in redox reactions.（给出第一个结论(假说)及较充分理由）

The second hypothesis indicates that CoO (contained in $Co_3O_4$) might also be the species responsible for the heterogeneous decomposition of Oxone. The results of Table 1 show that pure CoO released relatively high amounts of cobalt in solution and was shown not to react heterogeneously. CoO contained in $Co_3O_4$, on the other hand, might be the species contributing to the catalysis and, being bound to $Co_2O_3$, released minimal amounts of cobalt in solution as opposed to the case of pure CoO where there was no such inhibition. The difference in the reactivity observed might thus be due to the difference in the availability of CoO in the two oxides used; pure CoO and $Co_3O_4$.（给出第二个结论(假说)及充分理由）

Further insights on the mechanism of the heterogeneous catalysis, including exact cobalt speciation when cobalt transits from one oxidation state to the other and clear proof of the species responsible for the heterogeneity observed are currently under way.（提出对与本研究密切相关的未来研究工作的展望(今后的研究方向)）

该讨论“Discussion”部分，由四个自然段组成，但围绕“the species responsible for the heterogeneous catalysis with $Co_3O_4$”(属于催化机理范畴)这一个焦点展开讨论。本讨论内容与“Results”部分的论述不重复。作为“Discussion”单列的论文风格，论点专一，论述深入。与一般论文的“Results and discussion”中的讨论表述有所不同。

实例2：来自于《Water Research》上已发表论文的“Discussion”部分。该论文正文包括：Introduction、Materials and Methods、Results 和 Discussion 和 Conclusions 五部分。论文题目为：Hydrogen and Methane Production from Household Solid Waste in the Two-stage Fermentation Process。下面在文中，插入中文说明，进行简要剖析。阅读时，注意段落的开头、要素、时态、语态和动词选用。

Discussion

4.1. Comparison of two-stage and one-stage processes

The two-stage process was demonstrated as an optimal way which combined hydrogen（1st stage）and methane（2nd stage）production in this study.［首先引出结论(论点)］The short HRT in the first stage（2 d）was resulting in effective separation of hydrogen production from methane production（15 d），without the need of external additions. Figs. 2 and 3 demonstrate that two-stage process in this study worked very well. The stable hydrogen production yield was 43 mL $H_2$/g VS added or 250 mL $H_2$/g VSrem.（概述重要研究结果(论据)）It was higher than 165 mL $H_2$/g VSrem which was also produced from HSW at 37℃ by Valdez-Vazquez et al.（2005）. It fell in the hydrogen potential range（26.3-96 mL $H_2$/g VS added）of HSW reported by Okamoto et al.（2000）.（与已有结果比较，说明本结果的可靠和优点）

As shown in Fig. 3，for the methane production，two-stage process generated 7500 mL $CH_4$/d（or 500 mL $CH_4$/g VS），which was 21% higher than the methane（6200 mL $CH_4$/d or 413 mL $CH_4$/g VS）from one-stage process（Fig. 4）. This was consistent with VFA data. Total VFA value in the second-stage process（1.8 mM）was much lower than that of the one-stage process（3.5 mM）. It shows that more VFAs were converted to methane in two-stage process. It shall be also noticed that HRT was 17 days in total for two-stage process while 15 days for one-stage process. Mata-alverez et al.（1993）found 510 mL $CH_4$/g VS was achieved in two-stage process for HSW fermentation while 428 mL $CH_4$/g VS in one-stage process，resulting in 19% methane increase but without hydrogen production. Pavan et al.（2000）reported 83.5% VS removal in two-stage process，which was similar to ours（86%）（进一步进行比较说明，说明本结果的可靠和优点）.

4.2. Sparging effect on hydrogen production

As illustrated in Fig. 5，gas sparging resulted in significant increase of the hydrogen production（88%）.（重述重要研究结果）Mizuno et al.（2000）also reported that hydrogen production was increased 68% after sparging with $N_2$.（引用已有结果，作比较）This phenomenon can be directly explained by the decrease of hydrogen partial pressure and $CO_2$ concentration. The hydrogen partial pressure is an important factor in the hydrogen process. At high $H_2$ partial pressure，the hydrogen synthetic pathways shift to production of more reduced substrates such as lactate，ethanol，acetone or alanine (Adams, 1990). The $CO_2$ concentration can also affect the hydrogen synthetic pathway. High $CO_2$ concentration favors the production of fumarate or succinate, which consumes electrons, and therefore decreases hydrogen production. (对结果的一个解释)Another reason for it may be due to the removal of carbon monoxide in the system. Levin et al. (2004) reported that CO could influence bacterial metabolism away from hydrogen production towards solvent (i. e. ethanol) production. (对结果的另一个解释)

以上材料是一篇论文"Discussion"中的一部分，与论文引言中的研究目的等相呼应，都围绕一定的主题讨论问题。讨论中比较说明较多，这有助于说明研究结果的可靠性和研究的先进性。

实例3：来自在《Environ. Sci. Technol.》上一篇已发表论文的"Results and Discussion"部分。该论文正文包括：Introduction、Experimental Section、Results and Discussion三部分。即"Conclusions"部分省略。"Results and Discussion"下又分若干小部分，都有各自的标题。论文题目为：Pharmaceuticals and Endocrine Disrupting Compounds in U. S. Drinking Water。下面在文中，插入中文说明，进行简要剖析。阅读时，注意要素、时态、语态、句型和动词选用。

Results and Discussion

A summary of the occurrence of pharmaceuticals and EDCs in source, finished, and distribution system water is shown in Table 3. Thirty-four of the 51 targeted compounds were detected in at least one sample, while the remaining 17 compounds were not detected in any samples. Eleven compounds (atenolol, atrazine, carbamazepine, estrone, gemfibrozil, meprobamate, naproxen, phenytoin, sulfamethoxazole, tris (2-chloroethyl) phosphate (TCEP), and trimethoprim) were detected in more than half of the source waters, while only atrazine, meprobamate, and phenytoin were detected in more than half of finished waters or distribution systems. All raw data are available in Supporting Information Tables S2, S3, and S4. (介绍研究结果(这里的实验研究就是监测和分析)，作为下面分述的共同基础)

Occurrence of Pharmaceuticals and EDCs in Source Water Samples. Targeted compounds were detected most frequently in source waters as compared to treated drinking waters (Table 3 and Supporting Information Table S2). At least one compound was detected in all 19 source waters. The 11 compounds which were detected in greater than half of source waters were atenolol, atrazine, carbamazepine, estrone, gemfibrozil, meprobamate, naproxen, phenytoin, sulfamethoxazole, TCEP, and trimethoprim. (介绍实验研究结果)The three DWTPs (-10, -11, and -14) utilizing water from reservoirs with no direct input of wastewater and where no recreational use is permitted had the lowest numbers of individual compounds detected in source waters (3, 3, and 2 compounds detected). The four DWTPs (-16, -17, -18, and -19) utilizing water from reservoirs with no direct input of wastewater, but where recreational use is allowed, had similar numbers of individual compounds detected in their source waters as compared to those DWTPs withdrawing water from wastewater impacted sources. (对实验结果的中肯解释，可以看作"讨论")

Atrazine is a widely used herbicide and was detected in the source water of almost every DWTP, including those far removed from areas with no agricultural atrazine application. (交代实验结果)The frequent detection of atrazine suggests that it is a widespread environmental contaminant.

（由结果得出推论）For example，it was one of the most commonly detected compounds in the 22-month monitoring program at DWTP-6（see Supporting Information Tables S5 and S6），a plant located in an arid region of the U. S. where there is no known atrazine use（25）. However，atrazine has previously been detected in wastewater effluent influencing the reservoir of this DWTP（24）. Detectable levels of atrazine have been measured in some foods（26）which may explain the loading to this wastewater treatment plant and DWTP.（提供上述推论的佐证）（得出推论和提供佐证可看作"讨论"）In this study，the largest atrazine source water concentrations were measured in source waters proximate to agricultural areas where it is heavily applied.（介绍研究结果）

In some samples，pharmaceutical parent-metabolite compounds were detected together，whereas in others they were not. The hydroxylated metabolites of atorvastatin occurred when the parent compound，atorvastatin，was also detected. Norfluoxetine was not detected in any source waters，including the three in which its parent compound fluoxetine was detected. Neither simvastatin nor its metabolite，simvastatin hydroxy acid，was detected in any source waters.（介绍研究结果）

The most prescribed pharmaceutical in 2006 and 2007 in the U. S.（27），atorvastatin，was detected in only three source waters and was not detected in any finished or distribution waters. Conversely，the most frequently detected prescription pharmaceuticals（carbamazepine，gemfibrozil，meprobamate，sulfamethoxazole，and trimethoprim）were not included in the top 200 prescribed pharmaceuticals for 2006 or 2007. Only atenolol（ranked no. 99 in 2007）and phenytoin（ranked no. 128 in 2006 and no. 151 in 2007）were frequently detected in source water.（以上三句话提供的都是研究结果）Thus，prescription information alone is a poor proxy for source water occurrence because it does not take into account the dosage，pharmacokinetics，removal during wastewater treatment，or environmental fate.（基于上述结果得出的结论）

本摘录材料是一篇论文"Results and Discussion"的一部分，该论文的"Results and Discussion"首先集中介绍一些实验研究结果（见上），以提供后文分述的基础。每部分的叙述都有一个主题，各个分述相对独立。在本分述（主题）中可见，实验研究结果介绍充分而有分量，在介绍结果时有机进行讨论。行文有"夹叙夹议"的特点，讨论简明，注重采用佐证，英文表达规范、地道。

实例4：来自在《Applied Catalysis B：Environmental》上一篇已发表论文的"Results and discussion"部分。该论文正文包括：Introduction、Experimental（Materials and Methods）、Results and discussion 和 Conclusions 四部分。即，将 Results 和 Discussion 合并。"Results and Discussion"下又分若干部分，都有各自的标题。论文题目为：Iron-cobalt Mixed Oxide Nanocatalysts：Heterogeneous Peroxymonosulfate Activation，Cobalt Leaching，and Ferromagnetic Properties for Environmental Applications。下面在文中插入中文说明，进行简要剖析。阅读时，注意时态、语态、句型和动词选用。

Degradation of 2,4-DCP and cobalt leaching

The 2,4-DCP degradation profiles in the systems of 1Fe1Co catalysts，$Fe_2O_3$，$Co_3O_4$，and physical mixture of $Fe_2O_3$ and $Co_3O_4$（mole ratio of Fe/Co = 1）are shown in Fig. 7. The bulk $Fe_2O_3$ could not activate PMS since Fe（Ⅲ）in $Fe_2O_3$ does not act as an electron donor to activate PMS. In contrast，a relatively faster 2,4-DCP degradation kinetics was achieved in $Co_3O_4$ system with Co（Ⅱ）species. It is interesting that the 1Fe1Co catalysts are more effective than both the bulk $Co_3O_4$

and the physical mixture of $Fe_2O_3$ and $Co_3O_4$.（以上语句都是介绍实验结果）This implies the crucial role of Fe-Co interactions on the efficient PMS activation over the Fe-Co catalysts.（由实验结果得出推论，仍属于"Results"范畴）In our previous study, we addressed the effect of homogeneous reaction resulting from dissolved Co(Ⅱ) ions on heterogeneous PMS activation, and found that 36 μg/L of Co(Ⅱ) ions could only induce around 10% of 2,4-DCP degradation within 2 h under the identical experimental conditions used in this study.（提供已有研究结果）In addition, it should be noted that the cobalt leaching from 1Fe1Co catalysts was fairly low between 30 and 40 μg/L.（再介绍实验结果）This means the influence of homogeneous PMS activation due to the leached Co(Ⅱ) ions on the degradation of 2,4-DCP in the 1Fe1Co systems was not significant.（通过已有研究结果和本实验结果得出结论，这部分都属于"Discussion"范畴）It is quite worthwhile to note that the catalytic activity of 1Fe1Co700 is still high in spite of its extremely small $S_{BET}$ as compared with 1Fe1Co300 and 1Fe1Co500 (Fig. 7d).（介绍实验结果）As revealed by previous studies, Co(Ⅱ) is the most efficient species for PMS activation to generate highly oxidizing SRs while Co(Ⅲ) is unable to directly activate PMS because it needs to be reduced to Co(Ⅱ) with the expense of PMS consumption by forming much less reactive peroxymonosulfate radicals ($SO_5^-$).（提供已有研究结论）Cobalt in 1Fe1Co300 and 1Fe1Co500 exists as $Co_3O_4$ (i.e. $CoO \cdot Co_2O_3$) and thus Co(Ⅲ) is dominant, whereas 1Fe1Co700 contains significant content of $CoFe_2O_4$ where Co species is of Co(Ⅱ).（提供已有知识）The relatively higher concentration of Co(Ⅱ) in 1Fe1Co700 compensates its much low $S_{BET}$, which can explain its high catalytic activity on heterogeneous PMS activation.（对实验结果作解释，即提供实验结果的缘由）（以上在实验结果介绍后的叙述都属于"Discussion"范畴）After 2 h reaction, around 40% mineralization of 2,4-DCP could be achieved with the Fe-Co catalysts.（再介绍实验结果）Furthermore, a previous study in our group has clearly revealed that in the aqueous homogeneous $Co^{2+}$/PMS system, the major reaction intermediates during 2,4-DCP degradation by sulfate radical attack were 2,4,6-trichlorophenol, 2,3,5,6-tetrachloro-1,4-benzenediol, 1,1,3,3-tetrachloroacetone, pentachloroacetone, and carbon tetrachloride.（再提供已有研究结果）We expect that the major reaction intermediates in the present study should be similar to the above-mentioned species identified in the aqueous homogeneous $Co^{2+}$/PMS system; however, the degradation pathway of 2,4-DCP might be slightly different between the homogeneous ($Co^{2+}$/PMS) and heterogeneous (Fe-Co catalysts/PMS) systems.（对研究结果进行比较，指出异同点）Therefore, detailed studies will be performed to identify various reaction intermediates of 2,4-DCP degradation in heterogeneous Fe-Co catalysts/PMS systems in future.（指明今后的研究方向）（以上在实验结果介绍后的叙述都属于"Discussion"范畴）

As shown in Fig. 8, all the catalysts displayed fairly low Co leaching between 20 and 50 μg/L except $FeCo/TiO_2$. Furthermore, it should be noted that 7Fe3Co700 exhibited the lowest cobalt leaching mainly due to the efficient formation of $CoFe_2O_4$.（介绍实验结果）Given the effectiveness of Co(Ⅱ) on PMS activation, Anipsitakis et al. investigated the feasibility of using CoO for heterogeneous PMS activation. However, they observed extremely high concentration of dissolved cobalt (at neutral pH). On the contrary, $Co_3O_4$, where CoO is bound to $Co_2O_3$, showed a better performance for the heterogeneous PMS activation due to the intimate interactions between CoO and

$Co_2O_3$. Nonetheless, the presence of abundant Co(Ⅲ) species in $Co_3O_4$ is able to impair the overall efficiency for PMS activation because of the reaction between Co(Ⅲ) and PMS to form much less reactive $SO_5^-$.（提供已有研究结果和知识）The $CoFe_2O_4$ in FeCo700 catalysts successfully overcame the limitations above as Co species in $CoFe_2O_4$ are of Co(Ⅱ).（对实验结果的解释）（以上在实验结果介绍后的叙述都属于"Discussion"范畴）

可以看出，文中"夹叙夹议"，"结果"的介绍和"讨论"的展开有机结合，结果介绍依次、分步进行，以便对问题逐一进行分析和讨论。文中表述规范、清晰。做到"Results"、"Discussion"部分组成要素的充分显现，体现了一般"Results and Discussion"的写作风格。

## 5.5 结论部分

通常情况下，在英语科技论文中有关结论的内容都已出现在"结果与讨论"或"结果"和"讨论"部分。但在英语科技论文中往往将"结论"单独列为一个部分，紧跟在"讨论"部分(或"结果与讨论"部分)的后面，作为论文正文的最后一个重要组成部分。读者在看过摘要之后，紧接着就看结论部分，以了解研究工作的主要成果，再决定是否有必要更详细阅读。但也有一些期刊(包括很著名的英文环境类学术期刊)，在正文中省略"结论 Conclusions"。

### 5.5.1 功用

与引言相呼应，回答引言中提到的要解决的问题及预期目标，在论文中继摘要、引言之后，第三次重申问题的重要性和研究价值，是论文实质内容的浓缩。

### 5.5.2 组成要素

结论部分的语言严谨、精练、连贯，具有高度的概括性。其组成要素一般有以下几方面。但对于一篇具体论文而言，未必每一项都体现。

(1) 对引言中问题的回答

(2) 研究的主要结果及这些结果的内涵

表达比结果、讨论部分更精辟，不是已有表述的重复和简单组合。

(3) 研究成果的应用前景及局限性

(4) 进一步研究的设想或建议

结论部分不新增前文未曾提及的事实，不简单重复摘要、引言、结果或讨论中的表述，尤其不要重复其中的句子。应从结论的组成要素着力，重新组织句子。

### 5.5.3 时态、语态与常用句型

与其他各部分不同，英语科技论文的结论部分中现在时态(特别是一般现在时和现在完成时)使用频率很高，这是由于结论部分总结研究者到目前为止做了哪些工作，得出了什么结果，这些结果在现在有什么影响、意义和价值，可应用于什么场合，解决什么问题等。至于语态，使用较灵活，根据需要，主动和被动态兼用。常用句型如下：

(1) Two factors to influence A have been studied....

(2) Through the example of A, it has demonstrated that a simplified approach can be used....

(3) Overall, our study has revealed a variety of patterns....

(4) 用主动语态时，常用 this paper, this investigation, this study, this survey, the results, the analysis 或 we, I 等作主语。

This study (This investigation/The result) clearly demonstrates (discovers/reveals/has described/has shown/has proposed/discusses/represents) that...

(5) 作者有时对自己的研究结果采取慎重的态度，不采用肯定的说法“结果说明了(证明了)……”，而往往对动词加以修饰，以表示作者不太肯定或慎重的态度，或用以缓和语气，用情态动词是这种修饰的常用技巧。最常用的情态动词有 will，can，may 及其过去式 would，could，might 等，表示“可能”之意。

(a) A might provide additional useful information to be used in understanding B.

(b) A may be a natural and necessary phenomenon.

(c) A may also improve the effectiveness of student science projects.

(6) A appears to be B.

(7) A causes (seem to cause/may cause) B.

(8) 建议进一步研究或阐述实际应用时，动词常辅以 may，could，would 等情态动词。

(a) We recommend that these experiments be repeated using....

(b) Recommendation for further work are listed below.

(c) Experiments similar to those reported here should be conducted using....

(d) An investigation to study the effect of A should be carried out.

(e) It is (most) desirable that (to do)....

(f) It would be best if....

其他实际典型例句(单句或连续语句)如下所示：

(1) In the present study, efficient and environmental benign (i.e. minimum cobalt leaching and recoverable features) Fe-Co mixed oxide nanocatalysts have been developed for heterogeneous activation of peroxymonosulfate (PMS) to generate SRs targeting the decomposition of 2,4-DCP.

(2) It was clearly revealed that both Fe/Co molar ratio and calcination temperature were of crucial importance for the catalyst performance.

(3) Contrary to our expectation, the immobilized catalyst FeCo/$TiO_2$ exhibited very high cobalt leaching, which suggests that the presence of $TiO_2$ could hinder the formation of intimate Fe-Co interaction.

(4) The effectiveness of three representative inorganic, synthetic and naturally occurring chelating agents (pyrophosphate, EDDS and citrate, respectively) in stabilizing ferrous iron at neutral pH was evaluated.

(5) Activation of three common oxidants (PMS, PS and $H_2O_2$) by iron/chelating agent complexes for the degradation of a representative contaminant (4-CP) was investigated.

(6) In summary, PMS was found to be universally activated to a certain extent by all three iron/chelating agent systems.

(7) The results obtained in the present study showed that apart from the type of chelating agent used for iron stabilization, the nature of peroxide also plays a significant role in activation of oxidants for radical generation and ultimately contaminant degradation.

(8) In the present work, different cobalt catalysts were prepared by varying the support materials ($TiO_2$, $Al_2O_3$ and $SiO_2$) and cobalt precursors ($Co(NO_3)_2$, $CoCl_2$ and $CoSO_4$).

(9) The impact of these variables on the properties of the supported cobalt catalysts was evaluated by various characterization techniques.

(10) All the results shown in this study support that the coupling of PMS-0.001Co/$TiO_2$ with

UV-A is a good choice to simultaneously generate HRs and SRs for the rapid 2,4-DCP transformation with very low cobalt leaching.

(11) In the present study, sulfate radical-based environmentally friendly AOTs were investigated for PCB degradation in aqueous and sediment systems.

(12) In selected experiments using Fe(Ⅱ)/PMS, around 80% TOC removal was observed within 24 h indicating the effectiveness of these processes for complete mineralization of recalcitrant PCBs.

(13) During the course of the study, a significant decrease in degradation efficiency was observed with increasing initial pH, primarily due to the precipitation of iron at higher pH values.

(14) In this study, the applicability of solar radiation to drive sulfate radical-based AOPs was demonstrated for the first time.

(15) The results revealed that by using solar light as energy source for the photo-process, a significant enhancement in the reaction rate constant was achieved for the degradation of 2,4-D pesticide, a widely distributed, highly toxic and environmentally important pollutant in aquatic systems.

(16) Therefore, the Co/MgO catalyst prepared in this work could be potentially used in advanced oxidation technologies towards the removal of organic pollutants.

(17) Based on the present experimental results, the most effective activator was $CoCl_2$.

(18) According to experiments with various concentrations of PMS and Co(Ⅱ), increasing the molar ratio of PMS/Co(Ⅱ) increased the degradation of diesel.

(19) The experimental results showed that the use of this $Co^{2+}$/PMS oxidative process without any photopromotion is very beneficial to not only the complete decolorization of RBB but also to the degradation efficiency of its derivative aromatic fragments.

(20) Fortunately, this expected low dosage $Co^{2+}$/PMS oxidative process can be seen as a superior choice as a good pretreatment to selectively decompose the stubborn structure of benzenic and phenolic derivatives first although a low mineralization efficiency is the common drawback of a single use of the sulfite-bisulfite-pyrosulfite systems.

(21) This study was conducted to determine the interactive role of Pb, Cu, Ni and Zn on metal uptake, plant growth and anti-oxidative system of S.drummondii.

(22) The co-presence of metals resulted in a greater reduction in S.drummondii biomass than exposure to a single metal suggesting synergistic or additive response.

(23) Enhanced level of enzymatic and non-enzymatic anti-oxidants indicates that S.drummondii may have a detoxification mechanism to cope with different metals.

(24) In addition to acquiring the heritable resistance to the acute teratogenic effects of PAHs. ER fish appear to have concomitantly developed resistence to chronic effects, including cancer.

(25) The frequent consumption of food commodities with episodic presence of pesticides that are suspected to cause developmental and neurological effects in young children supports the need for further mitigation.

(26) Engineered nanomaterials can become airborne when mixed in solution by sonication, especially when nanomaterials are functionalized or in water containg NOM.

(27) This finding indicates that laboratory workers may be at increased risk of exposure to engineered nanomaterials.

### 5.5.4 实例简析

实例1：来自在《Applied Catalysis B：Environmental》发表的一篇论文"Conclusions"部分。论文题目为：Iron-cobalt Mixed Oxide Nanocatalysts：Heterogeneous Peroxymonosulfate Activation, Cobalt Leaching, and Ferromagnetic Properties for Environmental Applications。下面在文中，插入中文说明，进行简要剖析。阅读时，注意时态、语态、句型和动词选用。

Conclusions

In the present study, efficient and environmental benign (i.e.minimum cobalt leaching and recoverable features) Fe-Co mixed oxide nanocatalysts have been developed for heterogeneous activation of peroxymonosulfate (PMS) to generate SRs targeting the decomposition of 2,4-DCP.（与引言相呼应，回答了引言中提到的要解决的问题）It was clearly revealed that both Fe/Co molar ratio and calcination temperature were of crucial importance for the catalyst performance. In 7Fe3Co700 (Fe：Co=7：3, calcined at 700℃) catalyst, ferromagnetic $CoFe_2O_4$ composites effectively formed by thermal oxidation of a mixed phase of Fe and Co. It was found that this catalyst is the most promising for the efficient and environmentally friendly activation of PMS because of its exceptional physicochemical properties including (i) the cobalt species in $CoFe_2O_4$ are of Co(Ⅱ), unlike $Co_3O_4$ that shows some limitation of Co(Ⅲ) on PMS activation, (ii) $CoFe_2O_4$ possesses suppressed Co leaching properties due to strong Fe-Co interaction (Fe-Co linkages), and (iii) Fe-Co catalysts in form of $CoFe_2O_4$ are easy to recover due to the unique ferromagnetic nature of $CoFe_2O_4$.（概述重要研究结果及其缘由）Moreover, XPS results demonstrated that the conjunction of Co with Fe is beneficial for enhancing the content of hydroxyl groups on the catalyst surface, which is believed to facilitate the formation of Co(Ⅱ)-OH complexes that are vital for heterogeneous PMS activation.（概括另一重要研究结果及内涵）In addition, immobilization of Fe-Co catalyst on $TiO_2$ support was attempted for further suppressing the cobalt leaching. Contrary to our expectation, the immobilized catalyst $FeCo/TiO_2$ exhibited very high cobalt leaching, which suggests that the presence of $TiO_2$ could hinder the formation of intimate Fe-Co interaction（指出与预期设想及目标不一致的方面和原因）

对照引言，可以看出本结论首先说明该研究回答了引言中关切的问题（Consequently, it is critical to develop a highly efficient nanoscale metal catalyst for the heterogeneous PMS activation, where the adverse impact of metal leaching on aqueous environments is notably minimized and the catalyst is readily recovered after its application.）。然后概述主要的研究结果及其缘由或内涵，语言精练、严谨，对照摘要和论文正文其他部分的文字叙述，发现论文中的句子经重新组织，语言表达规范、地道。体现了"结论"部分的特定功用。

实例2：来自在《Journal of Hazardous Materials》发表的一篇论文"Conclusions"部分。论文题目为：Degradation of a Xanthene dye by Fe(Ⅱ)-Mediated Activation of Oxone Process。下面在文中，插入中文说明，进行简要剖析。阅读时，注意时态、语态、句型和动词选用。

Conclusions

The degradation of RhB was investigated using $SO_4^{\cdot-}$ generated by the coupling of Oxone and ferrous ions as oxidant and transition metal, respectively.（与引言相呼应，交代研究主题）

Experimental results showed that the performance of dye degradation was dramatically influenced by operating parameters, such as the molar ratio and the concentration of Fe( Ⅱ ) and Oxone, initial solution pH, and inorganic salts. Optimum molar ratio of Fe( Ⅱ )/Oxone was identified to be 1 : 1. Furthermore, the results indicated the higher the Oxone concentration, the faster the decay rate as well as the higher the performance. Optimum dye removal efficiency was observed at an initial solution pH of 3. 51 within the investigated pH range of 2. 67-10. 22.（概述研究结果）The overall dye degradation can be described as a two-stage reaction consisting of a rapid degradation stage followed by a retarded reaction stage under the optimum conditions. In addition, the presence of certain anions had a significant effect on the Fe ( Ⅱ )/Oxone process. It was found that $SO_4^{2-}$ demonstrated adverse effect in the process, whereas the existence of $Cl^-$ facilitated the transformation of RhB. The mineralization of RhB is feasible and the performance can be greatly improved by using stepwise Fe( Ⅱ ) and Oxone dosing approach.（说明研究结果的内涵）

本结论短小、紧凑。主要围绕研究结果和含义叙述。尽管论文在摘要、引言和结论中都提到研究主题，但写作中没有重复句子结构，结果的表述更加概括。

实例3：来自在《Water Research》发表的一篇论文“Conclusions”部分。论文题目为：Coupling Reverse Osmosis with Electrodialysis to Isolate Natural Organic Matter from Fresh Waters。下面在文中，插入中文说明，进行简要剖析。阅读时，注意时态、语态、句型和有关要素组合。

Conclusions

ED ( Electrodialysis ) experiments were conducted on synthetic freshwater samples whose chemical compositions most closely resemble those of un-desalted freshwater samples that have been concentrated by reverse osmosis. ED experiments were also conducted on RO( Reverse osmosis )-concentrated solutions of NOM ( Natural organic matter) from two rivers.（与引言相呼应，重述研究主题）These experiments have shown:

(1) The selective removal of $SO_4^{2-}$ and retention of TOC were best achieved at pH > 6 and conductivity>1m $Scm^{-1}$. If the ED process was conducted at pH>9, $H_4SiO_4$ could also be removed to a significant extent.

(2) When the RO and ED processes were applied in tandem to fresh waters using optimal conditions for ED, 102% of TOC was recovered, 79% of $SO_4^{2-}$ was removed, and 65% of $H_4SiO_4$ was removed. More importantly, the molar ratios of $SO_4^{2-}$ : TOC and $H_4SiO_4$ : TOC were reduced to 0. 0067 and 0. 025, respectively, thus surpassing the goal for removal of $SO_4^{2-}$ and almost achieving the goal for removal of $H_4SiO_4$.

(3) The ED process can lower the $SO_4^{2-}$ : TOC ratio in samples whose initial $SO_4^{2-}$ : TOC ratios are already far below the limit of 0. 008 used in this study.

(4) The degree to which $SO_4^{2-}$ and $H_4SiO_4$ can be removed depends on the amount of time used for the ED process, on chemical conditions of pH and conductivity, and an adequate supply of electric current.

(5) For all synthetic and natural samples tested in this study, the loss of TOC during the ED process is minimal, ranging from 0% to 12% with an average of 4%. Combined with an average TOC loss of 12% (or less) during the RO process, the combined RO/ED process yields an average TOC recovery of 84%.

（以上 1~5 条分述重要的实验研究结果）

Although some further enhancements are planned for the near future, the coupled RO/ED process that has been described in this study offers a fast, simple, chemically mild (relative to other methods), and reproducible method of isolation of large quantities of relatively unfractionated, low-ash NOM from fresh waters.（进一步研究的设想和本研究成果的价值）

本结论结构清晰，易读，研究主题表述明确，研究结果表达充分，语句结构和措辞避免了与论文其他部分的重复，作为论文实质内容的浓缩，注重与摘要的区别。结论的要素组合反映了一般结论的要求。英语表达规范、地道。

实例 4：来自在《Water Research》发表的一篇论文“Conclusions”部分。论文题目为：Simultaneous Adsorption and Biodegradation Processes in Sequencing Batch Reactor (SBR) for Treating Copper and Cadmium-containing Wastewater。下面在文中，插入中文说明，进行简要剖析。阅读时，注意时态、语态、句型和有关要素组合。

Conclusions

Based on the finds of this study, the following conclusions can be drawn:

(1) In the presence of 10 mg/L Cu(Ⅱ) and 30 mg/L Cd(Ⅱ), respectively, the simultaneous adsorption and biodegradation processes under SBR operation with a PAC dosage of 143 mg/L or 1.0 g/cycle was capable of removing more than 85% of COD as compared to around 60% for biodegradation process only under SBR operation.（实验结果）

(2) Cu(Ⅱ) was found to exert a more pronounced inhibitory effect on the bioactivity of the microorganisms compared to Cd(Ⅱ).（由实验结果提炼出的结论）

(3) The combined presence of Cu(Ⅱ) and Cd(Ⅱ) did not exert synergistic effects on the microorganisms and the combined effect was less severe than the summation of the two individual effects.（由实验结果提炼出的结论）

(4) The merit of PAC addition was in terms of minimizing the inhibitory effect of Cu(Ⅱ) and Cd(Ⅱ), respectively, on the bioactivity of microorganisms.（由实验结果提炼出的结论）

(5) The PAC-added biomass played the key role in the uptake of metals in the SBR reactors.（由实验结果提炼出的结论）

本结论是一个一览表式结论，结构紧凑，条理清楚，易读，表述重点突出，直指结果及其含义，语句精练。

# 参 考 文 献


1 Vesilind P A, Peirce J J, Weiner R F. *Environmental Engineering* (*$3^{rd}$ ed.*). Newton, MA: Butterworth-Heinemann, 1994

2 Theodore M K, Theodore L. *Major Environmental Issues Facing the $21^{st}$ Century.* New Jersey: Prentice Hall PTR, Prentice-Hall, Inc., 1996

3 Enger E D, Smith B F. *Environmental Science*(*$5^{th}$ ed.*). Dubuque, IA: Wm. C. Brown Publishers, 1995

4 Kiely G. *Environmental Engineering.* McGraw-Hill Publishing Company, 1997

5 John M Swales, Christine B Feak. *Academic Writing for Graduate Students: Essential Tasks and Skills*(*$2^{nd}$ edition*). The University of Michigan Press, 2004

6 钟似璇，王新英. 英语科技论文写作与发表. 天津：天津大学出版社，2004

7 任胜利. 英语科技论文撰写与投稿. 北京：科学出版社，2004

8 孙钰，Sheryl Holt. 英语论文成功写作. 北京：群言出版社，2006

9 秦荻辉. 实用科技英语写作技巧. 上海：上海外语教育出版社，2001

10 王亚光，李三喜. 科技英语教程. 北京：清华大学出版社，2008

11 戴浩中主编. 科技英语句子结构. 上海：上海科学技术出版社，1983

12 彭举威，王若竹，钱永梅. 新编环境科学与工程专业英语. 北京：化学工业出版社，2010